Preface to the fourth edition

There can be few scientific disciplines that have advanced as rapidly as molecular biology in the years since the last edition. Thus, our hope that the improvements made to the last edition would make future changes simpler was rather naïve. Our revisions and updates to the fourth edition have been major and have involved all Sections and Topics; such has been the rate of progress. The first two Sections of the third edition have been combined and simplified to reduce overlap with other titles in the *Instant Notes* series, whereas other Sections have been reorganized and restructured in a more logical fashion. This has created room for the consideration of advances in topics such as 'next-generation' DNA sequencing and genomics, global gene expression analysis, regulatory RNAs, proteomics, stem cells, systems biology, and many other areas. One difficulty has been judging what to leave out in order to accommodate new material. As knowledge expands and technology advances, and the old ways of doing things fall out of favor, the challenge is to keep the reader excited with new discoveries and with the power of new methods while keeping enough of the traditional background story to allow complete understanding of a topic. We are also well aware that this is an introductory textbook and so we have tried to avoid unnecessary complexity and detail. We hope we have achieved our aims. As always, we are most grateful to those many reviewers who made suggestions for improvements to the previous edition and are particularly indebted to Liz Owen and Vicki Noyes for their patience and understanding during the revision process.

Alexander McLennan, Andy Bates, Phil Turner, and Mike White

November 2011

BIOS INSTANT NOTES

Molecular Biology

FOURTH EDITION

Alexander McLennan, Andy Bates, and Phil Turner
Institute of Integrative Biology
University of Liverpool, Liverpool, UK

Mike White
Faculty of Life Sciences
University of Manchester, Manchester, UK

GS **Garland Science**
Taylor & Francis Group
NEW YORK AND LONDON

Garland Science
Vice President: Denise Schanck
Editor: Elizabeth Owen
Editorial Assistant: Vicky Noyes
Production Editor: Ioana Moldovan
Copyeditor: Alison Gibbs
Typesetting and illustrations: Phoenix Photosetting, Chatham, Kent
Proofreader: Dawn Booth
Printed by: T. J. International

ISBN 978-0-4156-8416-3

Library of Congress Cataloging-in-Publication

Molecular biology / Alexander McLennan ... [et al.]. — 4th ed.
   p. ; cm. — (BIOS instant notes)
Rev. ed. of: Molecular biology / Phil Turner ... [et al.]. 3rd ed. New York, NY : Taylor &
Francis, 2005.
Includes bibliographical references and index.
Summary: "Instant Notes in Molecular Biology, Fourth Edition is the perfect text for
undergraduates looking for a concise introduction to the subject, or a study guide to use
before examinations. Each topic begins with a summary of essential facts, an ideal revision
checklist followed by a description of the subject that focuses on core information, with
clear, simple diagrams that are easy for students to understand and recall in essays and
exams"—Provided by publisher.
ISBN 978-0-415-68416-3 (pbk.)
1. Molecular biology—Outlines, syllabi, etc. I. McLennan, Alexander G. II. Series: BIOS
instant notes.
[DNLM: 1. Molecular Biology—Outlines. QH 506]
QH506.I4815 2013
572.8—dc23   2012013042

Published by Garland Science, Taylor & Francis Group, LLC, an informa business,
711 Third Avenue, 8th Floor, New York NY 10017, USA, and 3 Park Square, Milton Park,
Abingdon, OX14 4RN, UK.

15 14 13 12 11 10 9 8 7 6 5 4 3 2 1

Visit our web site at http://www.garlandscience.com

Contents

Preface to the fourth edition   v

Section A – Informational macromolecules
    A1   Information processing and molecular biology   1
    A2   Nucleic acid structure and function   4
    A3   Protein structure and function   12
    A4   Macromolecular assemblies   22
    A5   Analysis of proteins   26

Section B – Properties of nucleic acids
    B1   Chemical and physical properties of nucleic acids   33
    B2   Spectroscopic and thermal properties of nucleic acids   37
    B3   DNA supercoiling   40

Section C – Prokaryotic and eukaryotic chromosome structure
    C1   Prokaryotic chromosome structure   45
    C2   Chromatin structure   48
    C3   Eukaryotic chromosome structure   53
    C4   Genome complexity   59

Section D – DNA replication
    D1   DNA replication: an overview   64
    D2   Bacterial DNA replication   69
    D3   Eukaryotic DNA replication   74

Section E – DNA damage, repair, and recombination
    E1   DNA damage   78
    E2   Mutagenesis   82
    E3   DNA repair   86
    E4   Recombination and transposition   91

Section F – Transcription in bacteria
    F1   Basic principles of transcription   96
    F2   *Escherichia coli* RNA polymerase   99
    F3   The *E. coli* σ^{70} promoter   102
    F4   Transcription initiation, elongation, and termination 105

Section G – Regulation of transcription in bacteria
    G1   The *lac* operon   110
    G2   The *trp* operon   114
    G3   Transcriptional regulation by alternative σ factors and RNA 119

Section H – Transcription in eukaryotes

H1	The three RNA polymerases: characterization and function	123
H2	RNA Pol I genes: the ribosomal repeat	126
H3	RNA Pol III genes: 5S and tRNA transcription	130
H4	RNA Pol II genes: promoters and enhancers	134
H5	General transcription factors and RNA Pol II initiation	137

Section I – Regulation of transcription in eukaryotes

I1	Eukaryotic transcription factors	141
I2	Examples of transcriptional regulation	148

Section J – RNA processing and RNPs

J1	rRNA processing and ribosomes	154
J2	tRNA and other small RNA processing	160
J3	mRNA processing, hnRNPs, and snRNPs	164
J4	Alternative mRNA processing	171

Section K – The genetic code and tRNA

K1	The genetic code	176
K2	tRNA structure and function	181

Section L – Protein synthesis

L1	Aspects of protein synthesis	188
L2	Mechanism of protein synthesis	192
L3	Initiation in eukaryotes	199
L4	Translational control and post-translational events	204

Section M – Bacteriophages and eukaryotic viruses

M1	Introduction to viruses	210
M2	Bacteriophages	213
M3	DNA viruses	218
M4	RNA viruses	222

Section N – Cell cycle and cancer

N1	The cell cycle	226
N2	Oncogenes	231
N3	Tumor suppressor genes	236
N4	Apoptosis	240

Section O – Gene manipulation

O1	DNA cloning: an overview	244
O2	Preparation of plasmid DNA	249
O3	Restriction enzymes and electrophoresis	253
O4	Ligation, transformation, and analysis of recombinants	258

Section P – Cloning vectors

P1	Design of plasmid vectors	265
P2	Bacteriophages, cosmids, YACs, and BACs	270
P3	Eukaryotic vectors	278

Section Q – Gene libraries and screening

Q1	Genomic libraries	283
Q2	cDNA libraries	286
Q3	Screening procedures	290

Section R – Analysis and uses of cloned DNA

R1	Characterization of clones	293
R2	Nucleic acid sequencing	297
R3	Polymerase chain reaction	303
R4	Analysis of cloned genes	309
R5	Mutagenesis of cloned genes	313

Section S – Functional genomics and the new technologies

S1	Introduction to the 'omics	317
S2	Global gene expression analysis	321
S3	Proteomics	328
S4	Cell and molecular imaging	333
S5	Transgenics and stem cell technology	337
S6	Bioinformatics	341
S7	Systems and synthetic biology	350

Further reading	356
Abbreviations	365
Index	368

A1 Information processing and molecular biology

Key Notes

The 'central dogma'	The central dogma is the original proposal that 'DNA makes RNA makes protein,' which happens via the processes of transcription and translation respectively. This is broadly correct, although a number of examples are known that contradict parts of it. Retroviruses reverse transcribe RNA into DNA, other viruses can replicate RNA directly into an RNA copy, whereas some RNAs can be edited after synthesis so that the resulting sequence is not directly specified by the DNA sequence.
Recombinant DNA technology	The ability to sequence and manipulate the genomes of microorganisms, animals, and plants has led to major advances in our understanding of cellular biology. In addition, transgenic organisms containing DNA from other sources have found many applications in medicine, agriculture, and industry. The ability to synthesize novel genomes will lead to even greater advances in these areas.
Related topics	(A2) Nucleic acid structure and function (A3) Protein structure and function (Section F) Transcription in bacteria	(Section H) Transcription in eukaryotes (Section K) The genetic code and tRNA (Section L) Protein synthesis

---

The central dogma

Molecular biology is the study of the molecular reactions and interactions that underpin biological function. This overlaps considerably with biochemistry and genetics, and so it is often deemed mainly to deal with the structural basis and control of information processing in the cell and the technologies required to investigate these. Through the pioneering experiments of Avery, MacLeod, and McCarty, and Hershey and Chase in the 1940s and 1950s, it became firmly established that the genetic instructions for creating a cell were held in the nucleus within the linear sequence of **bases** contained in the structure of a long chemical polymer, **deoxyribonucleic acid** (**DNA**). Then, in 1953, the famous double helical structure of DNA was proposed by Crick and Watson, which revealed exactly how this information was stored and passed on to subsequent generations. To explain how cells use the instructions encoded in this DNA **genome**, Crick suggested that there was a unidirectional flow of genetic information from DNA through an intermediary nucleic acid, **ribonucleic acid** (**RNA**), to **protein**, i.e. 'DNA makes RNA makes protein.' This became known as the **central dogma** of molecular biology, as it was proposed without much evidence for the individual steps. We now know that the broad thrust of the central dogma is correct, although a number of modifications have now been made to the original scheme. A diagrammatic version of this information flow is shown in Figure 1. The primary route remains from DNA to RNA to protein, and this is now known to include the DNA in the small, independent genomes of mitochondria and chloroplasts. In all cells, DNA is divided conceptually (though not physically) into discrete coding units (**genes**) that contain the information for individual proteins. This DNA is **transcribed** (Sections F and H) to yield RNA molecules (**messenger RNA, mRNA**) that contain the same sequence information as the DNA, and which can be regarded as working copies of the genes present in the master DNA blueprint. These mRNAs are then **translated** (Section L) into the amino acid sequences of proteins according to the **genetic code** (Section K1). The combination of all the processes that are required to decode the information in DNA to produce a functional molecule is called **gene expression**. We can also include DNA **replication** (Section D) in Figure 1, in which two daughter DNA molecules are formed by duplication of the information in the parent DNA, resulting in information flow and preservation from one generation to the next.

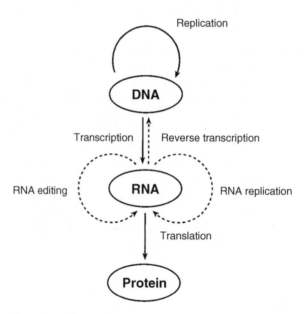

Figure 1. The flow of genetic information.

However, several exceptions to this basic scheme have been identified. Many RNA molecules are not translated into protein, but function as RNAs in their own right (Sections I2, L4, and J). Their genes are known as **RNA genes**. A number of classes of virus have no DNA but contain a genome consisting of one or more RNA molecules (Section M4). In the **retroviruses**, which include **human immunodeficiency virus (HIV)**, the causative agent of acquired immune deficiency syndrome (AIDS), the single-stranded RNA molecule is converted to a double-stranded DNA copy, which is then inserted into the genome of the host cell. This process has been termed **reverse transcription**. There are also a number of viruses known whose RNA genome is copied directly into RNA without the use of DNA as an intermediary (**RNA replication**) (Section M4). Examples include the influenza and hepatitis C viruses. As far as is known, there are no examples of a protein being 'reverse translated' to generate a specific RNA or DNA sequence, so the translation step of the central dogma does appear to be unidirectional. Finally, one fascinating exception to the dogma that RNA and protein sequences are faithfully encoded in the DNA is the process of **RNA editing**. Examples are known (mainly in eukaryotes) where the base sequence of an RNA is actually altered after it is transcribed from the DNA so that it, and any protein product if it is an mRNA, no longer correspond precisely to the DNA (Section J4).

Discussion of these systems in this book is based on the **three-domain system** of biological classification in which the **last universal common ancestor** (**LUCA**) of all life first split into the **bacteria** and the common progenitor of the **archaea** and **eukarya**, which split later. Bacteria and archaea are both **prokaryotic**, in that they lack a nucleus, but in many aspects of information processing, archaea have more in common with the nucleated **eukaryotes**. Most examples are drawn from bacteria and eukaryotes.

Recombinant DNA technology

Major advances in molecular biology became possible in the late 1970s with the development of **recombinant DNA technology** (**genetic engineering**). This has allowed genes to be isolated, sequenced, modified, and transferred from one organism to another and has been of paramount importance in advancing our understanding of how cells work. Furthermore, **transgenic** microorganisms produced in this way are now routinely used to produce human therapeutics on a large scale while transgenic animals and plants have huge potential to increase the range of useful products as well as leading to improved growth, disease resistance, and models for human disease, etc. (Section S5). The permanent correction of genetic disease by **gene therapy** is also now a realistic possibility. In recent years, the technology behind determining DNA base sequences has progressed and the costs have fallen so rapidly that it will soon become feasible to sequence an individual's entire genome to determine disease susceptibility as part of a routine health care program. New genes can now even be chemically synthesized and assembled into complete genomes. In 2010, J. Craig Venter and colleagues recreated the complete chromosomal DNA of a small mycoplasma bacterium and inserted it into an 'empty' cell, denuded of its own chromosome, thus recreating the living organism (Section S7). This DNA molecule also had some novel features, paving the way to the future possibility of creating truly **synthetic life** – 'designer' organisms with artificial genomes, able to carry out novel biochemical functions not seen in the natural world, with the aim of producing new medicines, fuels, and other products. Thus, molecular biology and the technologies that have been created around it have played a central role in the development of human and animal medicine, agriculture, and the biotechnology industry, and are now set to meet the challenges of global health, environmental change, and food security that we face in the 21st century.

A2 Nucleic acid structure and function

Key Notes	
Bases	In DNA, there are four heterocyclic bases: adenine (A) and guanine (G) are purines; cytosine (C) and thymine (T) are pyrimidines. In RNA, thymine is replaced by the structurally very similar pyrimidine, uracil (U).
Nucleosides	A nucleoside consists of a base covalently bonded to the 1′-position of a pentose sugar molecule. In RNA, the sugar is ribose and the compounds are ribonucleosides, or just nucleosides, whereas in DNA it is 2′-deoxyribose, and the nucleosides are named 2′-deoxyribonucleosides, or just deoxynucleosides. Base+sugar=nucleoside.
Nucleotides	Nucleotides are nucleosides with one or more phosphate groups covalently bound to the 3′-, 5′-, or, in some ribonucleotides, the 2′-position. Base+sugar+phosphate =nucleotide. The nucleoside 5′-triphosphates, NTPs and dNTPs, are the building blocks of polymeric RNA and DNA respectively.
Phosphodiester bonds	In nucleic acid polymers, the ribose or deoxyribose sugars are linked by a phosphate between the 5′-position of one sugar and the 3′-position of the next, forming a 3′,5′-phosphodiester bond. Hence, nucleic acids consist of a directional sugar-phosphate backbone with a base attached to the 1′-position of each sugar. The repeat unit is a nucleotide. Nucleic acids are highly charged polymers with a negative charge on each phosphate.
DNA/RNA sequence	The nucleic acid sequence is the sequence of bases A, C, G, T/U in the DNA or RNA chain. The sequence is conventionally written from the free 5′- to the free 3′-end of the molecule, for example 5′-ATAAGCTC-3′ (DNA) or 5′-AUAGCUUGA-3′ (RNA).
DNA double helix	DNA most commonly occurs as a double helix. Two separate and antiparallel chains of DNA are wound around each other in a right-handed helical (coiled) path, with the sugar-phosphate backbones on the outside and the bases, paired by hydrogen bonding and stacked on each other, on the inside. Adenine pairs with thymine; guanine pairs with cytosine. The two chains are complementary; one specifies the sequence of the other.

A, B, and Z helices	As well as the 'standard' DNA helix discovered by Watson and Crick, known as the B-form, and believed to be the predominant structure of DNA *in vivo*, nucleic acids can also form the right-handed A-helix, which is adopted by RNA sequences *in vivo* and the left-handed Z-helix, which only forms in specific alternating base sequences and is probably not a very important *in vivo* conformation.
RNA secondary structure	Most RNA molecules occur as a single strand, which may be folded into a complex conformation, involving local regions of intramolecular base pairing and other hydrogen bonding interactions. This complexity is reflected in the varied roles of RNA in the cell.
Modified nucleic acids	Covalent modifications of nucleic acids have specific roles in the cell. In DNA, these are mostly restricted to methylation of adenine and cytosine bases, but the range of modifications of RNA is much greater.
Nucleic acid function	DNA acts only as a carrier of expressible genetic information. However, the more versatile RNAs have numerous structural and functional roles in the mechanisms and regulation of information storage, flow, and processing.
Related topics	(B1) Chemical and physical properties of nucleic acids (B2) Spectroscopic and thermal properties of nucleic acids	(B3) DNA supercoiling (Section C) Prokaryotic and eukaryotic chromosome structure

Bases

The **bases** of DNA and RNA are heterocyclic (carbon- and nitrogen-containing) aromatic rings, with a variety of substituents (Figure 1). Adenine (A) and guanine (G) are **purines**, bicyclic structures with two fused rings, whereas cytosine (C), uracil (U), and thymine (T) are monocyclic **pyrimidines**. In DNA, the uracil base of RNA is replaced by thymine. Thymine differs from uracil only in having a methyl group at the 5-position, i.e. thymine is 5-methyluracil.

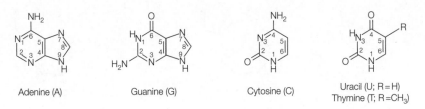

Figure 1. Nucleic acid bases.

Nucleosides

In nucleic acids, the bases are covalently attached to the 1′-position of a pentose sugar ring, to form a **nucleoside** (Figure 2). In RNA, the sugar is **ribose**, and in DNA it is **2′-deoxyribose**, in which the hydroxyl group at the 2′-position is replaced by a hydrogen. The point of attachment to the base is the 1-position (*N*-1) of the pyrimidines and the 9-position (*N*-9) of the purines (Figure 1). The numbers of the atoms in the ribose ring are designated 1′-, 2′-, etc., merely to distinguish them from the base atoms. The bond between the bases and the sugars is the **glycosylic** (**or glycosidic**) **bond**. If the sugar is ribose, the nucleosides (technically **ribonucleosides**) are adenosine, guanosine, cytidine, and uridine. If the sugar is deoxyribose (as in DNA), the nucleosides (**2′-deoxyribonucleosides**) are deoxyadenosine, etc. The terms thymidine and deoxythymidine may be used interchangeably.

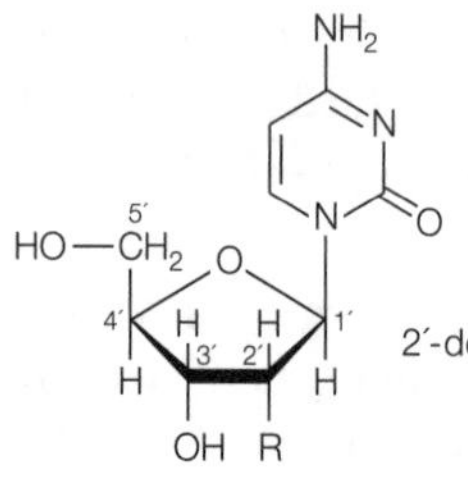

Ribonucleoside (R = OH; cytidine)
2′-deoxyribonucleoside (R = H; deoxycytidine)

Figure 2. Nucleosides.

Nucleotides

A **nucleotide** is a nucleoside with one or more phosphate groups bound covalently to the 3′-, 5′-, or (in some **ribonucleotides** only) the 2′-position. If the sugar is deoxyribose, then the compounds are termed **2′-deoxyribonucleotides**, or just **deoxynucleotides** (Figure 3). Chemically, the compounds are phosphate esters. In the case of the 5′-position, up to three phosphates may be attached, to form, for example, adenosine 5′-triphosphate, or deoxyguanosine 5′-triphosphate, commonly abbreviated to ATP and dGTP respectively. In the same way, we have deoxycytidine triphosphate (dCTP), uridine triphosphate (UTP) and deoxythymidine triphosphate (dTTP; also just called TTP). 5′-mono and -diphosphates are abbreviated as, for example, AMP and dGDP. Nucleoside 5′-triphosphates (NTPs), or deoxynucleoside 5′-triphosphates (dNTPs) are the building blocks of the polymeric nucleic acids. In the course of DNA or RNA synthesis, two phosphates are split off as pyrophosphate to leave one phosphate per nucleotide incorporated

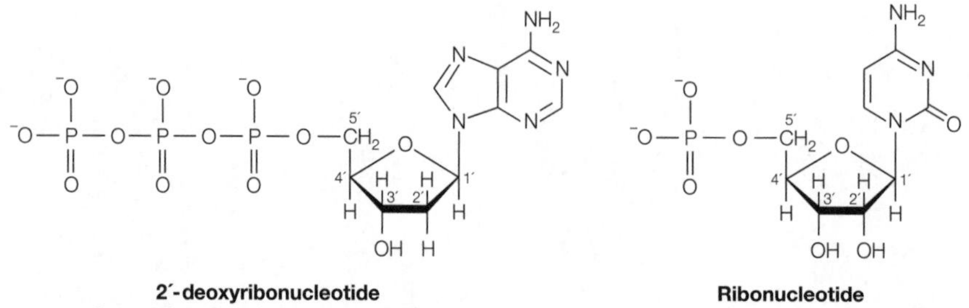

2′-deoxyribonucleotide
Deoxyadenosine 5′-triphosphate (dATP)

Ribonucleotide
Cytidine 5′-monophosphate (CMP)

Figure 3. Nucleotides.

into the nucleic acid chain (Sections D1 and F1). Hence, the repeat unit of a DNA or RNA chain is a nucleotide.

Phosphodiester bonds

In a DNA or RNA molecule, deoxyribonucleotides or ribonucleotides respectively are joined into a polymer by the covalent linkage of a phosphate group between the 5′-hydroxyl of one ribose and the 3′-hydroxyl of the next (Figure 4). This kind of bond or linkage is called a **phosphodiester bond**, since the phosphate is chemically in the form of a diester. Thus, a nucleic acid chain can be seen to have a direction, or **polarity**. Any nucleic acid chain, of whatever length (unless it is circular, Section B3), has a free 5′-end, which may or may not have any attached phosphate groups, and a free 3′-end, which is most likely to be a free hydroxyl group. At neutral pH, each phosphate group has a single negative charge. This is why nucleic acids are termed acids; they are the anions of strong acids. Nucleic acids are thus **highly negatively charged polymers**.

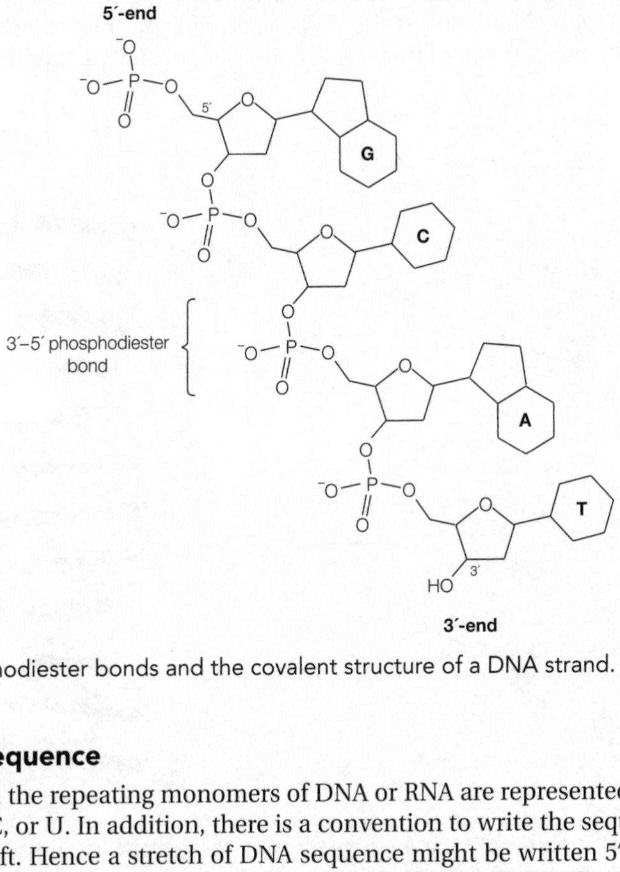

Figure 4. Phosphodiester bonds and the covalent structure of a DNA strand.

DNA/RNA sequence

Conventionally, the repeating monomers of DNA or RNA are represented by their single letters A, T, G, C, or U. In addition, there is a convention to write the sequences with the 5′-end to the left. Hence a stretch of DNA sequence might be written 5′-ATAAGCTC-3′, or even just ATAAGCTC. An RNA sequence might be 5′-AUAGCUUGA-3′. Note that the directionality of the chain means that, for example, ATAAG is not the same as GAATA.

DNA double helix

DNA most commonly occurs in nature as the well-known **double helix**. The basic features of this structure were deduced by James Watson and Francis Crick in 1953. Two separate chains of DNA are wound around each other, each following a helical (coiling) path, resulting in a **right-handed** double helix (Figure 5a). The negatively charged sugar-phosphate backbones of the molecules are on the outside, and the planar bases of each strand stack one above the other in the center of the helix (Figure 5b). Between the backbone strands run the **major** and **minor grooves**, which also follow a helical path. The strands are joined noncovalently by hydrogen bonding between the bases on opposite strands, to form **base pairs** (**bp**). There are around 10 bp/turn in the DNA double helix. The two strands are oriented in opposite directions (**antiparallel**) in terms of their 5′→3′ direction and, most crucially, the two strands are **complementary** in terms of sequence. This last feature arises because the structures of the bases and the constraints of the DNA backbone dictate that the bases hydrogen-bond (Section A3) to each other as purine–pyrimidine pairs, which have very similar geometry and dimensions (Figure 6). Guanine pairs with cytosine (three H-bonds) and adenine pairs with thymine (two H-bonds). Hence, any sequence can be accommodated within a regular double-stranded DNA structure. The sequence of one strand uniquely specifies the sequence of the other, and Watson and Crick were quick to realize that this fact implies an obvious mechanism for the replication of DNA (Section D1). Of course, it also underlies the mechanism of transcription of DNA sequence into RNA (Section F1).

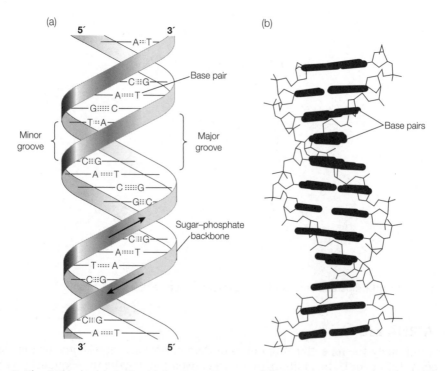

Figure 5. The DNA double helix. (a) A schematic view of the structure; (b) a more detailed structure, highlighting the stacking of the base pairs (in bold).

A, B, and Z helices

In fact, a number of different forms of nucleic acid double helix have been observed and studied, all having the basic pattern of two helically wound antiparallel strands. The structure identified by Watson and Crick, and described above, is known as **B-DNA** (Figure 7a), adenine : thymine   guanine : cytosine

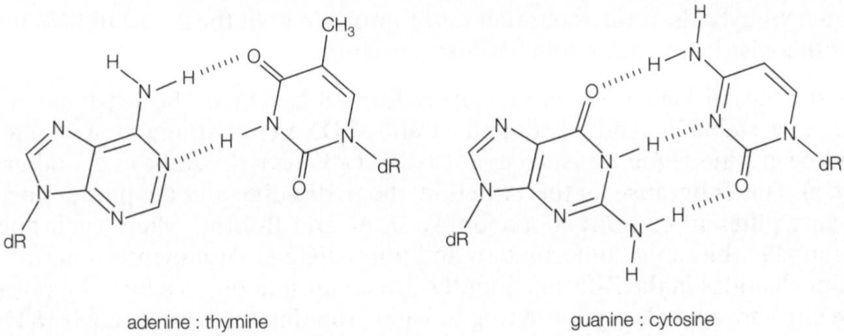

Figure 6. The DNA base pairs. Hydrogen bonds are shown as dashed lines; dR=deoxyribose.

and is believed to be the idealized form of the structure adopted by virtually all DNA *in vivo*. It is characterized by a helical repeat of 10 bp/turn, by the presence of base pairs lying on the helix axis and almost perpendicular to it, and by having well defined, deep major and minor grooves. Actually, real DNA sequences have a helical repeat closer to 10.5 bp/turn and have a variety of other structural distortions that depend on the exact base sequence.

DNA can be induced to form an alternative helix, known as the **A-form** (Figure 7b) under conditions of low humidity. The A-form is right-handed, like the B-form, but has a wider, more compressed structure in which the base pairs are tilted with respect to the helix axis, and actually lie off the axis (seen end-on, the A-helix has a hole down the middle). The helical repeat of the A-form is around 11 bp/turn. Although it may be that the A-form, or something close to it, is adopted by DNA *in vivo* under unusual circumstances, the major importance of the A-form is that it is the helix formed by RNA (see below), and

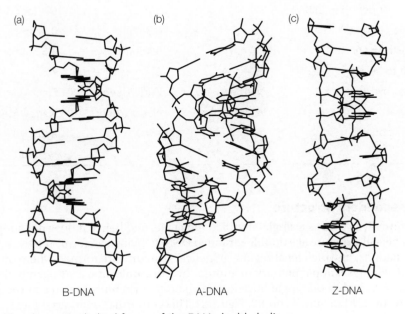

(a)   (b)   (c)

B-DNA   A-DNA   Z-DNA

Figure 7. The alternative helical forms of the DNA double helix.

by DNA–RNA hybrids; it turns out that it is impossible to fit the 2′-OH of RNA into what would otherwise be the more stable B-form structure.

A further unusual helical structure can be formed by DNA. The left-handed **Z-DNA** (Figure 7c) is stable in synthetic double-stranded DNA consisting purely of alternating pyrimidine–purine sequence (such as 5′-CGCGCG-3′, with the same in the other strand, of course). This is because, in this structure, the pyrimidine and the purine nucleotides adopt very different conformations, unlike in A- and B-form, where each nucleotide has essentially the same conformation and immediate environment. In particular, the purine nucleotides in the Z-form adopt the *syn* conformation, in which the purine base lies directly above the deoxyribose ring (imagine rotating the base through 180° around the glycosylic bond in Figure 3; the nucleotides shown there are in the alternative *anti* conformation). The pyrimidine nucleotides in Z-DNA and all nucleotides in the A- and B-forms adopt the *anti* conformation. The Z-helix has a zigzag appearance, with 12 bp/ turn, although it probably makes sense to think of it as consisting of six 'dimers of base pairs' per turn; the repeat unit along each strand is really a dinucleotide. Z-DNA does not easily form in normal DNA, even in regions of repeating CGCGCG, since the boundaries between the left-handed Z-form and the surrounding B-form would be very unstable. The Z-form is probably not a common feature of DNA (or RNA) *in vivo*, although it has been proposed to play a role in the dissipation of torsional stress (Section B3) in DNA during transcription in some specific cases. A comparison of the A, B, and Z helices is shown in Table 1.

Table 1. Summary of the major features of A, B, and Z nucleic acid helices

	A-form	B-form	Z-form
Helical sense	Right handed	Right handed	Left handed
Diameter	~2.6 nm	~2.0 nm	~1.8 nm
Base pairs per helical turn (*n*)	11	10	12 (6 dimers)
Helical twist per bp (= 360/*n*)	33°	36°	60° (per dimer)
Helix rise per bp (*h*)	0.26 nm	0.34 nm	0.37 nm
Helix pitch (= *nh*)	2.8 nm	3.4 nm	4.5 nm
Base tilt to helix axis	20°	6°	7°
Major groove	Narrow/deep	Wide/deep	Flat
Minor groove	Wide/shallow	Narrow/deep	Narrow/deep
Glycosylic bond	*anti*	*anti*	*anti* (pyr)
			syn (pur)

RNA secondary structure

RNA normally occurs as a single-stranded molecule, and hence it does not adopt a long regular helical structure like double-stranded DNA. RNA instead forms relatively globular conformations, in which local regions of helical structure are formed where one part of the RNA chain is complementary to another by **intramolecular** hydrogen bonding and base stacking within the single nucleic acid chain to form **hairpin** and **stem-loop** structures (Sections F1, Figure 3 and K2, Figure 2). This conformational variability is reflected in the more diverse roles of RNA in the cell, when compared with DNA (see below).

Modified nucleic acids

The chemical modification of bases or nucleotides in nucleic acids is widespread, and has a number of specific roles. In cellular DNA, the modifications are restricted to the methylation of the N-6 position of adenine, and the N-4 and 5-positions of cytosine (Figure 1), although more complex modifications occur in some phage DNAs. These methylations have a role in restriction modification (Section O3), base mismatch repair (Section E3) and eukaryotic genome structure and expression (Sections C2 and C3). A much more diverse range of modifications occurs in RNA after transcription, which again reflects the different roles of RNA in the cell. These are considered in more detail in Sections J3 and K2.

Nucleic acid function

DNA functions exclusively as a carrier of genetic information from generation to generation and, in that role, as a template for the synthesis of complementary RNA species. In contrast, although they can function as genomes or templates themselves (e.g. **RNA viruses** and **telomerase RNA**; Sections M4 and D3) and act as intermediates in the flow of information from DNA to protein (**mRNA**; Section A1), RNA molecules are less reliable as permanent stores of information due to their inherent chemical instability (Section B1). However, as they can achieve a wide range of tertiary structures and base-pair to DNA, many RNAs have additional functions similar to proteins. These highly abundant RNAs are called **noncoding RNAs (ncRNAs)** as they are not translated into protein, and their genes are known as **RNA genes**. Many are structural and functional components of the **pre-mRNA** processing and protein synthesis machineries (e.g. **snRNA, tRNA, rRNA**, and **7SL RNA**; Sections J and L), whereas some, known as **ribozymes**, have catalytic activity (e.g. **rRNA** again, and **RNase P**; Sections J2 and L2). Recently, it has become clear that a surprisingly large part of the eukaryotic genome encodes further ncRNAs that are essential for the control of gene expression. These are divided into **lncRNAs** (>200 nt), which are primarily involved in transcriptional control (Section I2), and the smaller (<200 nt) **miRNAs, siRNAs**, and **piRNAs** that are mainly involved in translational control (Section L4), although the size and functional distinctions are not absolute. As RNA has the ability to store genetic information and also catalyze and control chemical reactions, life based on RNA may have predated the existing system of life based on DNA, RNA, and proteins.

A3 Protein structure and function

Key Notes

Amino acid structure

The 20 common amino acids found in proteins have a chiral α-carbon atom linked to a proton, amino and carboxyl groups, and a specific side chain that confers different physical and chemical properties. These side chains may be basic (positively charged), acidic (negatively charged), hydrophobic (both aliphatic and aromatic) or possess other specific functional groups, e.g. hydroxyls, amides or thiols. They behave as zwitterions in solution. With two notable exceptions, nonstandard amino acids in proteins are formed by post-translational modification.

Protein sizes and shapes

Globular proteins, including most enzymes, behave in solution like compact, roughly spherical particles. Fibrous proteins have a high axial ratio and are often of structural importance, for example fibroin and keratin. Sizes range from a few thousand to several million Daltons. Some proteins have associated nonproteinaceous material, for example lipid or carbohydrate or small cofactors.

Primary structure

Amino acids are linked by peptide bonds between α-carboxyl and α-amino groups. The resulting polypeptide sequence has an N-terminus and a C-terminus. Polypeptides commonly have between 100 and 1500 amino acids linked in this way.

Noncovalent interactions

A large number of weak interactions maintain the three-dimensional structure of proteins. Charge–charge, charge–dipole and dipole–dipole interactions involve attractions between fully or partially charged atoms. Hydrogen bonds and hydrophobic interactions that exclude water are also important.

Secondary structure

Polypeptides can fold into a number of regular structures. The right-handed α-helix has 3.6 amino acids per turn and is stabilized by hydrogen bonds between peptide N–H and C=O groups three residues apart. Parallel and antiparallel β-pleated sheets are stabilized by hydrogen bonds between different portions of the polypeptide chain.

Tertiary structure

The different sections of secondary structure and connecting regions fold into a well-defined tertiary structure, with hydrophilic amino acids mostly on the surface and hydrophobic ones in the interior. The structure is stabilized by noncovalent interactions and, sometimes, disulfide

	bonds. Denaturation leads to loss of secondary and tertiary structure.
Quaternary structure	Many proteins have more than one polypeptide subunit. Hemoglobin has two α and two β chains. Large complexes such as microtubules are constructed from the quaternary association of individual polypeptide chains. Allosteric effects usually depend on subunit interactions.
Prosthetic groups	Some proteins have associated nonprotein molecules (prosthetic groups) that provide additional chemical functions to the protein. Small prosthetic groups include nicotinamide adenine dinucleotide (NAD$^+$), heme, and metal ions, for example Zn^{2+}.
Domains, motifs, families, and evolution	Domains form semi-independent structural and functional units within a single polypeptide chain. New proteins can evolve through new combinations of domains. Motifs are groupings of primary or secondary structural elements often found in related members of protein families. Protein families arise through gene duplication and subsequent divergent evolution of the new genes.
Protein function	Proteins have a wide variety of functions. They can act as enzymes, antibodies, structural components inside and outside the cell, receptors and transporters for chemical ligands, regulators, and nutritional stores.
Related topics	(A4) Macromolecular assemblies (Section L) Protein synthesis

Amino acid structure

Proteins are polymers of L-amino acids. Apart from **proline**, all of the 20 amino acids found in proteins have a common structure in which a carbon atom (the α-carbon) is linked to a carboxyl group, a primary amino group, a proton and a **side chain** (R) that is different in each amino acid (Figure 1). Except in **glycine**, the α-carbon atom is asymmetric – it has four chemically different groups attached. Thus, amino acids can exist as pairs of optically active stereoisomers (D- and L-). However, only the L-isomers are found in proteins. As glycine, the simplest amino acid, has a hydrogen atom in place of a side chain, it is optically inactive. Amino acids are dipolar ions (**zwitterions**) in aqueous solution and behave as both acids and bases (they are **amphoteric**). The side chains differ in size, shape, charge, and chemical reactivity, and are responsible for the differences in the properties of different proteins (Figure 2). Many proteins also contain nonstandard

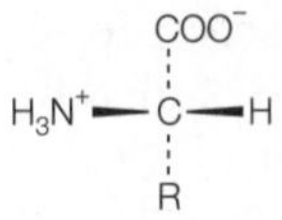

$$\text{H}_3\text{N}^+ \!\!—\!\! \overset{\displaystyle \text{COO}^-}{\underset{\displaystyle \text{R}}{\text{C}}} \!\!—\!\! \text{H}$$

Figure 1. General structure of an L-amino acid. The R group is the side chain.

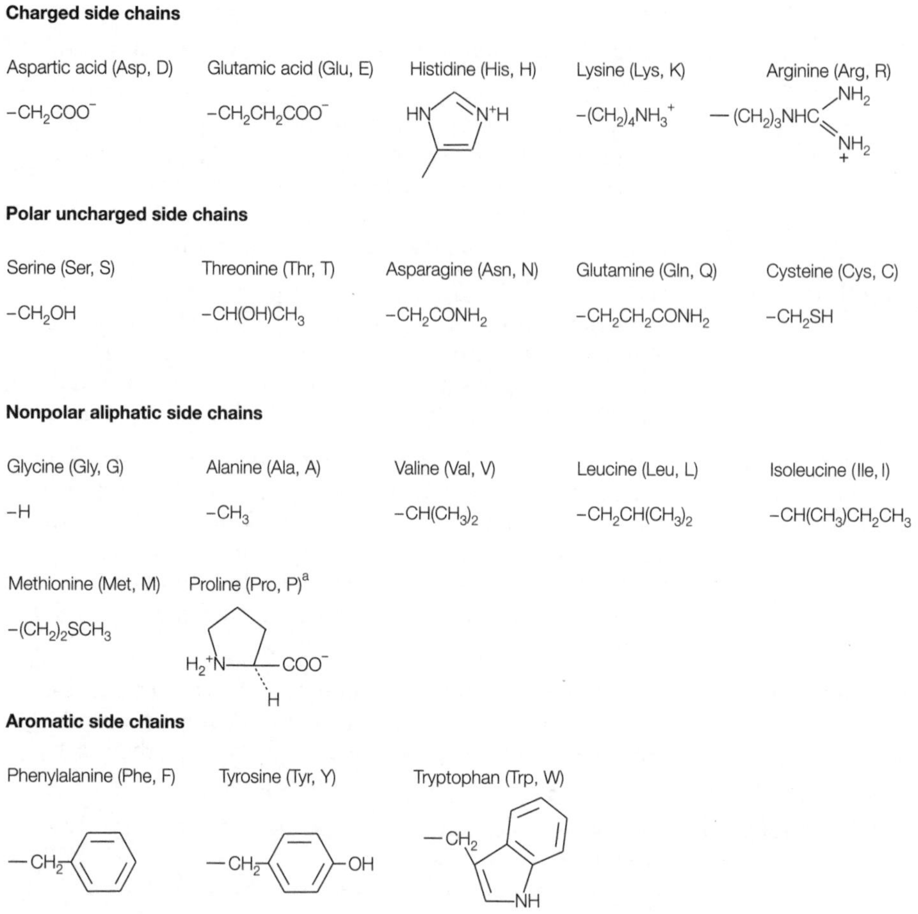

Charged side chains

| Aspartic acid (Asp, D) | Glutamic acid (Glu, E) | Histidine (His, H) | Lysine (Lys, K) | Arginine (Arg, R) |

$-CH_2COO^-$ $-CH_2CH_2COO^-$ $-(CH_2)_4NH_3^+$

Polar uncharged side chains

| Serine (Ser, S) | Threonine (Thr, T) | Asparagine (Asn, N) | Glutamine (Gln, Q) | Cysteine (Cys, C) |

$-CH_2OH$ $-CH(OH)CH_3$ $-CH_2CONH_2$ $-CH_2CH_2CONH_2$ $-CH_2SH$

Nonpolar aliphatic side chains

| Glycine (Gly, G) | Alanine (Ala, A) | Valine (Val, V) | Leucine (Leu, L) | Isoleucine (Ile, I) |

$-H$ $-CH_3$ $-CH(CH_3)_2$ $-CH_2CH(CH_3)_2$ $-CH(CH_3)CH_2CH_3$

Methionine (Met, M)   Proline (Pro, P)[a]

$-(CH_2)_2SCH_3$

Aromatic side chains

Phenylalanine (Phe, F)   Tyrosine (Tyr, Y)   Tryptophan (Trp, W)

Figure 2. Side chains (R) of the 20 common amino acids. The standard three-letter abbreviations and one-letter code are shown in brackets. [a]The full structure of proline is shown as it is a secondary amino acid.

amino acids, such as 4-hydroxyproline and 5-hydroxylysine in collagen. These are mostly formed by **post-translational modification** of the parent amino acids, e.g. proline and **lysine**, in the newly synthesized protein (Section L4). However, **selenocysteine** (found in a number of enzymes), in which Se replaces the S of cysteine, and **pyrrolysine** (a modified lysine found only in certain archaeal proteins) are both incorporated into growing protein chains by a subtle manipulation of the genetic code (Section K1) and are regarded by some as the 21st and 22nd 'standard' amino acids.

Taking pH 7 as a reference point, several amino acids have ionizable groups in their side chains that provide an extra positive or negative charge at this pH. The 'acidic' amino acids, **aspartic acid** and **glutamic acid**, have additional carboxyl groups that are usually ionized (negatively charged). The 'basic' amino acids have positively charged groups – **lysine** has a second amino group attached to the ε-carbon atom while **arginine** has a guanidino group. The imidazole group of **histidine** has a pK_a near neutrality. Reversible protonation of this group under physiological conditions contributes to the catalytic mechanism of many enzymes. Together, acidic and basic amino acids can form important salt bridges in proteins.

Polar, uncharged side chains contain groups that form hydrogen bonds with water. Together with the charged amino acids, they are often described as **hydrophilic** ('water-loving'). **Serine** and **threonine** have hydroxyl groups that can be reversibly phosphorylated by protein kinases (see below) while **asparagine** and **glutamine** are the amide derivatives of aspartic acid and glutamic acid. **Cysteine** has a **thiol** (sulfhydryl) group, which often oxidizes to **cystine**, in which two cysteines form a structurally important disulfide bond.

Phenylalanine, **tyrosine** (which can also be phosphorylated), and **tryptophan** have bulky **hydrophobic** ('water-hating') side chains that participate in **hydrophobic interactions** in protein structure (see below). The aromatic structures of tyrosine and tryptophan account for most of the ultraviolet (UV) absorbance of proteins, which absorb maximally at 280 nm. The phenolic hydroxyl group of tyrosine can also form hydrogen bonds. Other nonpolar, hydrophobic side chains include the aliphatic alkyl groups of **alanine**, **valine**, **leucine**, **isoleucine**, and **methionine**, which contains a **sulfur** atom in a thioether link, and the cyclic ring of proline, which is unusual in being a secondary amino (or **imino**) acid.

Protein sizes and shapes

Two broad classes of protein may be distinguished. **Globular proteins** are folded compactly and behave in solution more or less as spherical particles; most enzymes are globular in nature. **Fibrous proteins** have very high axial ratios (length/width) and are often important structural proteins, for example silk fibroin and keratin in hair and wool. Molecular masses can range from a few thousand Daltons (Da), e.g. the hormone insulin with 51 amino acids and a molecular mass of 5734 Da (5.7 kiloDaltons, kDa), to nearly 4 million Da (4 MDa) in the case of the muscle protein titin. Some proteins contain bound **nonproteinaceous** material, either in the form of small **prosthetic groups**, which may act as cofactors in enzyme reactions, or as large associations (e.g. the lipids in **lipoproteins** or the carbohydrate in **glycoproteins**, Section A4).

Primary structure

The α-carboxyl group of one amino acid is covalently linked to the α-amino group of the next amino acid by an amide bond, commonly known as a **peptide bond** when in proteins. When two amino acid **residues** are linked in this way the product is a **dipeptide**. Many amino acids linked by peptide bonds form a **polypeptide** (Figure 3). The repeating sequence of α-carbon atoms and peptide bonds provides the structural **backbone** of the polypeptide while the different amino acid **side chains** confer functionality on the protein. The amino acid at one end of a polypeptide has an unattached α-amino group while the one at the other end has a free α-carboxyl group. Hence, polypeptides are directional, with an **N-terminus** and a **C-terminus**. Sometimes the N-terminus is **blocked** with, for example, an acetyl group. The sequence of amino acids from the N- to

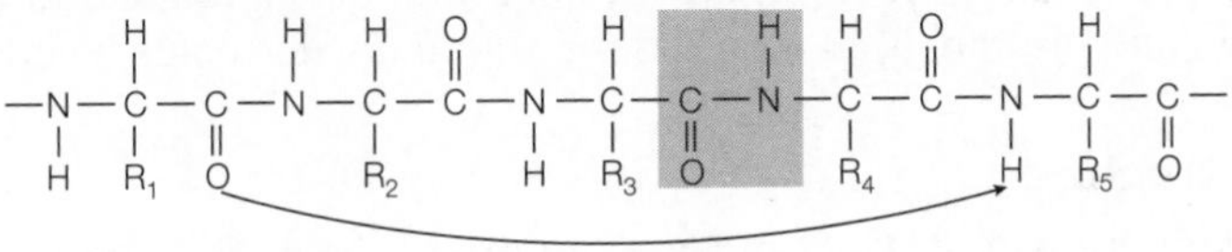

Figure 3. Section of a polypeptide chain. The peptide bond is boxed. In the α-helix, the CO group of amino acid residue n is hydrogen-bonded to the NH group of residue $n+4$ (arrowed).

the C-terminus is the **primary structure** of the polypeptide. Typical sizes for single poly-peptide chains are within the range 100–1500 amino acids, although shorter and longer ones exist, e.g. titin has around 34,000.

Noncovalent interactions

The three-dimensional structure of proteins and large, protein-containing assemblies (Section A4) is maintained by many different noncovalent interactions. Electrostatic **charge–charge** interactions (**salt bridges**) operate between ionizable groups of opposite charge at physiological pH, e.g. between positive lysine and arginine side chains and negative glutamic acid and aspartic acid side chains or the negative phosphates of DNA in DNA-binding proteins such as histones (Section C2). **Charge–dipole** and **dipole–dipole** interactions are weaker and form when either or both of the participants is a dipole because of the asymmetric distribution of charge in the molecule (Figure 4a). Even uncharged groups like methyl groups can attract each other weakly through transient dipoles arising from the motion of their electrons (**dispersion forces**).

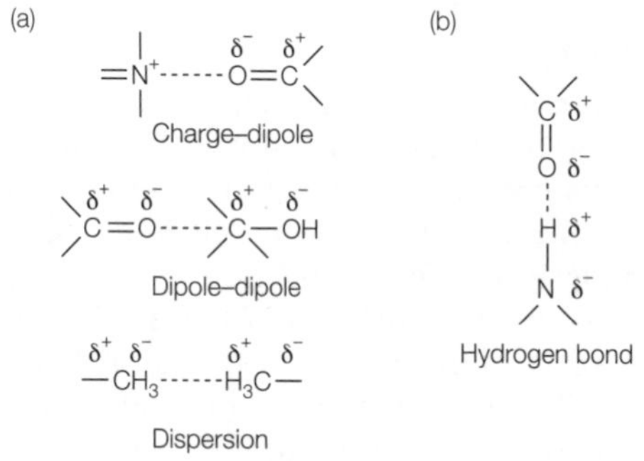

Figure 4. Examples of (a) van der Waals forces and (b) a hydrogen bond.

Noncovalent associations between electrically neutral molecules are known collectively as **van der Waals forces**. **Hydrogen bonds** are of great importance. They form between a covalently bonded hydrogen atom on a donor group (e.g. –O-H or –N-H) and a pair of nonbonding electrons on an acceptor group (e.g. :O=C– or :N–) (Figure 4b). Hydro-gen bonds and other interactions involving dipoles are directional in character and so help define macromolecular shapes and the specificity of molecular interactions. The presence of uncharged and nonpolar substances, e.g. lipids, in an aqueous environ-ment tends to force a highly ordered structure on the surrounding water molecules. This is energetically unfavorable, as it reduces the entropy of the system. Hence, nonpolar molecules and functional groups such as the aliphatic and aromatic amino acid side chains tend to clump together, reducing the overall surface area exposed to water. This attraction is termed a **hydrophobic** interaction and is a major stabilizing force in protein–protein and protein–lipid interactions and in nucleic acids (Section A2).

Secondary structure

The highly polar nature of the C=O and N–H groups of the peptide bonds gives the C–N bond partial double bond character. This makes the peptide bond unit rigid and planar, though there is free rotation between adjacent peptide bonds. This polarity also favors hydrogen bond formation between appropriately spaced and oriented peptide bond units. Thus, polypeptide chains are able to fold into a number of regular structures that are held together by these hydrogen bonds. The best-known **secondary structure** is the **α-helix** (Figure 5a). The polypeptide backbone forms a right-handed helix with 3.6 amino acid residues per turn such that each peptide N–H group is hydrogen bonded to the C=O group of the peptide bond three residues away (Figure 3). Sections of α-helical secondary structure are often found in globular proteins and in some fibrous proteins. The rarer **3₁₀-helix** is similar, but with different dimensions. The **β-pleated sheet** (β-sheet) is formed by hydrogen bonding of the peptide bond N–H and C=O groups to the complementary groups of another section of the polypeptide chain (Figure 5b). Several sections of polypeptide chain may be involved side-by-side, giving a sheet structure with the side chains (R) projecting alternately above and below the sheet. If these sections run in the same direction (e.g. N-terminus→C-terminus), the sheet is **parallel**; if they alternate N→C and C→N, then the sheet is **antiparallel**. **Mixed β-sheets** comprising both orientations are also found. β-Sheets are strong and rigid and are important in structural proteins, for example silk fibroin. The connective tissue protein **collagen** has an unusual **triple helix** secondary structure in which three polypeptide chains are intertwined, making it very strong.

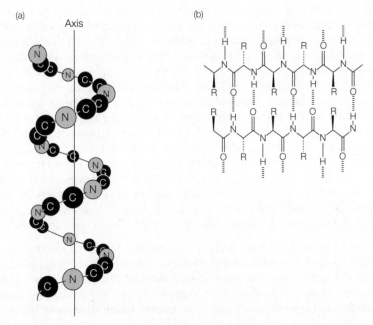

Figure 5. (a) α-Helix secondary structure. Only the α-carbon and peptide bond carbon and nitrogen atoms of the polypeptide backbone are shown for clarity. (b) Section of a β-sheet secondary structure.

Tertiary structure

The way in which the different sections of α-helix, β-sheet, other minor secondary structures and connecting, unstructured loops fold in three dimensions is the **tertiary structure** of the polypeptide (Figure 6). The nature of the tertiary structure is inherent in the primary structure and, given the right conditions, most polypeptides will fold

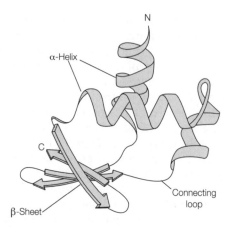

Figure 6. Schematic diagram of a section of protein tertiary structure.

spontaneously into the correct tertiary structure, as it is generally the lowest energy conformation for that sequence. However, proper folding *in vivo* is usually promoted by proteins called **chaperones,** which help prevent misfolding of new polypeptides before their synthesis (and primary structure) is complete. Folding is such that amino acids with hydrophilic side chains locate mainly on the exterior of the protein, where they can inter-act with water or solvent ions, whereas the hydrophobic amino acids become buried in the interior, from which water is excluded. This gives overall stability to the structure. Various types of noncovalent interaction between side chains hold the tertiary struc-ture together: van der Waals forces, hydrogen bonds, electrostatic salt bridges between oppositely charged groups, and hydrophobic interactions between nonpolar side chains. In addition, covalent disulfide bonds can form between two cysteine residues that may be far apart in the primary structure but close together in the folded tertiary structure. Disruption of secondary and tertiary structure by heat or extremes of pH leads to **dena-turation** of the protein and formation of a **random coil** conformation. Any associated biological activity is usually lost and the denatured proteins tend to clump into insoluble aggregates, as their exposed hydrophobic interiors interact to exclude water.

The importance of correct protein folding is illustrated by the fact that many neurode-generative disorders such as **Alzheimer's disease** are associated with the accumulation of insoluble protein aggregates in neurons called **amyloid fibrils**, which contain extensive β-sheet structure. Also, the fatal diseases **scrapie** (sheep), **bovine spongiform enceph-alopathy** (**BSE**, 'mad cow disease') and **Creutzfeldt–Jakob disease** (**CJD**, humans) are caused by an infectious agent called a **prion,** which consists solely of a misfolded protein. When the prion enters a neuron it binds to a related cellular protein causing it to misfold, thus setting off a chain reaction of misfolding and amyloid formation, leading to loss of cell function.

Quaternary structure

Many proteins are composed of two or more polypeptide chains (**subunits**) form-ing **oligomers** – a few subunits, or **multimers** – many subunits. The subunits may be identical (**homomers**), or different (**heteromers**). For example, **hemoglobin** has two α-globin and two β-globin chains ($\alpha_2\beta_2$). The same forces that stabilize tertiary structure hold subunits together, including in some cases disulfide bonds between cysteines on separate polypeptides. This level of organization is known as the **quaternary structure** and has certain consequences. First, it allows very large protein structures to be made, e.g. the **microtubules** of the cytoskeleton (Section A4, Figure 1). Secondly, it can provide greater functionality to a protein by combining different activities into a single entity, as in DNA polymerase III holoenzyme and the replisome (Section D2). Often, the interactions between the subunits are modified by the binding of small molecules and this can lead to the **allosteric** effects seen in enzyme regulation. There are also many examples of transient protein complexes, particularly in cell signaling pathways, where a post-translational modification (such as phosphorylation, Section L4) of one protein causes it to briefly associate with another, often resulting in a conformational change in the second protein that switches its function on or off.

Prosthetic groups

Many proteins contain covalently or noncovalently attached small molecules called **prosthetic groups** that give structural or chemical functionality to the protein that the amino acid side chains cannot provide. Many of these are **cofactors** in enzyme-catalyzed reactions. Examples are nicotinamide adenine dinucleotide (NAD^+) in many dehydrogenases, pyridoxal phosphate in transaminases, heme in hemoglobin and cytochromes, metal ions such as Zn^{2+}, and fatty acyl groups that can anchor proteins in cell membranes through hydrophobic interactions. Such proteins are termed **conjugated** proteins and the protein without its prosthetic group is known as an **apoprotein**. Other conjugated proteins contain associated macromolecules in large complexes such as carbohydrate (**glycoproteins**), lipid (**lipoproteins**), or nucleic acid (**nucleoproteins**) (Section A4).

Domains, motifs, families, and evolution

Many individual polypeptides are composed of structurally independent units, or **domains**, that are connected by sections with limited higher order structure within the same polypeptide. The connections can act as hinges to permit the individual domains to move in relation to each other, and breakage of these connections by limited proteolysis can often separate the domains, which can then behave like independent globular proteins. The active site of an enzyme is sometimes formed in a groove between two domains, which wrap around the substrate. Domains can also have a specific function such as binding a commonly used molecule, for example ATP. When such a function is required in many different proteins, the same domain structure is often found. In eukaryotes, domains are often encoded by discrete parts of genes called **exons** (Section J3). Therefore, it has been suggested that during evolution, new proteins were created by the duplication and rearrangement of domain-encoding exons in the genome to produce new combinations of binding sites, catalytic sites, and structural elements in the resulting new polypeptides. In this way, the rate of evolution of new functional proteins may have been greatly increased.

Structural motifs (also known as **supersecondary structures**) are groupings of secondary structural elements that frequently occur in globular proteins. They often have functional significance and can represent the essential parts of binding or catalytic sites that have been conserved during the evolution of protein families from a common ancestor. Alternatively, they may represent the best solution to a structural–functional requirement that has been arrived at independently in unrelated proteins. A common example is the βαβ **motif**, in which the connection between two consecutive parallel strands of a β-sheet is an α-helix (Figure 7). Two overlapping βαβ motifs (βαβαβ) form a dinucleotide (e.g. NAD^+) binding site in many otherwise unrelated proteins. **Sequence motifs** consist

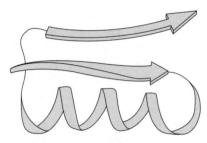

Figure 7. Representation of a βαβ motif. The α-helix is shown as a coiled ribbon and the β-sheet segments as flat arrows.

of linear sequences of conserved, functionally important amino acids, i.e. primary structure, rather than supersecondary structure. They can also represent binding or catalytic sites.

Protein families arise through successive duplications and subsequent **divergent evolution** of an ancestral gene. Myoglobin, the oxygen-carrying protein in muscle, the α- and β-globin chains and the minor δ-(delta) chain of adult hemoglobin and the γ- (gamma), ε- (epsilon) and ζ- (zeta) globins of embryonic and fetal hemoglobins are all related polypeptides within the **globin family** (Figure 8). Their genes, and the proteins, are said to be **homologs**. Family members in different species that have retained the same function and carry out the same biochemical role (e.g. rat and mouse myoglobin) are **orthologs** while those that have evolved different but often related functions (e.g. α-globin and β-globin) are **paralogs** (Section S6). The degree of similarity between the amino acid sequences of orthologous members of a protein family in different organisms depends on how long ago the two organisms diverged from their common ancestor and on how important conservation of the sequence is for the function of the protein. This function, whether structural or catalytic, is inherently related to its structure. As indicated above, similar structures and functions can also be achieved by **convergent evolution** whereby unrelated genes evolve to produce proteins with similar structures or catalytic activities. A good example is provided by the proteolytic enzymes **subtilisin** (bacterial) and **chymotrypsin** (animal).

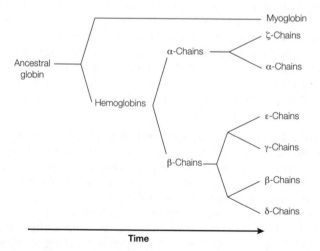

Figure 8. Evolution of globins from an ancestral globin gene.

Even though their amino acid sequences are very different and they are composed of different structural motifs, they have evolved the same spatial orientation of the **catalytic triad** of active site amino acids – serine, histidine, and aspartic acid – and use exactly the same catalytic mechanism to hydrolyse peptide bonds. Such proteins are termed **functional analogs.** Where similar structural motifs have evolved independently, the resulting proteins are **structural analogs**.

Protein function

- **Enzymes:** Apart from a few catalytically active RNA molecules (Section J2), all enzymes are proteins. These can enhance the rate of biochemical reactions by several orders of magnitude. Binding of the **substrate** involves various noncovalent interactions with specific amino acid side chains, including van der Waals forces, hydrogen bonds, salt bridges, and hydrophobic forces. Specificity of binding can be extremely high, with only a single substrate binding (e.g. glucose oxidase binds only glucose), or it can be group-specific (e.g. hexokinase binds a variety of hexose sugars). Side chains can also be directly involved in catalysis, for example by acting as nucleophiles, or proton donors, or abstractors.

- **Signaling:** Receptor proteins in cell membranes can bind **ligands** (e.g. hormones) from the extracellular medium and, by virtue of the resulting conformational change, initiate reactions within the cell in response to that ligand. Ligand binding is similar to substrate binding but the ligand usually remains unchanged. Some hormones are themselves small proteins, such as insulin and growth hormone. **Protein kinases**, which modify the properties of other proteins by adding a phosphoryl group from ATP to them, are extremely important enzymes in intracellular signaling.

- **Transport and storage: Hemoglobin** transports oxygen in the red blood cells while **transferrin** transports iron to the liver. Once in the liver, iron is stored bound to the protein **ferritin**. Dietary fats are carried in the blood by **lipoproteins**. Many other molecules and ions are transported and stored in a protein-bound form. This can enhance solubility and reduce reactivity until they are required.

- **Structure and movement: Collagen** is the major protein in skin, bone, and connective tissue, whereas hair is made mainly from **keratin**. There are also many structural proteins within the cell, for example in the **cytoskeleton**. The major muscle proteins **actin** and **myosin** form sliding filaments, which are the basis of muscle contraction.

- **Nutrition: Casein** and **ovalbumin** are the major proteins of milk and eggs, respectively, and are used to provide the amino acids for growth of developing offspring. Seed proteins also provide nutrition for germinating plant embryos.

- **Immunity: Antibodies**, which recognize and bind to bacteria, viruses and other foreign material (the **antigen**), are proteins.

- **Regulation: Transcription factors** bind to and modulate the function of DNA. Many other proteins modify the functions of other molecules by binding to them.

A4 Macromolecular assemblies

Key Notes

Large protein complexes

Proteins can associate with each other in extremely large structures. The eukaryotic cytoskeleton consists of various such complexes including microtubules (made of tubulin), microfilaments and muscle fibers (containing actin and myosin), cilia, and flagella. These organize the shape and movement of cells and subcellular organelles.

Conjugated proteins

Glycoproteins and proteoglycans (mucoproteins) are proteins with covalently attached carbohydrate and are generally found on extracellular surfaces and in extracellular spaces. Lipoproteins are used to transport lipids in aqueous environments. Mixed macromolecular complexes such as these provide a wider range of functions than the component parts.

Nucleoproteins

Bacterial 70S ribosomes comprise a large 50S subunit, with 23S and 5S ribosomal RNA (rRNA) molecules and 31 proteins, and a small 30S subunit, with a 16S rRNA molecule and 21 proteins. Eukaryotic 80S ribosomes have 60S (28S, 5.8S, and 5S rRNAs) and 40S (18S rRNA) subunits. Chromatin contains DNA and the basic histone proteins. Viruses are also nucleoprotein complexes.

Membranes

Membrane phospholipids and sphingolipids form bilayers with the polar groups on the exterior surfaces and the hydrocarbon chains in the interior. Membrane proteins may be peripheral or integral and act as receptors, enzymes, transporters, or mediators of cellular interactions.

Related topics

(A3) Protein structure and function

(C2) Chromatin structure

(J1) rRNA processing and ribosomes

(Section M) Bacteriophages and eukaryotic viruses

Large protein complexes

Few macromolecules work in isolation as monomers. They generally associate with other macromolecules of the same or a different class in stable or weak, transient complexes to carry out their functions. Some of these can be extremely large. For example, many of the major structural and locomotory elements of the cell consist of large protein complexes. The **cytoskeleton** is an array of protein filaments that organizes the shape and motion of cells and the intracellular distribution of subcellular organelles. **Microtubules** are 200- to 25,000-nm-long polymers of tubulin, a 110-kDa globular protein, which is itself a dimer of distinct α and β subunits (Section A3) (Figure 1). These are a major component of the cytoskeleton, the **mitotic spindle** (Section C3), and of eukaryotic **cilia** and **flagella**, the hair-like structures on the surface of many cells that whip to move the cell or to move fluid across the cell surface. Cilia also contain the proteins nexin and dynein. **Microfilaments** consisting of the protein **actin** form huge contractile assemblies with the protein **myosin** to cause cytoplasmic motion. Actin and myosin are also major components of muscle fibers.

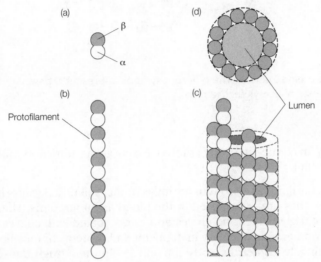

Figure 1. The structure of a microtubule: (a) tubulin consists of α- and β-subunits; (b) a tubulin protofilament consisting of many adjacent subunits; (c) the microtubule is formed from 13 protofilaments aligned in parallel; (d) cross-section of the hollow microtubule. From Hames and Hooper (2011) *Instant Notes Biochemistry*, 4th edn, Garland Science.

Conjugated proteins

Some conjugated proteins (Section A3) comprise associations of proteins with one or more of the other major classes of large biomolecules – carbohydrates, lipids, and nucleic acids. This can greatly increase the functionality or structural capabilities of the resulting complex.

Glycoproteins contain both protein and carbohydrate (between <1% and >90% of the weight) components; **glycosylation** is the commonest form of **post-translational modification** of proteins (Section L4). The carbohydrate is always covalently attached to the surface of the protein, never the interior, and is often variable in composition, causing microheterogeneity (Figure 2). This has made glycoproteins difficult to study. Glycoproteins have functions that span the entire range of protein activities, and are usually found extracellularly, either as secreted proteins or embedded in the plasma membrane (see below) where they can mediate cell–cell recognition or function as **receptors**. **Antibodies** and several protein hormones are glycoproteins.

Proteoglycans (**mucoproteins**) are large complexes (>10^7 Da) of protein and **mucopolysaccharides** (**glycosaminoglycans**) and are important components of the **extracellular matrix**, the material that binds and organizes cells in tissues. Their sugar units often have sulfate groups (e.g. **chondroitin sulfate**, **heparan sulfate**), which makes them highly charged and hydrated. This, coupled with their lengths (>1000 units), produces solutions

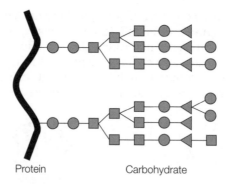

Protein   Carbohydrate

Figure 2. Glycoprotein structure. The different symbols represent different monosaccharide units (e.g. galactose, *N*-acetylglucosamine).

of high viscosity. In connective tissue, proteoglycans act as lubricants and shock absorbers in the extracellular space.

In **lipoproteins**, the lipids and proteins are linked noncovalently. Because lipids are poorly soluble in water, they are transported in the blood as lipoproteins. These are basically large particles of triglycerides and cholesterol esters coated with a layer of phospholipids, cholesterol and protein and vary in diameter and protein composition from 500 nm and 1–2% protein (**chylomicrons**) to 10 nm and 35% protein (**high-density lipoprotein, HDL**). The structures of the **apolipoproteins** (protein part without lipid) are such that their hydrophobic amino acids face towards the lipid interior of the particles while the charged and polar amino acids (Section A3) face outwards into the aqueous environment. This renders the particles soluble.

Nucleoproteins

Nucleoproteins, comprising both nucleic acid and protein, provide particularly important examples of conjugated proteins in molecular biology. Small **ribonucleoproteins** include **telomerase** (Sections C3 and D3) and the ribozyme **ribonuclease P** (Section J2), both of which contain RNA. **Ribosomes** are much larger ribonucleoprotein complexes and are the sites of protein synthesis in the cytoplasm (Section L). Bacterial 70S ribosomes have large (50S) and small (30S) subunits with a total mass of 2.5×10^6 Da. (The **S value**, e.g. 50S, is the numerical value of the **sedimentation coefficient**, s, and describes the rate at which a macromolecule or particle sediments in a centrifugal field. It is determined by both the mass and shape of the molecule or particle; hence S values are not additive.) The 50S subunit contains 23S and 5S ribosomal RNA (**rRNA**) molecules and 31 different proteins while the 30S subunit contains a 16S rRNA and 21 proteins. Under the correct conditions, mixtures of the rRNAs and proteins will self-assemble in a precise order into functional ribosomes *in vitro*. Thus, all the information for ribosome structure is inherent in the structures of the components. The rRNAs are not simply frameworks for the assembly of the ribosomal proteins, but participate in both the binding of the messenger RNA and in the catalysis of peptide bond synthesis (Section L2).

Chromatin is the material from which eukaryotic chromosomes are made. It is a **deoxyribonucleoprotein** complex made up of roughly equal amounts of DNA and small, basic proteins called **histones** that together form a discrete, repeating unit called a **nucleosome** (Section C2). Histones neutralize the repulsion between the negative charges of the DNA sugar–phosphate backbone and allow the DNA to be tightly packaged within the chromosomes but are also of great functional importance in the control of gene expression. **Bacteriophages** and **viruses** are another example of nucleoprotein complexes. They are discussed in Section M.

Membranes

When placed in an aqueous environment, phospholipids and sphingolipids naturally form a **lipid bilayer** with the polar groups on the outside and the nonpolar hydrocarbon chains on the inside. This is the structural basis of all biological membranes. Such membranes form cellular and organellar boundaries and are selectively permeable to uncharged molecules. The precise lipid composition varies from cell to cell and from organelle to organelle. Proteins are also a major component of cell membranes (Figure 3). **Peripheral** membrane proteins are loosely bound to the outer surface or are anchored via a lipid or **glycosyl phosphatidylinositol** anchor and are relatively easy to remove. **Integral membrane proteins** are embedded in the membrane and cannot be removed without destroying the membrane. Some protrude from the outer or inner surface of the membrane while **transmembrane proteins** span the bilayer completely and have both extracellular and intracellular **domains** (Section A3). The transmembrane regions of these proteins contain predominantly hydrophobic amino acids. Many membrane proteins are also glycoproteins and have a variety of functions, for example:

- Receptors for signaling molecules such as hormones and neurotransmitters

- Enzymes for degrading extracellular molecules before uptake of the products

- Pores or channels for the selective transport of small, polar ions and molecules

- Mediators of cell–cell interactions (mainly glycoproteins)

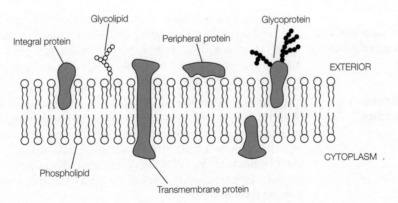

Figure 3. Schematic diagram of a plasma membrane showing the major macromolecular components.

A5 Analysis of proteins

Key Notes

Recombinant protein production

Purification of a protein from its native source for analysis is often inefficient and laborious and so they are now generally produced by inserting the gene encoding the protein into an expression vector and expressing in a suitable host cell. Sequence tags added to the recombinant protein simplify purification and detection, and aid solubility. Correct post-translational modifications may require a eukaryotic host–vector system.

Protein sequencing

After breaking a polypeptide down into smaller peptides using specific proteases, the peptides can be sequenced by chemical methods or by mass spectrometry. The original sequence is recreated from the overlaps produced by cleavage with proteases of different specificities. However, prediction of a protein sequence from the sequence of its gene or complementary DNA (cDNA) is simpler.

Biophysical methods

Many proteins can be crystallized and their three-dimensional structures determined by X-ray diffraction. The structures of small proteins in solution can also be determined by multi-dimensional nuclear magnetic resonance, particularly if the normal ^{12}C and ^{14}N are substituted by ^{13}C and ^{15}N. Many other techniques are available to study protein structures and interactions.

Mass determination and mass spectrometry

Approximate molecular masses can be obtained by gel electrophoresis in the presence of sodium dodecyl sulfate. Mass spectrometry using electrospray ionization gives masses that are accurate to within 0.01%. Mass spectrometry also detects post-translational modifications.

Applications for antibodies

Antibodies are proteins produced by the immune system of vertebrates in response to a foreign agent (the antigen), such as an injected protein. Their high binding affinities and specificities for the protein antigens make them useful laboratory tools for the detection and analysis of proteins by immunofluorescence, western blotting and immunoprecipitation.

Functional analysis

Functional analysis of a protein involves its isolation and study in vitro combined with a study of the behavior of a mutant organism in which the protein has been rendered nonfunctional by mutation or deletion of its gene. The function of a new protein can sometimes be predicted by comparing its sequence and structure to those of known proteins.

Related topics	(A2) Nucleic acid structure and function (A3) Protein structure and function	(S3) Proteomics (S6) Bioinformatics

Recombinant protein production

A typical eukaryotic cell may contain thousand of different proteins, some abundant and some present in only a few copies. In order to analyze the structure and function of a protein, it must be separated from other proteins and nonprotein molecules and purified in sufficient quantity. Classical protein purification methods include ion-exchange, gel filtration, hydrophobic interaction, and affinity chromatography, as well as various forms of electrophoresis, but these are beyond the scope of this book. Such methods tend to give low yields and <100% purity, and so isolation of a protein from its native source has now been largely supplanted by **recombinant protein** production. A full appreciation of this process requires an understanding of the cloning procedures described in Sections O and P.

Briefly, the gene or complementary DNA (cDNA) encoding the protein is inserted into an **expression vector**, usually a modified plasmid or viral DNA (Section P), which is then introduced into a host cell such as the bacterium *Escherichia coli*, which uses its transcription and translation machinery to synthesize the protein. By engineering the appropriate control sequences into the vector, such as a strong promoter (Section F3), the recombinant protein can be synthesized to form up to 30% of the total protein of the cell. Purification can be simplified by adding a **purification tag** sequence, such as six histidine codons (Section K1), to the 5'- or 3'-end of the cloned gene. In this case, the protein is synthesized with six histidine residues at the N- or C-terminus, which allows one-step purification on an **affinity column** containing immobilized metal ions such as Ni^{2+} or Co^{2+}, which bind the histidines. Untagged proteins pass straight through the column and the pure, recombinant protein can then be eluted from the column with a solution of histidine. Fortunately, the tag usually has little effect on the structure and function of the protein.

Often, a eukaryotic protein will not express well in *E. coli* because of **codon usage bias** (Section K1) and so problematic codons may need to be changed to the favored *E. coli* codons by **site-directed mutagenesis** (Section R5) or a specialized host strain of *E. coli* used that expresses a more eukaryotic pattern of transfer RNAs (tRNAs). 'Foreign' proteins expressed in *E. coli* are frequently insoluble, either because they lack the required post-translational modifications, or because the host chaperones are overloaded or inappropriate (Sections A3 and L4). They may form cytoplasmic aggregates called **inclusion bodies**, from which they can sometimes be resolubilized using urea. Solubility can often be improved by expressing the recombinant as a **fusion protein** with a more soluble partner. This involves placing the coding sequences of the two proteins next to each other in the vector and deleting the termination codon of the upstream partner so that both are translated as a single polypeptide (Section L1). Common fusion partners are **thioredoxin, maltose-binding protein** (**MBP**) and the enzyme **glutathione-S-transferase** (**GST**). GST and MBP can also be used as purification tags using affinity columns containing immoblized **glutathione**, the tripeptide substrate of GST, or maltose for MBP. Other tags that may be included in the protein by encoding them in the vector include an **epitope tag** for antibody-based detection (see below) and a tag encoding the recognition sequence for a proteolytic enzyme such as Factor Xa, which cleaves proteins specifically after the sequence IleGluGlyArg. If placed between the two partners of a fusion protein, this tag allows them to be separated by proteolysis after purification.

Sometimes expression in *E. coli* just fails, and many eukaryotic proteins require expression in a eukaryotic host–vector system to achieve the correct folding and post-translational modifications (Sections P3 and L4). As well as producing proteins for laboratory analysis, recombinant methods are now routinely used to produce therapeutic proteins such as insulin and blood clotting factors, with over 130 different proteins currently approved for treatment.

Protein sequencing

An essential requirement for understanding how a protein works is a knowledge of its primary structure. The amino acid composition of a protein can be determined by hydrolyzing all the peptide bonds with strong acid and separating the resulting amino acids by chromatography. This indicates how many glycines and serines, etc. there are but does not give the actual sequence. Early methods of sequence determination involved splitting the protein into a number of smaller peptides using specific proteolytic enzymes or chemicals that break only certain peptide bonds. For example, trypsin cleaves only after lysine or arginine, and V8 protease only after glutamic acid. Cyanogen bromide cleaves polypeptides after methionine residues. Each peptide is then subjected to sequential **Edman degradation** in an automated protein sequencer. Phenylisothiocyanate reacts with the N-terminal amino acid, which, after acid treatment, is released as the phenylthiohydantoin (PTH) derivative, leaving a new N-terminus. The PTH-amino acid is identified by chromatography by comparison with standards and the cycle repeated to identify the next amino acid, and so on. The order of the peptides in the original protein can be deduced by sequencing peptides produced by proteases with different specificities and looking for the overlapping sequences. However, this method is both laborious and expensive, and it is now much easier to 'sequence' proteins indirectly by sequencing the DNA of the gene or cDNA (Section R2) and deducing the protein sequence using the genetic code (Sections K1 and S6). However, this misses post-transcriptional (e.g. mRNA editing, Section J4) and post-translational (Section L4) modifications, and so there is still a need for direct protein sequencing. This is now achieved by mass spectrometry (see below). In organisms whose genome has not been fully sequenced, partial protein sequencing can be used to provide information for the construction of an oligonucleotide **probe** (Section Q3), which is then used to find the corresponding gene or cDNA, from which the full protein sequence can then be deduced.

Biophysical methods

Several methods are available to determine the secondary and tertiary structures, and the physical properties of proteins. **Circular dichroism**, a form of UV spectroscopy, determines the relative proportions of different secondary structures and is useful for measuring changes in protein conformation (shape) under different conditions. Because they have such well-defined tertiary structures, many globular proteins have been crystallized and so the tertiary structure can be determined by **X-ray crystallography**. X-rays interact with the electrons in the matter through which they pass. By measuring the pattern of diffraction of a beam of X-rays as it passes through a crystal, the positions of the atoms in the crystal can be calculated. By crystallizing an enzyme in the presence of its substrate, the precise intermolecular interactions responsible for binding and catalysis can be seen. The power and resolution of modern X-ray crystallography using high intensity **synchrotron radiation** is such that the detailed structures of large macromolecular assemblies like nucleosomes and ribosomes have now been determined.

The structures of small globular proteins in solution can also be determined by two- or three-dimensional **nuclear magnetic resonance** (**NMR**) **spectroscopy**. In NMR, the relaxation of protons is measured after they have been excited by the radiofrequencies in a strong magnetic field. The properties of this relaxation depend on the relative positions of the protons in the molecule. The multi-dimensional approach is required for proteins to spread out and resolve the overlapping data produced by the large number of protons. Substituting ^{13}C and ^{15}N for the normal isotopes ^{12}C and ^{14}N in the protein also greatly improves data resolution by eliminating unwanted resonances. In this way, the detailed structures of proteins up to about 40 kDa in size can be deduced, while partial information on larger proteins can also be obtained. NMR is particularly useful for proteins that do not crystallize readily because they contain unstructured, flexible regions. **Solid state NMR** is used to analyze insoluble membrane proteins. Where both X-ray and NMR methods have been used to determine the structure of a protein, the results usually agree well. This suggests that the measured structures are the true *in vivo* structures.

Cryo-electron microscopy is performed at extremely low temperatures and has lower resolution than X-ray and NMR techniques but can provide structures of much larger entities, e.g. viruses and organelles, whereas **isothermal titration calorimetry** and **surface plasmon resonance** are used to measure the thermodynamic and kinetic parameters of protein–ligand interactions, including protein–protein and protein–DNA.

Mass determination and mass spectrometry

The mass of individual polypeptide chains can be determined by electrophoresis through a polyacrylamide gel in the presence of the ionic detergent sodium dodecyl sulfate (**SDS polyacrylamide gel electrophoresis**, **SDS-PAGE**). SDS binds to, denatures, and imparts a negative charge to polypeptides, so all move towards the anode during electrophoresis at a rate that depends on their mass. Masses are determined by reference to known standards. Denaturation disrupts quaternary structure, so multimeric proteins are split into individual subunits. SDS-PAGE is cheap and easy though not particularly accurate (5–20% error). **Mass spectrometry** (**MS**) offers an extremely accurate method. A **mass spectrometer** consists of an **ion source** that generates characteristic, multiply-charged ionic fragments in the gas phase from the sample molecule, a **mass analyzer** that measures the mass-to-charge ratio (*m/z*) of the ionized sample, and a **detector** that counts the numbers of ions of each *m/z* value. For small molecule analysis, samples are traditionally vaporized and ionized by a beam of Xe or Ar atoms. The degree of deflection of the various ions in an electromagnetic field is mass dependent and can be measured, giving a **mass spectrum** (or '**fingerprint**') that identifies the original molecule. However, such methods have an upper mass limit of only a few kDa and are too destructive for protein analysis, so non-destructive ionization techniques are necessary to extend the mass range.

An **electrospray ionization** (**ESI**) ion source creates positively charged (protonated) ions of individual protein molecules by creating then vaporizing a fine spray of highly charged droplets of the protein solution, whereas a **matrix-assisted laser desorption/ionization** (**MALDI**) ion source generates gas-phase ions by the laser vaporization of the sample contained in a solid bed of one of several chemicals (the 'matrix'). These ion sources can be coupled to a variety of different mass analyzers and detectors, each suited to a different purpose. An ESI source is commonly attached to a **quadrupole** analyzer. This uses a set of four parallel rods to produce a time-varying electric field that filters ions of different *m/z* values, allowing them to arrive and be counted individually at the detector. A MALDI source is commonly coupled to a **time-of-flight** (**TOF**) analyzer. The ions are accelerated along a flight tube and separate according to their velocities. The detector then counts the different ions as they arrive. A system set up in this way is called **MALDI-TOF** and is most commonly used for protein identification by **peptide mass fingerprinting** of individual peptides generated by prior proteolysis of the protein of interest (Section S3).

For mass determination, the highest quality data are obtained from an **ESI-Q-TOF** mass spectrometer with combined quadrupole and TOF mass analyzers. Because proteins have multiple sites to carry a proton (all lysine, arginine, and histidine residues), they acquire multiple positive charges, and in a slightly unpredictable fashion. Thus a protein with 20 basic residues might be charged with between eight and 18 protons. Because mass spectrometers analyze ions according to the *m/z* ratio, each differently protonated variant creates a different signal (the mass, *m*, stays almost the same but the charge, *z*, increases in integral values of +1, +2, etc.). From the profile of *m/z* values, it is possible to calculate the molecular weight of the protein. The precision of this method is around 0.01%, so the measured mass of a 50-kDa protein would be accurate to about ±5 Da. This method has also been used to study protein complexes in the MDa range.

Applications for antibodies

Antibodies are useful molecular tools for investigating protein structure and function. Antibodies are themselves glycoproteins and are generated by the immune system of higher animals when they are injected with a macromolecule (the **antigen**) such as a protein that is not native to the animal. Their physiological function is to bind to antigens on the surface of invading viruses and bacteria as part of the animal's response to kill and eliminate the infectious agent. Antibodies fall into various classes, but all have the same Y-shaped structure comprising two **heavy chains** and two **light chains**, linked by disulfide bonds (Figure 1). The most useful are the immunoglobulin G (IgG) class, produced as soluble proteins by B lymphocytes. IgGs have a very high affinity and specificity for the corresponding antigen and can be used to detect and quantitate the antigen in cells and cell extracts. The specificity lies in the variable region of the molecule, which is generated by recombination (Section E4). This region recognizes and binds to a short sequence of

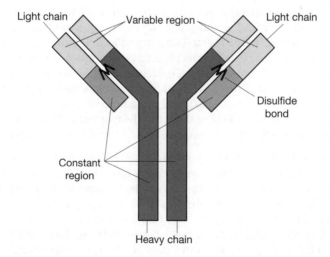

Figure 1. Structure of an antibody molecule.

5-8 amino acids on the surface of the antigen (an **epitope**). Usually, a single antigen elicits the production of several different antibodies by different B cell clones, each of which recognizes a different epitope on the antigen (a **polyclonal** antibody mixture). However, it is possible to isolate the B cell clones (usually from mice) and grow them in culture after fusion to cancer cells; the resulting **hybridoma** cell lines secrete **monoclonal** antibodies, which bind individual epitopes and so tend to be more specific than polyclonals. Antibodies can be used to detect specific proteins in cells by **immunocytochemical** techniques, particularly **immunofluorescence** (Section S4). They can also be used to detect proteins in cell extracts after separation by SDS-PAGE, which separates polypeptides according to their molecular mass (Figure 2). (Note that the detergent SDS splits multimeric proteins into their individual polypeptide subunits.) After separation, the polypeptides are transferred from the gel to a membrane in a procedure similar to Southern blotting (Section R1). This so-called **western blot** is incubated with an antibody that is specific for and binds to the polypeptide of interest (the **primary** antibody), followed by a **second antibody,** several molecules of which recognize and bind the first antibody (the use of a labeled second antibody has several practical advantages, including signal amplification). As the second antibody has attached to it an enzyme or chemical that can generate a color or a light signal, the position of the polypeptide on the blot can be visualized using a suitable detector. This can show, for example, if a protein is present or absent, or increases or decreases in a cell under particular conditions, e.g. hormone stimulation. If it moves position, this could indicate a modification, perhaps partial degradation.

In theory, detection of different proteins requires a different antibody in each case, However, recombinant proteins can be engineered and expressed with an **epitope tag** at one end. This is a short sequence of extra amino acids (an epitope) that is recognized by just one antibody, thus allowing this same antibody to be used to detect any protein

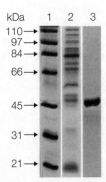

Figure 2. Sodium dodecyl sulfate polyacrylamide gel electrophoresis (SDS-PAGE) and western blot. A set of molecular weight standards (lane 1) and a sample of rat liver extract (lane 2) were boiled in buffer containing SDS, which denatures the proteins and imparts a mass-dependent negative charge to them. The proteins were then separated according to their mass by electrophoresis in a polyacrylamide gel and visualized by staining the gel with the dye Coomassie Blue. The masses of the standards are show in kiloDaltons (kDa). An unstained liver sample identical to lane 2 was blotted on to a nitrocellulose membrane and incubated with a rabbit IgG antibody specific for the enzyme nucleoside triphosphatase (NTPase). This was then incubated with a goat anti-rabbit immunoglobulin G (IgG), which had previously been covalently linked to the enzyme peroxidase. Finally, the blot was treated with a peroxidase substrate that generates a visible color. The presence of the 46-kDa NTPase in the liver extract can clearly be seen (lane 3).

containing the appropriate tag (Section S4). This saves a great deal of time and expense. Another important use for antibodies is in the detection of protein–DNA and protein–protein interactions by **immunoprecipitation** (Sections S2 and S3).

Functional analysis

As a result of high throughput structural analysis (Section A3), the three-dimensional structures of many proteins have now been determined. This structural information is of great value in the rational development of new drugs designed to bind specifically and with high affinity to target proteins and, in conjunction with functional analysis, permits **structure–function relationships** to be established. However, structural determination still lags well behind the availability of new protein primary structures predicted from genome sequencing projects (Sections R2 and S6). Thus, there is great interest in computational methods that will allow the prediction of both structure and possible function from simple amino acid sequence information. These methods involve both the mapping of new protein sequences on to the known three-dimensional structures of template proteins with related amino acid sequences and also *ab initio* **protein modeling** that attempts to build three-dimensional structures from first principles without relying on known template structures (Section S6). However, at least for the moment, understanding the true function of a protein still requires both biochemical and genetic analysis. Recombinant methods allow the production of virtually any protein that has been identified from its DNA sequence so that its properties can be studied *in vitro* (see above and Section P1). Identification of all the other proteins with which a protein interacts in the cell is another important aspect of functional analysis (Section S3). If the gene for the protein can be inactivated by mutagenesis or deleted by recombinant DNA techniques, then the **phenotype** of the resulting mutant can be studied. In conjunction with the biochemical information, the altered behavior of the mutant cell can help to pinpoint the function of the protein *in vivo*. All 6000 or so protein-coding genes of the yeast *Saccharomyces cerevisiae* have been individually deleted to produce a set of mutants that should help to define the role of all the proteins in this relatively simple eukaryote. Similarly, a set of over 20,000 mutant strains of the model plant *Arabidopisis thaliana*, each defective in a different gene, has been created by **transposon mutagenesis** (Section E4), whereas the use of **transgenic** and **knockout mice** will eventually help to define the function of all mammalian proteins (Section S5). Finally, a simple, alternative approach to mutagenesis is the suppression of individual gene expression using **siRNA**, which can yield 'phenotypic' mutants in virtually any biological system (Sections J2 and S2).

B1 Chemical and physical properties of nucleic acids

Key Notes

Stability of nucleic acids	Although it might seem obvious that DNA double strands and RNA structures are stabilized by hydrogen bonding, this is really not the case. H-bonds help to determine the specificity of the base pairing, but the stability of a nucleic acid helix is mainly the result of hydrophobic and dipole–dipole interactions between the stacked base pairs.
Effect of acid	Highly acidic conditions may hydrolyze nucleic acids to their components: bases, sugar, and phosphate. Moderate acid causes the hydrolysis of the purine base glycosylic bonds to yield apurinic acid. More complex chemistry can be used to remove particular bases.
Effect of alkali	High pH denatures DNA and RNA by altering the tautomeric state of the bases and disrupting specific hydrogen bonding. RNA is also susceptible to hydrolysis at high pH, by participation of the 2′-OH in intramolecular cleavage of the phosphodiester backbone.
Chemical denaturation	Some chemicals, such as urea and formamide, can denature DNA and RNA at neutral pH by disrupting the hydrophobic forces between the stacked bases.
Viscosity	DNA is very long and thin, and DNA solutions have a high viscosity. Long DNA molecules are susceptible to cleavage by shearing in solution – this process can be used to generate DNA of a specific average length.
Buoyant density	DNA has a density of around 1.7 g cm^{-3}, and can be analyzed and purified by its ability to equilibrate at its buoyant density in a cesium chloride density gradient formed in a centrifuge. The exact density of DNA is a function of its G+C content, and this technique may be used to analyze DNAs of different composition.
Related topics	(A2) Nucleic acid structure and function (B2) Spectroscopic and thermal properties of nucleic acids

---

Stability of nucleic acids

At first sight, it might seem that the double helices of DNA and RNA secondary structure are stabilized by the hydrogen bonding between base pairs. In fact, this is not the case. As in proteins (Section A3), the presence of H-bonds within a structure does not normally confer stability. This is because one must consider the *difference* in energy between, in the case of DNA, the single-stranded random coil state, and the double-stranded conformation. H-bonds between base pairs in double-stranded DNA (dsDNA) merely replace what would be equally strong and energetically favorable H-bonds with water molecules in free solution, if the DNA were single-stranded (ssDNA). Hydrogen bonding, along with the shapes of the bases, contributes to the specificity required for base pairing in a double helix (dsDNA will only form if the strands are complementary; Section A2), but it does not contribute to the overall stability of that helix. The root of this stability lies elsewhere, in the **stacking interactions** between the base pairs (Section A2, Figure 5b). The flat surfaces of the aromatic bases cannot hydrogen-bond to water when they are in free solution, in other words they are **hydrophobic**. The hydrogen-bonding network of bulk water becomes destabilized in the vicinity of a hydrophobic surface, since not all the water molecules can participate in full hydrogen bonding interactions, and they become more ordered. Hence it is energetically favorable to exclude water altogether from pairs of such surfaces by stacking them together; more water ends up in the bulk hydrogen-bonded network. This also maximizes the interaction between **charge dipoles** (Section A3) on the bases. Even in ssDNA, the bases have a tendency to stack on top of each other, but this stacking is maximized in dsDNA, and the **hydrophobic effect** ensures that this is the most energetically favorable arrangement. In fact, this is a simplified discussion of a rather complex phenomenon; a fuller explanation is beyond the scope of this book.

Effect of acid

In strong acid and at elevated temperatures, for example perchloric acid ($HClO_4$) at more than 100°C, nucleic acids are **hydrolyzed** completely to their constituents: bases, ribose or deoxyribose, and phosphate. In more dilute acid, for example at pH 3–4, the most easily hydrolyzed bonds are selectively broken. These are the glycosylic bonds attaching the purine bases to the ribose ring, and hence the nucleic acid becomes **apurinic** (Section E1). More complex chemistry can remove bases with some degree of specificity, and this formed the basis for an early method of DNA sequencing; however, this has now been replaced by better methods (Section R2).

Effect of alkali

DNA

Increasing pH above the physiological range (pH 7–8) has more subtle effects on DNA structure. The effect of alkali is to change the tautomeric state of the bases. This effect can be seen with reference to the model compound, cyclohexanone (Figure 1a). The molecule is in equilibrium between the tautomeric keto and enol forms (1) and (2). At neutral pH, the compound is predominantly in the keto form (1). Increasing the pH causes the molecule to lose a proton and shift to the enolate form (3), since the negative charge is most stably accommodated on the electronegative oxygen atom. In the same way, the structure of guanine (Figure 1b) is also shifted to the enolate form at increased pH, and analogous shifts take place in the structures of the other bases as the pH becomes higher. This affects the specific hydrogen bonding between the base pairs and introduces negative charges in the hydrophobic environment of the stacked bases, with the result that the double-stranded structure of the DNA breaks down; that is the DNA becomes **denatured** (Figure 1c).

RNA

In RNA, the same denaturation of helical regions will take place at higher pH, but this effect is overshadowed by the susceptibility of RNA to **hydrolysis** in alkali. This comes

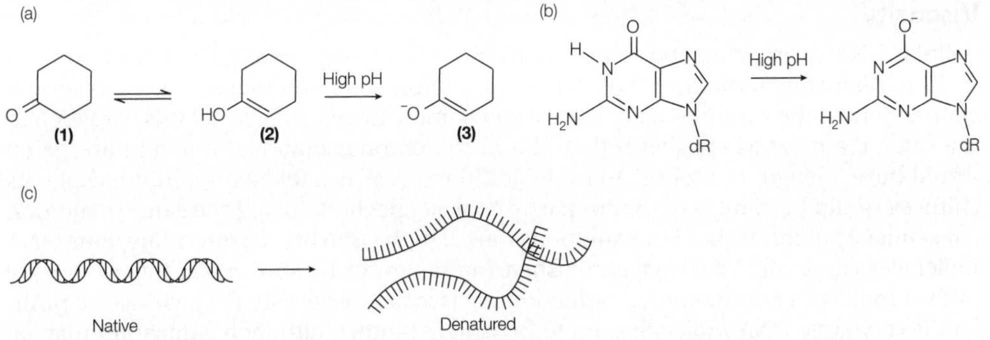

Figure 1. The denaturation of DNA at high pH. (a) Alkali shifts the tautomeric ratio to the enolate form; (b) the tautomeric shift of deoxyguanosine; (c) the denaturation of double-helical DNA.

about because of the presence of the 2′-OH group in RNA, which is perfectly positioned to participate in the cleavage of the RNA backbone by intramolecular attack on the phosphate of the phosphodiester bond (Figure 2). This reaction is promoted by high pH, since -OH acts as a general base, tending to remove the proton from the 2′-OH and making it more nucleophilic. The products are a free 5′-OH and a **2′,3′-cyclic phosphodiester**, which is subsequently hydrolyzed to either the 2′- or 3′-monophosphate. Even at neutral pH, RNA is much more susceptible to hydrolysis than DNA, which of course lacks the 2′-OH. This is a plausible reason why DNA has evolved to incorporate 2′-deoxyribose, since its function requires extremely high stability.

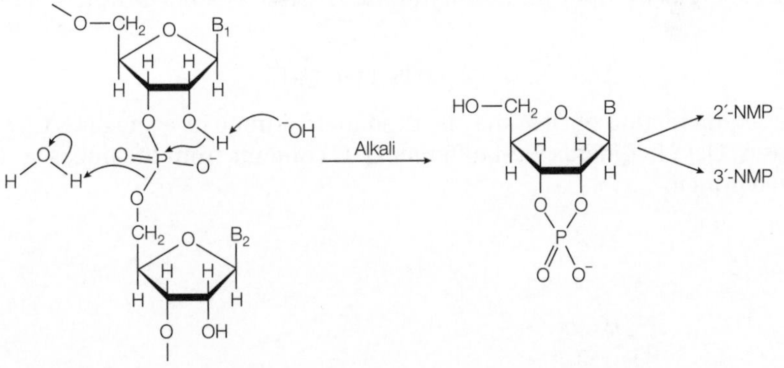

Figure 2. Intramolecular cleavage of RNA phosphodiester bonds in alkali.

Chemical denaturation

A number of chemical agents can cause the denaturation of DNA or RNA at neutral pH, the best known examples being **urea** (H_2NCONH_2) and **formamide** ($HCONH_2$). A relatively high concentration of these agents (several molar) has the effect of disrupting the hydrogen bonding of the bulk water solution. This means that the energetic stabilization of the nucleic acid secondary structure, caused by the exclusion of water from between the stacked hydrophobic bases, is lessened and the strands become denatured.

Viscosity

Cellular DNA is very long and thin; technically, it has a high **axial ratio**. DNA is around 2 nm in diameter, and may have a length of micrometers, millimeters, or even several centimeters in the case of eukaryotic chromosomes. To give a flavor of this, if DNA had the same diameter as spaghetti, then the *E. coli* chromosome (4.6 million base pairs) would have a length of around 1 km. In addition, DNA is a relatively stiff molecule; its stiffness would be similar to that of partly cooked spaghetti, using the same analogy. A consequence of this is that DNA solutions have a **high viscosity**. Furthermore, long DNA molecules can easily be damaged by **shearing** forces, or by **sonication** (high-intensity ultrasound), with a concomitant reduction in viscosity. Sensitivity to shearing is a problem if very large DNA molecules are to be isolated intact, although sonication may be used to produce DNA of a specified average length (Section C4). Note that neither shearing nor sonication denatures the DNA; they merely reduce the length of the double-stranded molecules in the solution.

Buoyant density

Analysis and purification of DNA can be carried out according to its density. In solutions containing high concentrations of a high molecular weight salt, for example 8 M **cesium chloride** (CsCl), DNA has a similar density to the bulk solution, around 1.7 g cm^{-3}. If the solution is centrifuged at very high speed, the dense cesium salt tends to migrate down the tube, setting up a **density gradient** (Figure 3). Eventually the DNA sample will migrate to a sharp band at a position in the gradient corresponding to its own **buoyant density**. This technique is known as **equilibrium density gradient centrifugation** or **isopycnic** (Greek for 'same density') **centrifugation**. Since, under these conditions, RNA pellets at the bottom of the tube and protein floats, this can be an effective way of purifying DNA away from these two contaminants (Section O2). However, the method is also analytically useful, since the precise buoyant density of the DNA (ρ) is a linear function of its **G+C content**:

$$\rho = 1.66 + 0.098 \times \mathrm{Frac(G+C)}$$

Hence, the sedimentation of DNA may be used to determine its average G+C content or, in some cases, DNA fragments with different G+C contents from the bulk sequence can be separated from it.

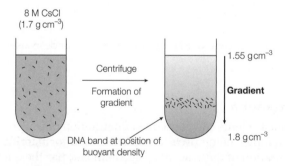

Figure 3. Equilibrium density gradient centrifugation of DNA.

B2 Spectroscopic and thermal properties of nucleic acids

Key Notes

UV absorption	The aromatic bases of nucleic acids absorb light with a λ_{max} of 260 nm.
Absorbance and structure	The extinction coefficient of nucleic acid bases depends on their environment. The absorbance of isolated nucleotides is greater than that of RNA and ssDNA, which is in turn greater than that of dsDNA.
Quantitation of nucleic acids	The absorbance at 260 nm is used to determine the concentration of nucleic acids. At a concentration of 1 mg ml^{-1} and 1 cm pathlength, dsDNA has A_{260}=20. RNA and ssDNA have A_{260}=25. The values for RNA and ssDNA depend on base composition and secondary structure.
Purity of DNA	The A_{260}/A_{280} ratio of a dsDNA sample can be used to assess its purity. For pure DNA, the value is 1.8. Values above 1.8 suggest RNA contamination and those below 1.8 suggest protein contamination.
Thermal denaturation	Increased temperature can bring about the denaturation of DNA and RNA. RNA denatures gradually on heating, but dsDNA 'melts' cooperatively to give single strands at a defined temperature, T_m, which is a function of the G+C content of the DNA. Denaturation may be detected by the change in A_{260}.
Hybridization	DNA denaturation is reversed on cooling, but fully double-stranded native DNA will only form (renature) if the cooling is sufficiently slow to allow the complementary strands to anneal, or hybridize.
Related topics	(A2) Nucleic acid structure and function (C4) Genome complexity (B1) Chemical and physical properties of nucleic acids

---

UV absorption

Nucleic acids absorb UV light due to the conjugated aromatic nature of the bases; the sugar-phosphate backbone does not contribute appreciably to absorption. The wavelength of maximum absorption of light by both DNA and RNA is 260 nm (λ_{max}=260 nm), which is conveniently distinct from the λ_{max} of protein (280 nm) (Section A3). The absorption properties of nucleic acids can be used for detection, quantitation, and assessment of purity.

Absorbance and structure

Although the λ_{max} for DNA or RNA bases is constant, the extinction coefficient and hence the absorption depends on the environment of the bases. The absorbance at 260 nm (A_{260}) is greatest for isolated nucleotides, intermediate for ssDNA or RNA, and least for dsDNA. This effect is caused by the fixing of the bases in a hydrophobic environment by stacking (Sections A2 and B1). There is a classical, although perhaps rather archaic, term for this change in absorbance – **hypochromicity**, i.e. dsDNA is **hypochromic** (from the Greek for 'less colored') relative to ssDNA. Alternatively, ssDNA may be said to be **hyperchromic** when compared with dsDNA.

Quantitation of nucleic acids

It is not generally convenient to consider the molar extinction coefficient (ε) of a nucleic acid, since its value will depend on the length of the molecule in question; instead, extinction coefficients are normally quoted in terms of concentration in mg ml^{-1}: 1 mg ml^{-1} dsDNA has an A_{260} of 20. The corresponding value for RNA or ssDNA is approximately 25. The values for ssDNA and RNA are approximate for two reasons: they arise from the sum of absorbances contributed by the different bases (purines have a higher extinction coefficient than pyrimidines), and hence are sensitive to the base composition of the molecule; and dsDNA has equal numbers of purines and pyrimidines, and so does not show this effect. The absorbance values also depend on the amount of secondary structure (double-stranded regions) in a given molecule, due to the effect of base stacking. In the case of a short oligonucleotide, where secondary structure will be minimal, it is usual to calculate the extinction coefficient based on the sum of values for individual nucleotides in the single strand.

Purity of DNA

The approximate purity of dsDNA preparations (Section O2) may be estimated by determination of the ratio of absorbance at 260 and 280 nm (A_{260}/A_{280}). The shape of the absorption spectrum (Figure 1), as well as the extinction coefficient, varies with the environment of the bases such that pure dsDNA has an A_{260}/A_{280} of 1.8, and pure RNA one of around 2.0. Protein, of course, with λ_{max}=280 nm has a A_{260}/A_{280} ratio of less than 1 (actually around

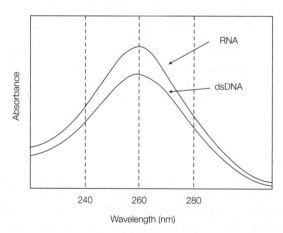

Figure 1. The ultraviolet absorption spectra of RNA and double-stranded DNA solutions at equal concentration (mg ml^{-1}).

0.5). Hence, if a DNA sample has A_{260}/A_{280} greater than 1.8, this suggests RNA contamination, whereas a value less than 1.8 suggests protein in the sample.

Thermal denaturation

A number of chemicals can bring about the denaturation of nucleic acids (Section B1). Heating also leads to the destruction of double-stranded hydrogen-bonded regions of DNA and RNA, as thermal motion in the strands increases. The process of denaturation can be observed conveniently by the increase in absorbance as double-stranded nucleic acids are converted to single strands (Figure 2). The thermal behaviors of dsDNA and RNA are very different. As the temperature is increased, the absorbance of an RNA sample gradually and erratically increases as the stacking of the bases in double-stranded regions is reduced. Shorter regions of base pairing will denature before longer regions, since they are more thermally mobile. In contrast, the **thermal denaturation**, or **melting**, of dsDNA is **cooperative**. The denaturation of the ends of the molecule, and of more mobile AT-rich internal regions, will destabilize adjacent regions of helix, leading to a progressive and concerted melting of the whole structure at a well-defined temperature corresponding to the mid-point of the smooth transition, and known as the **melting temperature** (T_m) (Figure 2a). The melting is accompanied by a 40% increase in absorbance. T_m is a function of the G+C content of the DNA sample, and ranges from 80°C to 100°C for long DNA molecules.

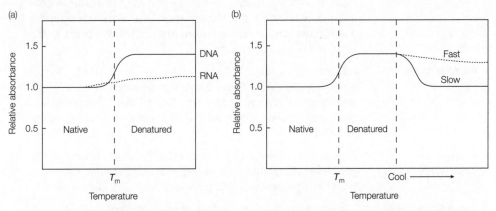

Figure 2. (a) Thermal denaturation of double-stranded DNA (cooperative) and RNA; (b) renaturation of DNA by fast and slow cooling.

Hybridization

The thermal denaturation of DNA may be reversed by cooling the solution. In this case, the rate of cooling has an influence on the outcome. Rapid cooling allows only the formation of local regions of dsDNA, formed by the base pairing or **annealing** of short regions of complementarity within or between DNA strands; hence the decrease in A_{260} is rather small (Figure 2b). On the other hand, slow cooling allows time for the wholly complementary DNA strands to find each other, and the sample can become fully double-stranded, with the same absorbance as the original native sample. This annealing or **renaturation** of regions of complementarity between different nucleic acid strands is known as **hybridization**. Hybridization is crucial to the analysis of specific sequences of DNA, since just about the only way to detect a specific DNA sequence is by using the corresponding complementary sequence as a 'probe' (Sections Q3 and S2).

B3 DNA supercoiling

<table>
<tr><td colspan="2">Key Notes</td></tr>
<tr><td>Closed-circular DNA</td><td>DNA frequently occurs in nature as closed-circular molecules, where the two single strands are each circular and linked together. The number of links is known as the linking number (Lk).</td></tr>
<tr><td>Supercoiling</td><td>Supercoiling is the coiling of the DNA axis upon itself, caused by a change in the linking number from $Lk°$, the value for a relaxed closed circle. Most natural DNA is negatively supercoiled, that is the DNA is deformed in the direction of unwinding of the double helix.</td></tr>
<tr><td>Topoisomer</td><td>A circular dsDNA molecule with a specific value of linking number is a topoisomer. Its linking number may not be changed without first breaking one or both strands.</td></tr>
<tr><td>Twist and writhe</td><td>Supercoiling is partitioned geometrically into a change in twist, the local winding up or unwinding of the double helix, and a change in writhe, the coiling of the helix axis upon itself. Twist and writhe are interconvertible according to the equation $\Delta Lk = \Delta Tw + \Delta Wr$.</td></tr>
<tr><td>Intercalators</td><td>Intercalators such as ethidium bromide bind to DNA by inserting themselves between the base pairs (intercalation), resulting in the local untwisting of the DNA helix. If the DNA is closed-circular, then there will be a corresponding increase in writhe.</td></tr>
<tr><td>Energy of supercoiling</td><td>Negatively supercoiled DNA has a high torsional energy, which facilitates the untwisting of the DNA helix and can help to drive processes that require the DNA to be unwound.</td></tr>
<tr><td>Topoisomerases</td><td>Topoisomerases alter the level of supercoiling of DNA by transiently breaking one or both strands of the DNA backbone. Type I enzymes change Lk by ±1; type II by ±2. DNA gyrase introduces negative supercoiling using the energy derived from ATP hydrolysis, and topoisomerase IV and eukaryotic topoisomerase IIs unlink (decatenate) daughter chromosomes.</td></tr>
<tr><td>Related topics</td><td>(C1) Prokaryotic chromosome structure
(C2) Chromatin structure
(C3) Eukaryotic chromosome structure
(D2) Bacterial DNA replication</td></tr>
</table>

Closed-circular DNA

Many DNA molecules in cells consist of **closed-circular** double-stranded molecules, for example bacterial plasmids and chromosomes, and many viral DNA molecules. This means that the two complementary single strands are each joined into circles, 5′ to 3′, and are twisted around one another by the helical path of the DNA. The molecule has no free ends, and the two single strands are linked together a number of times corresponding to the number of double-helical turns in the molecule. This number is known as the **linking number** (**Lk**).

Supercoiling

A number of properties arise from this circular constraint of a DNA molecule. A good way to imagine these is to consider the DNA double helix as a piece of rubber tubing with a line drawn along its length to enable us to follow its twisting. The tubing may be joined by a connector into a closed circle (Figure 1). If we imagine a twisting of the DNA helix (tubing) followed by the joining of the ends, then the deformation so formed is locked into the system (Figure 1). This deformation is known as **supercoiling**, since it manifests itself as a coiling of the DNA axis around itself in a higher-order coil, and corresponds to a change in linking number from the simple circular situation. If the twisting of the DNA is in the same direction as that of the double helix, i.e. the helix is twisted up before closure, then the supercoiling formed is **positive**; if the helix is untwisted, then the supercoiling is **negative**. Almost all DNA molecules in cells are on average negatively supercoiled. This is true even for linear DNAs such as eukaryotic chromosomes, which are constrained into large loops by interaction with a protein scaffold (Sections C2 and D3).

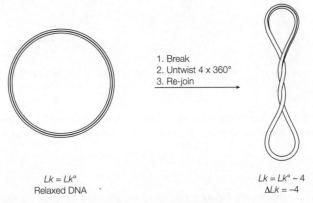

1. Break
2. Untwist 4 x 360°
3. Re-join

$Lk = Lk°$
Relaxed DNA

$Lk = Lk° - 4$
$\Delta Lk = -4$

Figure 1. A rubber tubing model for DNA supercoiling, which illustrates the change in DNA conformation. The changes in linking number are indicated (see text for details).

The level of supercoiling may be quantified in terms of the change in linking number (ΔLk) from that of the unconstrained (**relaxed**) closed-circular molecule ($Lk°$). This corresponds to the number of 360° twists introduced before ring closure. DNA isolated from cells is commonly negatively supercoiled by around six turns per 100 turns of helix (1000 bp), i.e. $\Delta Lk/Lk° = -0.06$.

Topoisomer

The linking number of a closed-circular DNA is a **topological** property, i.e. one that cannot be changed without breaking one or both of the DNA backbones. A molecule of a given linking number is known as a **topoisomer**. Topoisomers differ from each other only in their linking number.

Twist and writhe

The conformation (geometry) of the DNA can be altered while the linking number remains constant. Two extreme conformations of a supercoiled DNA topoisomer may be envisaged (Figure 2), corresponding to the partition of the supercoiling (ΔLk) completely into **writhe** (Figure 2a) or completely into **twist** (Figure 2c). The line on the rubber tubing model helps to keep track of local twisting of the DNA axis. The equilibrium situation lies between these two extremes (Figure 2b), and corresponds to some change in both twist and writhe induced by supercoiling. This partition may be expressed by the equation:

$$\Delta Lk = \Delta Tw + \Delta Wr.$$

ΔLk must be an integer, but ΔTw and ΔWr need not be. The topological change in supercoiling of a DNA molecule is partitioned into a conformational change of twist and/or a change of writhe.

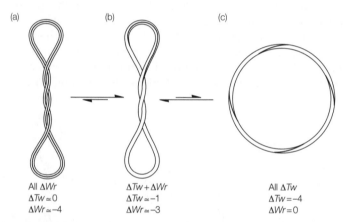

(a)   (b)   (c)

All ΔWr $\Delta Tw + \Delta Wr$ All ΔTw
$\Delta Tw \simeq 0$ $\Delta Tw \simeq -1$ $\Delta Tw = -4$
$\Delta Wr \simeq -4$ $\Delta Wr \simeq -3$ $\Delta Wr = 0$

Figure 2. Conformational flexibility in supercoiled DNA as modeled by rubber tubing. The changes in twist and writhe at constant linking number are shown (see text for details).

Intercalators

The geometry of a supercoiled molecule may be altered by any factor that affects the intrinsic twisting of the DNA helix. For example, an increase in temperature reduces the twist, and an increase in ionic strength may increase the twist. One important factor is the presence of an **intercalator**. The best-known example of an intercalator is **ethidium bromide** (Figure 3a). This is a positively charged polycyclic aromatic compound that binds to DNA by inserting itself between the base pairs (**intercalation**; Figure 3b). In addition to a large increase in fluorescence of the ethidium bromide molecule on binding, which is the basis of its use as a stain of DNA in gels (Section O3), the binding causes a local unwinding of the helix by around 26°. This is a reduction in twist, and hence results in a corresponding increase in writhe in a closed-circular molecule. In a negatively supercoiled molecule, this corresponds to a decrease in the (negative) writhe of the molecule, and the shape of the molecule will be altered in the direction a→b→c in Figure 2.

Energy of supercoiling

Supercoiling involves the introduction of **torsional stress** into DNA molecules. Supercoiled DNA hence has a higher energy than relaxed DNA. For negative supercoiling, this

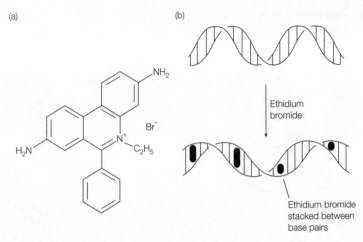

Figure 3. (a) Ethidium bromide; (b) the process of intercalation, illustrating the lengthening and untwisting of the DNA helix.

energy makes it easier for the DNA helix to be locally untwisted, or unwound. Negative supercoiling may thus facilitate processes that require the unwinding of the helix, such as **transcription initiation** or **replication** (Sections F4 and D2).

Topoisomerases

Enzymes exist which regulate the level of supercoiling of DNA molecules; these are termed **topoisomerases**. To alter the linking number of DNA, the enzymes must transiently break one or both DNA strands, which they achieve by the attack of a tyrosine residue on a backbone phosphate, resulting in a temporary covalent attachment of the enzyme to one of the DNA ends via a **phosphotyrosine bond**. There are two classes of topoisomerase. **Type I** enzymes break one strand of the DNA, and change the linking number in steps of ±1 by allowing rotation at the break point (Figure 4a) or by passing the other strand through the break (Figure 4b). **Type II** enzymes, which require the hydrolysis of ATP, break both strands of DNA and change the linking number in steps of ±2, by the transfer of another double-stranded segment through the break (Figure 4c). Most topoisomerases reduce the level of positive or negative supercoiling, i.e. they operate in the energetically favorable direction. However, **DNA gyrase**, a bacterial type II enzyme, uses the energy of ATP hydrolysis to introduce negative supercoiling, which maintains the level of negative supercoiling in bacterial DNA and allows it to efficiently remove positive supercoiling generated during replication (Section D2). Since their mechanism involves the passing of one double strand through another, type II topoisomerases are also able to unlink DNA molecules, such as daughter molecules produced in replication, which are linked (**catenated**) together (Section D2). In bacteria, this function is carried out by **topoisomerase IV** and in eukaryotes by topoisomerase II. Topoisomerases are essential enzymes in all organisms, being involved in replication, recombination, and transcription. The formation of a transient break in the DNA during the topoisomerase reaction is a potentially dangerous event that can be exploited to kill cells, for example by drugs that act to make the transient break permanent. If not repaired, double-stranded breaks in DNA lead to cell death, in eukaryotes by the process of apoptosis (Sections E3 and N4). DNA gyrase and topoisomerase IV are the targets of **anti-bacterial drugs**, such as the

quinolones, which operate by this mechanism, and both type I and type II enzymes are the target of **anti-cancer agents**, such as **etoposide** and **camptothecin** in humans.

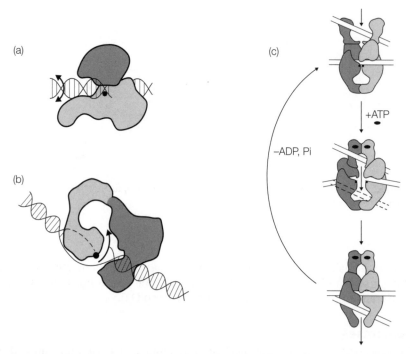

Figure 4. Cartoons of the mechanisms of type I and type II topoisomerases. Type I enzymes can operate by allowing DNA to rotate at the break point (a) or by passing one strand through a transient break in the other (b). (c) Type II enzymes hydrolyse ATP to adenosine 5′ disphophate (ADP) and phosphate and pass one double helix through a break in another.

C1 Prokaryotic chromosome structure

Key Notes

The prokaryotic chromosome	The single *E. coli* chromosome is a closed-circular DNA of length 4.6 million base pairs that resides in a region of the cell called the nucleoid. In normal growth the DNA is being replicated continuously. Many prokaryotes also contain plasmid DNAs while some have multiple chromosomes.
DNA domains	The *E. coli* genome is organized into 50–100 large loops or domains of 50–100 kb in length, which are constrained by binding to a membrane–protein complex.
Supercoiling of the genome	The genome is negatively supercoiled. Individual domains may be topologically independent and so are able to support different levels of supercoiling.
DNA-binding proteins	The DNA domains are compacted by wrapping around nonspecific DNA-binding proteins such as HU and H-NS (histone-like proteins). These proteins constrain about half of the supercoiling of the DNA. Other molecules such as integration host factor, RNA polymerase and mRNA may help to organize the nucleoid.
Related topics	(B3) DNA supercoiling (C4) Genome complexity (C2) Chromatin structure

The prokaryotic chromosome

Prokaryotic genomes vary in size from the 0.16 Mb of the tiny insect **endosymbiont** *Carsonella ruddi* to the 13 Mb of the soil bacterium *Sorangium cellulosum*. Most of these have one **chromosome** containing a single, closed circular (Section B3) DNA molecule. The 4.6-Mb *E. coli* chromosome is fairly typical. This DNA is packaged into a region of the cell known as the **nucleoid**, although the nucleoid is not bounded like the eukaryotic nucleus. This region has a very high DNA concentration, perhaps 30–50 mg mL^{-1}, as well as containing all the proteins associated with DNA, such as polymerases, transcription factors, and others (see below). A fairly high DNA concentration in the test tube would be 1 mg mL^{-1}. In normal growth, the DNA is being replicated continuously and there may be on average around two copies of the genome per cell, when growth is at the maximal rate (Section D2).

Many prokaryotes also contain single or multiple (up to a few hundred) copies of additional small circular DNA molecules called **plasmids**, generally between 2 and 200 kb in size, which contain a small number of genes. Some 'plasmids,' such as the 1 Mb circle found in *Vibrio cholerae* in addition to its 3-Mb chromosome, are large enough to be considered second chromosomes. Finally, linear chromosomes can also be found – some strains of *Agrobacterium tumefaciens* have a 2.9-Mb circle and a 2.1-Mb linear DNA as well as 0.5-Mb and 0.2-Mb circular plasmids. The latter **Ti plasmid** is used as a cloning vector in plants (Section P3).

DNA domains

Experiments in which DNA from *E. coli* is carefully isolated free of most of the attached proteins and observed under the electron microscope reveal one level of organization of the nucleoid. The DNA consists of 50–100 **domains** or **loops**, the ends of which are constrained by binding to a structure that probably consists of proteins attached to part of the cell membrane (Figure 1). The loops are about 50–100 kb in size. It is not known whether the loops are static or dynamic, but one model suggests that the DNA may spool through sites of polymerase or other enzymic action at the base of the loops.

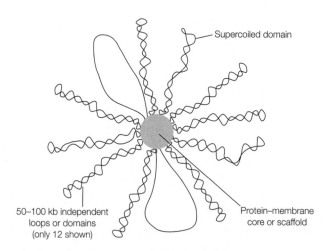

Figure 1. A schematic view of the structure of the *E. coli* chromosome (4600 kb) as visualized by electron microscopy. The thin line is the DNA double helix.

Supercoiling of the genome

The *E. coli* chromosome as a whole is negatively supercoiled ($\Delta Lk/Lk^\circ$=–0.06; Section B3), although there is some evidence that individual domains may be supercoiled independently. Electron micrographs indicate that some domains may not be supercoiled, perhaps because the DNA has become broken in one strand (Section B3), whereas other domains clearly do contain supercoils (Figure 1). The attachment of the DNA to the protein–membrane scaffold may act as a barrier to rotation of the DNA, such that the domains may be **topologically independent**. There is no real biochemical evidence for major differences in the level of supercoiling in different regions of the chromosome *in vivo*, although more local variations in supercoiling may occur as a result of transcription at particular locations (Section F3).

DNA-binding proteins

The looped DNA domains of the chromosome are constrained further by interaction with a number of DNA-binding proteins. The most abundant of these are **protein HU**, a small basic (positively charged) dimeric protein, which binds DNA nonspecifically by the wrapping of the DNA around the protein, and **H-NS**, a monomeric neutral protein that also binds DNA nonspecifically in terms of sequence, but which seems to have a preference for AT-rich regions of DNA that are intrinsically bent. These and other related proteins are sometimes known as **histone-like** proteins (Section C2), and have the effect of compacting the DNA, which is essential for the packaging of the DNA into the nucleoid and for stabilizing and constraining the supercoiling of the chromosome. This means that, although the chromosome when isolated has a $\Delta Lk/Lk°=-0.06$, i.e. approximately one supercoil for every 17 turns of DNA helix, roughly half of this is **constrained** as permanent wrapping of DNA around proteins such as HU (actually a form of writhing; Section B3). Only about half the supercoiling is **unconstrained** in the sense of being able to adopt the twisting and writhing conformations described in Section B3. It has also been suggested that RNA polymerase and mRNA molecules, as well as site-specific DNA-binding proteins such as **FIS**, **integration host factor** (a homolog of HU that binds to specific DNA sequences and bends DNA through 140°), and transcription factors may be important in the organization of the DNA domains. It is likely that the organization of the nucleoid is fairly complex, although highly ordered DNA–protein complexes such as nucleosomes (Section C2) have not been detected.

C2 Chromatin structure

Key Notes

Chromatin

Eukaryotic chromosomes each contain a long linear DNA molecule, which must be packaged into the nucleus. The name chromatin is given to the highly ordered DNA–protein complex that makes up the eukaryotic chromosomes. The chromatin structure serves to package and organize the chromosomal DNA, and is able to alter its level of packing at different stages of the cell cycle.

Histones

The major protein components of chromatin are the histones; small, basic (positively charged) proteins that bind tightly to DNA. There are four families of core histone, H2A, H2B, H3, H4, and a further family, H1, which has some different properties, and a distinct role. Individual species have a number of variants of the different histone proteins.

Nucleosomes

The nucleosome core is the basic unit of chromosome structure, consisting of a protein octamer containing two each of the core histones, with 146 bp of DNA wrapped 1.8 times in a left-handed fashion around it. The wrapping of DNA into nucleosomes accounts for virtually all of the negative supercoiling in eukaryotic DNA.

The role of H1

A single molecule of H1 stabilizes the DNA at the point at which it enters and leaves the nucleosome core, and organizes the DNA between nucleosomes. A nucleosome core plus H1 is known as a chromatosome. In some cases, H1 is replaced by a variant, H5, which binds more tightly, and is associated with DNA which is inactive in transcription.

Linker DNA

The linker DNA between the nucleosome cores varies between less than 10 and more than 100 bp, but is normally around 55 bp. The nucleosomal repeat unit is hence around 200 bp.

The 30-nm fiber

Chromatin is organized into a larger structure, known as the 30-nm fiber or solenoid, thought to consist of a left-handed helix of nucleosomes with approximately six nucleosomes per helical turn. Most chromatin exists in this form.

Higher order structure

On the largest scale, chromosomal DNA is organized into loops of up to 100 kb in the form of the 30-nm fiber, constrained by a protein scaffold, the nuclear matrix. The overall structure somewhat resembles that of the organizational domains of prokaryotic DNA.

Related topics	(B3) DNA supercoiling	(C3) Eukaryotic chromosome
	(C1) Prokaryotic	structure
	chromosome structure	(C4) Genome complexity

Chromatin

The total length of DNA in a eukaryotic cell depends on the species, but it can be thousands of times as much as in a prokaryotic genome, and is made up of a number of discrete bodies called **chromosomes** (46 in humans). The DNA in each chromosome is a single linear molecule, which can be up to several centimeters long (Section C3). All this DNA must be packaged into the **nucleus**, a space of approximately the same volume as a bacterial cell; in fact, in their most highly condensed forms, the chromosomes have an enormously high DNA concentration of perhaps 200 mg mL^{-1} (Section C1). This feat of packing is accomplished by the formation of a highly organized complex of DNA and protein, known as **chromatin**, a nucleoprotein complex (Section A4). More than 50% of the mass of chromatin is protein. Chromosomes greatly alter their level of compactness as cells progress through the cell cycle (Sections C3 and N1), varying between highly condensed chromosomes at **metaphase** (just before cell division), and very much more diffuse structures in **interphase**. This implies the existence of different levels of organization of chromatin.

Histones

Most of the protein in eukaryotic chromatin consists of **histones**, of which there are five families, or classes: H2A, H2B, H3, and H4, known as the **core histones**, and H1. The core histones are small proteins, with masses between 10 and 20 kDa, and H1 histones are a little larger at around 23 kDa. All histone proteins have a large positive charge; between 20 and 30% of their sequences consist of the basic amino acids, lysine, and arginine (Section A3). This means that histones will bind very strongly to the negatively charged DNA in forming chromatin.

Members of the same histone class are very highly conserved between relatively unrelated species, for example between plants and animals, which testifies to their crucial role in chromatin. Within a given species, there is normally a number of closely similar variants of a particular class, which may be expressed in different tissues, and at different stages in development. There is not much similarity in sequence between the different histone classes, but structural studies have shown that the classes do share a similar tertiary structure (Section A3), suggesting that all histones are ultimately evolutionarily related.

H1 histones are somewhat distinct from the other histone classes in a number of ways: in addition to their larger size, there is more variation in H1 sequences both between and within species than in the other classes. Histone H1 is more easily extracted from bulk chromatin, and seems to be present in roughly half the quantity of the other classes, of which there are very similar amounts. These facts suggest a specific and distinct role for histone H1 in chromatin structure.

Nucleosomes

A number of studies in the 1970s pointed to the existence of a basic unit of chromatin structure. Nucleases are enzymes that hydrolyze the phosphodiester bonds of nucleic acids. Exonucleases release single nucleotides from the ends of nucleic acid strands, whereas endonucleases cleave internal phosphodiester bonds. Treatment of chromatin with **microccocal nuclease**, an endonuclease that cleaves dsDNA, led to the isolation of DNA fragments with discrete sizes, in multiples of approximately 200 bp. It was discovered that each 200-bp fragment is associated with an **octamer** core of histone proteins, $(H2A)_2(H2B)_2(H3)_2(H4)_2$, which is why these are designated the **core histones**, and more loosely with one molecule of H1. The proteins protect the DNA from the action of micrococcal nuclease. More prolonged digestion with nuclease leads to the loss of H1 and yields a very resistant structure consisting of **146 bp** of DNA associated very tightly with the histone octamer. This structure is known as the **nucleosome core**, and is structurally very similar whatever the source of the chromatin.

The structure of the nucleosome core particle is now known in considerable detail, from structural studies culminating in X-ray crystallography (Section A5). The histone octamer forms a wedge-shaped disk, around which the 146 bp of DNA is wrapped in 1.8 turns in a left-handed direction. Figure 1 shows the basic features and the dimensions of the structure. The left-handed wrapping of the DNA around the nucleosome corresponds to negative supercoiling, i.e. the turns are **superhelical turns** (technically writhing; Section B3). Although eukaryotic DNA is negatively supercoiled to a similar level as that of prokaryotes, on average virtually all the supercoiling is accounted for by wrapping in nucleosomes and there is no unconstrained supercoiling (Section C1).

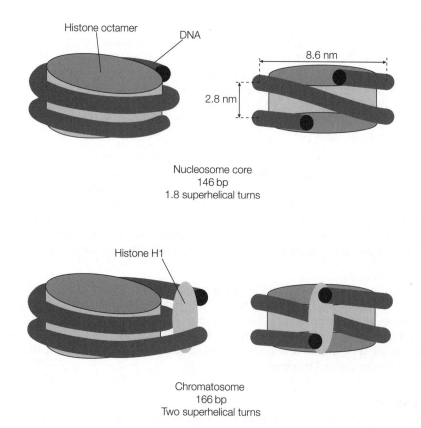

Figure 1. A schematic view of the structure of a nucleosome core and a chromatosome.

The role of H1

One molecule of histone H1 binds to the nucleosome, and acts to stabilize the point at which the DNA enters and leaves the nucleosome core (Figure 1). In the presence of H1, a further 20 bp of DNA is protected from nuclease digestion, making 166 bp in all, corresponding to two full turns around the histone octamer. A nucleosome core plus H1 is known as a **chromatosome**. The larger size of H1 compared with the core histones is due to the presence of an additional C-terminal tail, which serves to stabilize the DNA between the nucleosome cores. As stated above, H1 is more variable in sequence than the other histones and, in some cell types, it may be replaced by an extreme variant called histone H5, which binds chromatin particularly tightly, and is associated with DNA which is transcriptionally repressed (Section C3).

Linker DNA

In electron micrographs of nucleosome core particles on DNA under certain conditions, an array, sometimes called the '**beads on a string**' structure is visible. This comprises globular particles (the nucleosomes) connected by thin strands of DNA. This **linker DNA** is the additional DNA required to make up the 200-bp nucleosomal repeat apparent in

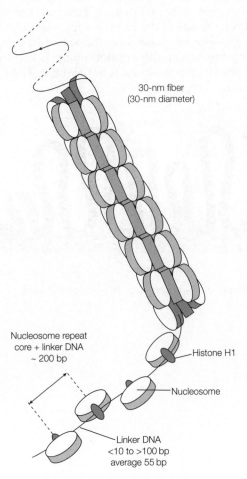

Figure 2. An array of nucleosomes separated by linker DNA, and the 30-nm fiber.

the micrococcal nuclease experiments (see above). The average length of linker DNA between core particles is 55 bp, but the length varies between species and tissues from almost nothing to more than 100 bp (Figure 2).

The 30-nm fiber

The presence of histone H1 increases the organization of the 'beads on a string' to show a zigzag structure in electron micrographs. With a change in the salt concentration, further organization of the nucleosomes into a **fiber** of 30-nm diameter takes place. Detailed studies of this process have suggested that the nucleosomes are wound into a higher order left-handed helix called a **solenoid**, with around six nucleosomes per turn (Figure 2). However, there is still some conjecture about the precise organization of the fiber structure, including the path of the linker DNA and the way in which different linker lengths might be incorporated into what seems to be a very uniform structure. Most chromosomal DNA *in vivo* is packaged into the 30-nm fiber (Section C3).

Higher order structure

The organization of chromatin at the highest level seems rather similar to that of prokaryotic DNA (Section C1). Electron micrographs of chromosomes that have been stripped of their histone proteins show a looped domain structure similar to that illustrated in Section C1, Figure 1. Even the size of the loops is approximately the same, up to around 100 kb of DNA, although there are many more loops in a eukaryotic chromosome. The loops are constrained by interaction with a protein complex known as the **nuclear matrix** (Sections C3 and D3). The DNA in the loops is in the form of 30-nm fiber, and the loops form an array about 300 nm across (Figure 3).

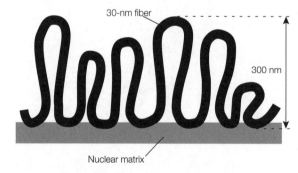

Figure 3. The organization of 30-nm fiber into chromosomal loops.

C3 Eukaryotic chromosome structure

Key Notes

The mitotic chromosome

The classic picture of paired sister chromatids at mitosis represents the most highly condensed state of chromatin. The linear DNA traces a single path from one tip of the chromosome to the other, in successive loops of up to 100 kb of 30-nm fiber anchored to the nuclear matrix in the core.

The centromere

The centromere is the region where the two chromatids are joined and is also the site of attachment, via the kinetochore, to the mitotic spindle, which pulls apart the sister chromatids at anaphase. Centromeres are characterized by specific short DNA sequences although, in mammalian cells, there may be an involvement of satellite DNA.

Telomeres

The ends of the linear chromosomal DNA are protected from degradation and gradual shortening by the telomeres, which are short repeating sequences synthesized by a specific enzyme, telomerase, independently of normal DNA replication. These sequences adopt a characteristic G-quadruplex structure.

Interphase chromosomes

In interphase, the chromosomes adopt a much more diffuse structure, although the chromosomal loops remain attached to the nuclear matrix.

Heterochromatin

Heterochromatin is a portion of the chromatin in interphase which remains relatively compacted and is transcriptionally inactive.

Euchromatin

Euchromatin is the more diffuse region of the interphase chromosome, consisting of inactive regions in the 30-nm fiber form, and actively transcribed regions where the fiber has dissociated and individual nucleosomes may be replaced by transcription initiation proteins.

DNase I hypersensitivity

Active regions of chromatin, or regions where the 30-nm fiber is interrupted by the binding of a specific protein to the DNA or by ongoing transcription, are characterized by hypersensitivity to deoxyribonuclease I (DNase I).

CpG methylation

5'-CG-3' (CpG) sequences in mammalian DNA are normally methylated on the cytosine base; however, 'islands' of unmethylated CpG occur near the promoters of frequently transcribed genes, and form regions of particularly high DNase I sensitivity. Patterns of CpG methylation can persist

	through multiple cell divisions, and through generations, resulting in the possibility of epigenetic inheritance.	
Histone variants and modification	The control of the degree of condensation of chromatin operates at least in part through the chemical modification of histone proteins, which changes their charge and their interaction with other proteins during the cell cycle, or through the use of histone variants in particular cell types or during development.	
Related topics	(C2) Chromatin structure (C4) Genome complexity	(D3) Eukaryotic DNA replication

The mitotic chromosome

The familiar picture of a chromosome (Figure 1) is actually that of its most highly condensed state at **mitosis**. As the daughter chromosomes are pulled apart by the **mitotic spindle** at cell division (Section N1), the fragile centimeters-long chromosomal DNA would certainly be sheared by the forces generated, were it not in this highly compact state. The structure in Figure 1 actually illustrates two identical **sister chromatids**, the products of replication of a single chromosome, joined at their **centromeres**. The tips of

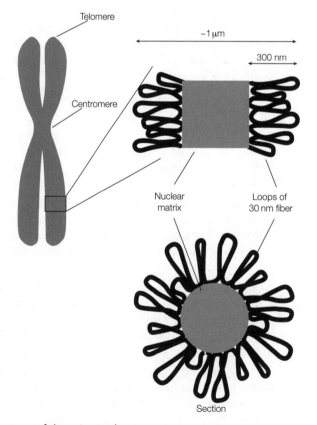

Figure 1. The structure of the mitotic chromosome.

the chromosomes are the **telomeres**, which are also the ends of the DNA molecule; the DNA maps in a linear fashion along the length of the chromosome, albeit in a very convoluted path. The structure of a section of a mitotic chromatid is shown in Figure 1. The chromosomal loops (Section C2) fan out from a central scaffold or **nuclear matrix** region that consists of protein. The loops comprise chromatin in the 30-nm fiber form (Section C2). Consecutive loops may trace a helical path along the length of the chromosome.

The centromere

The **centromere** is the constricted region where the two sister chromatids are joined in the metaphase chromosome. This is the site of assembly of the **kinetochore**, a protein complex which attaches to the **microtubules** (Section A4) of the mitotic spindle. The microtubules act to separate the chromatids at **anaphase**. The DNA of the centromere has been shown in yeast to consist merely of a short AT-rich sequence of 88 bp flanked by two very short conserved regions, although in mammalian cells centromeres seem to consist of rather longer sequences and are flanked by a large quantity of repeated DNA, known as **satellite DNA** (Section C4).

Telomeres

Telomeres are specialized DNA sequences that form the ends of the linear DNA molecules of the **eukaryotic** chromosomes. A telomere consists of up to hundreds of copies of a short repeated sequence (5′-TTAGGG-3′ in vertebrates), which is synthesized by the enzyme **telomerase** (a **ribonucleoprotein**, Section J1) in a mechanism independent of normal DNA replication. The telomeric DNA forms a special folded secondary structure, called a G-quadruplex, with three square planar arrays of G-residues. The functions of the telomere are to protect the ends of the chromosome from degradation and from end-to-end joining (Section E3). Independent synthesis of the telomere acts to counteract the gradual shortening of the chromosome that would result from the inability of normal replication to copy the very end of a linear DNA molecule (Section D3).

Interphase chromosomes

In interphase (Section N1), the genes on the chromosomes are being transcribed and DNA replication takes place. During this time, which comprises most of the cell cycle, the chromosomes adopt a much more diffuse structure and cannot be visualized individually. It is believed, however, that the chromosomal loops are still present, attached to the nuclear matrix (Section C2, Figure 3).

Heterochromatin

Heterochromatin comprises a portion of the chromatin in interphase that remains highly compacted, although not so compacted as at metaphase. It can be visualized under the microscope as dense regions at the periphery of the nucleus, and probably consists of closely packed regions of 30-nm fiber. It is transcriptionally inactive. **Constitutive heterochromatin** is always compacted and consists mainly of the repeated satellite DNA sequences close to the centromeres of the chromosomes (Section C4), although in some cases entire chromosomes can remain as heterochromatin, e.g. one of the two X chromosomes in female mammals. **Facultative heterochromatin** is chromatin that may be compacted in some tissues where its component genes are not expressed, but which can become part of the active **euchromatin** fraction (see below) in other tissues.

Euchromatin

The bulk of the chromatin that is not visible as heterochromatin is known historically by the catch-all name of **euchromatin**, and is the region where all transcription takes place. Euchromatin is not homogeneous however, and consists of relatively inactive regions comprising chromosomal loops compacted in 30-nm fibers, and regions (perhaps 10% of the whole) where genes are actively being transcribed or are destined to be transcribed in that cell type. In these regions the 30-nm fiber has been dissociated to the 'beads on a string' structure (Section C2). Parts of these regions may be depleted of nucleosomes altogether, particularly within promoters, to allow the binding of transcription factors and other proteins (Figure 2) (Section H5).

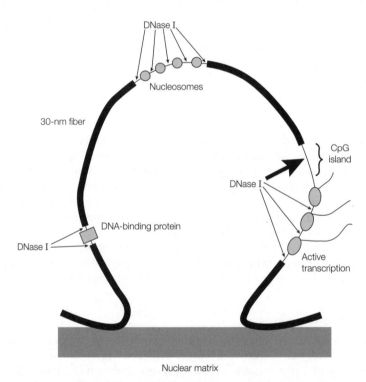

Figure 2. Euchromatin, showing active and inactive regions (see text for details) and sites of DNase I hypersensitivity indicated by large and small arrows.

DNase I hypersensitivity

The sensitivity of chromatin to the nuclease deoxyribonuclease I (**DNase I**), which cuts the backbone of DNA unless the DNA is protected by bound protein, has been used to map the regions of transcriptionally active chromatin in cells. Short regions of DNase I hypersensitivity (readily attacked by DNase I) are thought to represent stretches of 'naked' DNA where the 30-nm fiber is interrupted by the binding of a sequence-specific regulatory protein (Figure 2). Longer regions of sensitivity represent sequences where transcription is taking place. These regions vary between different cell types, and correspond to the sites of genes that are expressed specifically in those cells.

CpG methylation

An important chemical modification that may be involved in signaling the appropriate level of chromosomal packing at the sites of expressed genes in mammalian cells is the methylation of C-5 in the cytosine base of 5′-CG-3′ sequences, commonly known as **CpG methylation**. CpG sites are normally methylated in mammalian cells, and are relatively scarce throughout most of the genome. This is because 5-methylcytosine spontaneously deaminates to thymine and, since this error is not always repaired, methylated CpG mutates fairly rapidly to TpG (Section E1). The methylation of CpG is associated with transcriptionally inactive regions of chromatin. However, there exist throughout the genome '**islands**' of unmethylated CpG, where the proportion of CG dinucleotides is consequently much higher than average. These islands are commonly around 2 kb long, and are coincident with regions of particular sensitivity to DNase I (Figure 3). The CpG islands surround the promoter regions of genes that are expressed in almost all cell types, so called **housekeeping genes**, and may be largely free of nucleosomes. The pattern of CpG methylation is preserved through DNA replication and cell division by the action of a 'maintenance' **methyltransferase**, which methylates **hemimethylated** sites after replication (cf. hemimethylation in mismatch repair, Section E3). The pattern of global gene activation and repression may therefore be inherited through the generations by an **epigenetic** effect, i.e. a heritable change brought about by something other than a change in DNA sequence.

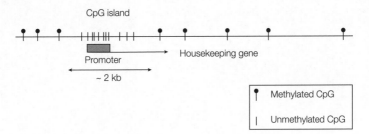

Figure 3. CpG islands and the promoters of housekeeping genes.

Histone variants and modification

The major mechanisms for the condensing and decondensing of chromatin are believed to operate directly through the histone proteins that carry out the packaging (Section C2). Short-term changes in chromosome packing during the cell cycle seem to be modulated by chemical modification of the histone proteins. For example, actively transcribed chromatin is associated with the acetylation and methylation of lysine residues in the N-terminal regions of the core histones (Sections C2 and I1), whereas the condensation of chromosomes at mitosis is accompanied by the phosphorylation of histone H1. These changes alter the positive charge on the histone proteins and may affect the stability of the various chromatin conformations, for example the 30-nm fibers or the interactions between them. Chemical changes may also influence the recruitment of other proteins, such as transcription factors, to specific regions of the chromosome, thus influencing patterns of gene expression (Sections H5 and I1). The precise effects are dependent on the type of modification and its position in the histone protein and attempts have been made to derive a so-called 'histone code' for the effects of multiple chemical modifications.

Longer-term differences in chromatin condensation are associated with changes that occur during development and differentiation into specific tissue types. These changes are associated with the utilization of alternative histone variants, which may also act by altering the stability of chromatin conformations. Histone H5 is an extreme example of this effect, replacing H1 in some very inactive chromatin, for example in avian red blood cells (Section C2).

C4 Genome complexity

Key Notes

Noncoding DNA

A large proportion of the DNA in eukaryotic cells does not code for protein. Most genes have large intron sequences, and genes or gene clusters are separated by long stretches of DNA, a considerable part of which has no known function. Much of this DNA consists of multiple repeats of a few relatively short sequence elements, some of which exist in thousands or hundreds of thousands of copies.

Unique sequence DNA

In multicellular organisms, substantially less than half the genomic DNA is of unique sequence. This essentially comprises single-copy protein-coding genes, or those repeated a few times. In complex genomes, microRNA genes also form part of this fraction. In contrast, *E. coli* DNA is essentially all unique sequence.

Tandem gene clusters

Part of the moderately repetitive DNA fraction is made up of genes or gene clusters that are tandemly repeated up to a few hundred times. Examples include rRNA genes and histone gene clusters.

Dispersed repetitive DNA

Much of the moderately repetitive DNA consists of families of DNA sequences a few hundred base pairs long (SINES, or short interspersed elements), or between one and five thousand base pairs long (LINES, or long interspersed elements), each repeated more than 100,000 times and scattered throughout the genome. The most prominent examples in humans are the *Alu* and the *L1* elements, which may be parasitic DNA sequences that replicate themselves by transposition.

Satellite DNA

Satellite DNA, which occurs mostly near the centromeres of chromosomes and which may be involved in attachment of the mitotic spindle, consists of huge numbers of short (up to 30 bp) tandem repeat sequences. Hypervariability in satellite DNA is the basis of the DNA fingerprinting technique.

Genetic polymorphism

Mutations in a gene or a chromosomal locus can create multiple forms (polymorphs) of that locus, which is then said to show genetic polymorphism. The term can describe different alleles of a single copy gene in a single individual as well as the different sequences present in different individuals in a population. Common types are single nucleotide polymorphisms (SNPs) and simple sequence length polymorphisms (SSLPs). Where SNPs create or destroy the sequence recognized by a restriction enzyme, a

	detectable restriction fragment length polymorphism (RFLP) will result. High throughput methods for polymorphism detection based on sequencing or MS are being developed.	
Related topics	(B2) Spectroscopic and thermal properties of nucleic acids	(H2) RNA Pol I genes: the ribosomal repeat

Noncoding DNA

Whilst some eukaryotes have genomes no bigger than the largest prokaryotes (e.g. the 12.5 Mb of the green alga *Ostreococcus tauri*), the genomes of many complex multicellular eukaryotic organisms can contain more than 1000 times as much DNA as *E. coli*. It is clear that much of this DNA does not code for protein since there are, for example, probably only five times as many genes in humans as in *E. coli*. The coding regions of most eukaryotic genes are interrupted by **intron** sequences (Section J3) and so genes may take up many kilobases of sequence. Furthermore, genes are by no means **contiguous** (adjacent) along the genome; they are separated by long stretches of DNA sequence whose function, if any, is often obscure. Biophysical methods such as **reassociation kinetics**, which directly measures the concentration of similar DNA sequences, and more recently **DNA sequencing**, have shown that much of this **noncoding** DNA consists of multiple repeats of similar or identical copies of a few different types of sequence. These sequences may be classified as **moderately** or **highly repetitive DNA**. These copies may follow one another directly (**tandemly repeated**), such as **satellite DNA**, or they may occur as multiple copies scattered throughout the genome (**interspersed**), such as the *Alu* **elements**. However, some noncoding DNA is classified as **unique sequence** (see below).

Unique sequence DNA

Unique sequence DNA corresponds to the coding regions of genes that occur in only one or a few copies per haploid genome (i.e. the great majority), and any unique intervening sequence. In the *E. coli* genome and in those of other bacteria and many unicellular eukaryotes, most of the DNA is unique sequence, since it consists predominantly of more or less contiguous single-copy genes. In the last 10 years, it has been discovered that, in addition to unique protein-coding genes, multicellular eukaryotes also possess a large number of single-copy genes encoding both long noncoding RNAs and microRNAs (Sections A2, I2, and L4), both of which are involved in the regulation of gene expression.

Tandem gene clusters

Moderately repetitive DNA consists of a number of types of repeated sequence. At the lower end of the repeat scale are genes that occur as clusters of multiple repeats. These are genes whose products are required in unusually large quantities, for example those encoding rRNA (**rDNA**). The gene encoding the 45S precursor of the 18S, 5.8S, and 28S rRNAs is repeated in **arrays** containing from around 10 to 10,000 copies depending on the species (Section H2, Figure 1). In humans, the 45S gene occurs in arrays on five separate chromosomes, each containing around 40 copies. In interphase, these regions are spatially located together in the **nucleolus**, a dense region of the nucleus which is a factory for rRNA production and modification (Sections H2 and J1). A second example of tandem gene clusters is given by the histone genes, whose products are produced in large quantities during S-phase (Sections C2 and N1). The five histone genes occur together in a **cluster**, which is directly repeated up to several hundred times in some species.

Dispersed repetitive DNA

In many species, most of the moderately repetitive DNA consists of sequences from a few hundred base pairs (**SINES**, or **short interspersed elements**) to between one and five thousand base pairs (**LINES**, or **long interspersed elements**) in length. Collectively, these are known as **dispersed repetitive DNA** sequences as they are repeated many thousands of times and are scattered throughout the whole genome. The commonest such sequence in humans is the ***Alu* element**, a 300-bp SINE that occurs between 300,000 and 500,000 times. The copies are all 80–90% identical, and most contain the *AluI* restriction site (Section O3), hence the name.

A second dispersed sequence is the ***L1* element** and together, copies of *Alu* and *L1* make up almost 10% of the human genome, occurring between genes and in introns. For many years *Alu* elements and other such families were regarded as parasitic, **selfish DNA** sequences with no actual function, duplicating themselves randomly within the genome by **transposition** (Section E4). However, there is evidence that *Alu* transcripts, an example of **long noncoding RNAs** (**lncRNAs**), can act as global regulators of gene expression (Section I2).

Satellite DNA

In eukaryotic genomes, highly repetitive DNA consists of very short sequences from 2 bp to 20–30 bp in length, in tandem arrays of many thousands of copies. Such arrays are known as **satellite DNA**, since they were first identified as satellite bands with a buoyant density different from that of **bulk DNA** in CsCl density gradients of chromosomal DNA fragments. Since they consist of repeats of such short sequences, they have non-average G+C content (Section B1). Satellite DNAs are divided into **minisatellites (variable number tandem repeats, VNTRs)** and **microsatellites (simple tandem repeats, STRs)** according to the length of the repeating sequence, microsatellites having the shortest repeats. Figure 1 shows an example of a *Drosophila* satellite DNA sequence that occurs millions of times in the insect's genome. As with dispersed repetitive sequences, satellite DNA has no conclusively demonstrated function, although it has been suggested that some of these arrays, i.e. those **heterochromatic** regions concentrated near the centromeres of chromosomes, may have a role in the binding of kinetochore components to the centromere (Section C3). Minisatellites (VNTRs) are more frequently found near chromosome ends whereas microsatellites (STRs) are more evenly distributed along the chromosomes.

Minisatellite repeats are the basis of the **DNA fingerprinting** technique used to unambiguously identify individuals and their familial relationships. The numbers of repeats in the arrays of some satellite sequences are **hypervariable**, i.e. they vary significantly

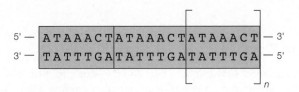

Figure 1. *Drosophila* satellite DNA repeat.

between individuals. These variations are examples of genetic polymorphism (see below). If Southern blots (Section R1) of genomic DNA from members of a family are digested with a restriction enzyme (Section O3) and hybridized with a probe that detects one of these types of repeats, each sample will show a set of bands of varying lengths (the length of the repeats between the two flanking restriction enzyme sites). One hybridizing locus (pair of alleles) is shown in Figure 2. Some of these bands will be in common with those of the mother and some with those of the father and the pattern of bands will be different for an unrelated individual. The different patterns in individuals at each of these kinds of simple repeat sites mean that, by using a small number of probes, the likelihood of two individuals having the same pattern becomes vanishingly small. Figure 2 also shows how DNA fingerprinting can be carried out on small DNA samples such as a blood spot or hair follicle left at a crime scene. Instead of digesting the DNA with restriction enzyme E at each end of the VNTR and Southern blotting, a pair of polymerase chain reaction (PCR) primers can be designed based on the unique sequences flanking the repeats (shown as arrows in Figure 2a). The VNTRs can thus be amplified and directly visualized by staining after agarose gel electrophoresis. The same techniques can also be used to show pedigree in commercially bred animals and to discover mating habits in wild animals.

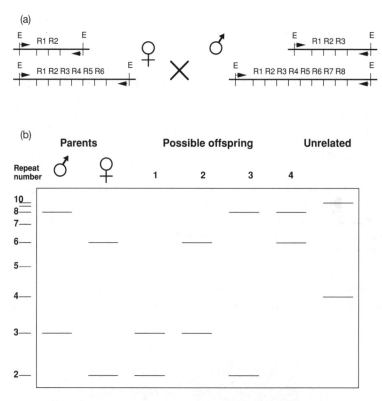

Figure 2. DNA fingerprinting showing how two VNTR alleles might be inherited. (a) Parental VNTR alleles; R is the repeat and E the flanking restriction enzyme sites. (b) Agarose gel analysis of VNTR alleles.

Genetic polymorphism

When the same region, or locus, of a chromosome has a slightly different DNA sequence in two paired chromosomes in a diploid individual or in the corresponding chromosomes in different individuals of the same species, these are described as **polymorphs** and the locus is said to show (**genetic**) **polymorphism**. Genetic polymorphism is caused by mutation (Section E2). For example, in a single individual, the two **alleles** (different forms) of a particular single copy gene could differ by just one nucleotide, and so the gene would be described as polymorphic. Similarly, in a population of individuals, if other alternative DNA sequences exist for the same gene, there would be many different alleles of that gene in the whole population. The mutational events that create these **single nucleotide polymorphisms** (**SNPs**) can also create length variations in arrays of repeated sequences. Each different length of the repeat is a different form and each is an example of a **simple sequence length polymorphism** (**SSLP**).

SNPs can occur within the short sequences that are recognized by restriction enzymes (Section O3) and so the length of a fragment generated by cutting a DNA molecule with a restriction enzyme could be different for each allele. This is known as a **restriction fragment length polymorphism** (**RFLP**). When a particular RFLP is associated with the allele responsible for a genetic disease, it can be used as a marker in clinical diagnosis. RFLPs can also be used to help create genetic maps of chromosomes and so help to work out the order of DNA sequences in genome sequencing projects (Section R2). It is possible to use a form of gel electrophoresis (Section O3) to detect SNPs in DNA fragments of the same length such as restriction fragments or PCR products (Section R3). To do this, the DNA fragments must be denatured so the two separated strands can adopt specific conformations that depend on their nucleotide sequence as they migrate through the gel. This technique is called **single stranded conformational polymorphism** (**SSCP**) analysis. Much more efficient methods for detecting polymorphisms, particularly SNPs, have recently been developed based on **high throughput sequencing** (Section R2) and **MS**.

Analysis of SNPs and other polymorphisms can be used to predict an individual's carrier status for recessive monogenic disease or their susceptibility to adult-onset disorders such as **Huntington's** disease and some inherited cancers. Predicting the lifetime risk of more complex multigenic diseases such as cardiovascular disease is less reliable and is surrounded by ethical issues but is likely to figure strongly in future medical practice, particularly when whole genome sequencing becomes so routine as to replace individual polymorphism analysis.

D1 DNA replication: an overview

Key Notes

Semi-conservative mechanism	During replication, the strands of the double helix separate and each acts as a template to direct the synthesis of a complementary daughter strand using deoxyribonucleoside 5′-triphosphates as precursors. Thus, each daughter cell receives one of the original parental strands. This mechanism can be demonstrated experimentally by density labeling experiments.
Replicons, origins, and termini	Small chromosomes, such as those of bacteria and viruses, replicate as single units called replicons. Replication begins from a unique site, the origin, and proceeds, usually bidirectionally, to the terminus. Large eukaryotic chromosomes contain multiple replicons, each with its own origin, which fuse as they replicate. Origins tend to be AT-rich to make opening easier.
Semi-discontinuous replication	At each replication fork, the leading strand is synthesized as one continuous piece while the lagging strand is made discontinuously as short fragments in the reverse direction. These fragments are 1000–2000 nt long in bacteria and 100–200 nt long in eukaryotes. They are joined by DNA ligase. This mechanism exists because DNA can only be synthesized in a 5′→3′ direction.
RNA priming	The leading strand and all lagging strand fragments are primed by synthesis of a short piece of RNA, which is then elongated with DNA. The primers are removed and replaced by DNA before ligation. This mechanism helps to maintain high replicational fidelity.
Related topics	(Section C) Prokaryotic and eukaryotic chromosome structure (D2) Bacterial DNA replication	(D3) Eukaryotic DNA replication (N1) The cell cycle

---

Semi-conservative mechanism

The key to the mechanism of DNA replication is the fact that each strand of the DNA double helix carries the same information – their base sequences are complementary (Section A2). Thus, during replication, the two **parental strands** separate and each acts as a **template** to direct the enzyme-catalyzed synthesis of a new complementary **daughter strand** following the normal base-pairing rules (A with T; G with C). The two new double-stranded molecules then pass to the two daughter cells at cell division (Figure 1a). The point at which separation of the strands and synthesis of new DNA takes place is known as the **replication fork**. This **semi-conservative** mechanism was demonstrated experimentally in 1958 by Meselson and Stahl (Figure 1b). *E. coli* cells were grown for several generations in the presence of the stable heavy isotope ^{15}N so that their DNA became fully **density labeled** (both strands ^{15}N labeled: $^{15}N/^{15}N$). The cells were then transferred to medium containing only normal ^{14}N and, after each cell division, DNA was prepared from a sample of the cells and analyzed on a CsCl gradient using the technique of **equilibrium** (**isopycnic**) **density gradient centrifugation**, which separates molecules according to differences in buoyant density (Section B1). After the first cell division, when the DNA had replicated once, it was all of hybrid density, appearing in a position on the gradient half way between fully labeled $^{15}N/^{15}N$ DNA and unlabeled $^{14}N/^{14}N$ DNA. After the second cell generation growing in ^{14}N, half of the DNA was hybrid density and half fully light ($^{14}N/^{14}N$). After each subsequent generation, the proportion of $^{14}N/^{14}N$ DNA increased but some DNA of hybrid density (i.e. $^{15}N/^{14}N$) persisted. These results could only be interpreted in terms of a semi-conservative model. All DNA molecules replicate in this manner. The template strands are read in the $3'{\rightarrow}5'$ direction (Section A2) while the new strands are synthesized $5'{\rightarrow}3'$.

The substrates for DNA synthesis are the deoxynucleoside 5′-triphosphates (dNTPs): **dATP, dGTP, dCTP**, and **dTTP** (Section A2). The reaction mechanism involves nucleophilic

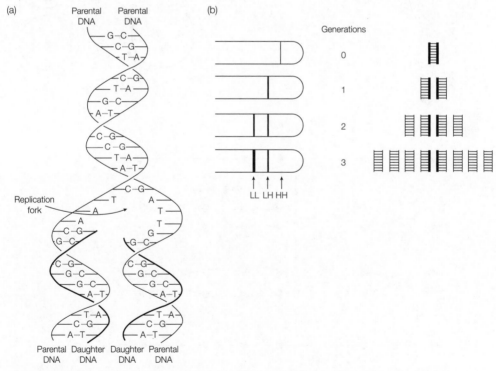

Figure 1. (a) Semi-conservative replication of DNA at the replication fork; (b) proof of semi-conservative mechanism. The centrifuge tubes on the left show the positions of fully ^{15}N-labeled (HH), hybrid density (LH) and fully ^{14}N-labeled (LL) DNA in CsCl density gradients after each generation and the diagram on the right shows the corresponding DNA molecules. The original parental ^{15}N molecules are shown as thick lines and all ^{14}N daughter DNA as thin lines.

attack of the 3′-OH group of the nucleotide at the 3′-end of the growing strand on the α-phosphorus of the dNTP complementary to the next template base, with the elimination of pyrophosphate. The energy for polymerization comes from the hydrolysis of the dNTPs and of the resulting pyrophosphate to two phosphates.

Replicons, origins, and termini

Any piece of DNA that replicates as a single unit is called a **replicon**. The **initiation** of DNA replication within a replicon always occurs at a fixed point known as the **origin** (Figure 2a). Usually, two replication forks proceed **bidirectionally** away from the origin and the strands are copied as they separate until the **terminus** is reached. Most prokaryotic chromosomes and many bacteriophage and viral DNA molecules are circular (Section C1) and comprise single replicons. Thus, there is a single termination site roughly 180° opposite the unique origin (Figure 2a). Linear viral DNA molecules usually have a single origin, but this is not necessarily in the center of the molecule. In all these

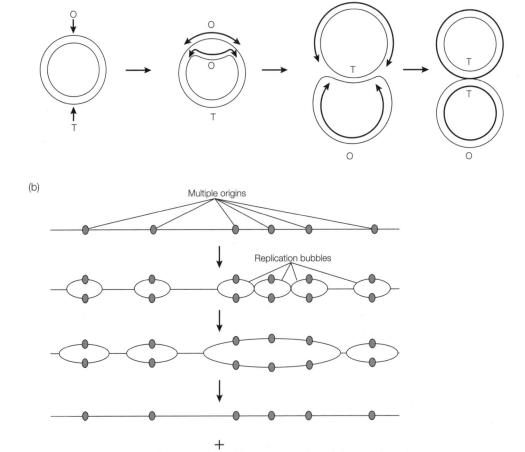

Figure 2. (a) Bidirectional replication of a circular bacterial replicon. Two replication forks proceed away from the origin (O) towards the terminus (T). Daughter DNA is shown as a thick line. (b) Multiple eukaryotic replicons. Each origin is marked (●).

cases, the origin is a complex region where the initiation of DNA replication and the control of the growth cycle of the organism are regulated and coordinated. In contrast, the long, linear DNA molecules of eukaryotic chromosomes consist of multiple replicons, each with its own origin. A typical mammalian cell has 50 000–100 000 replicons with a size range of 40–200 kb. Where replication forks from adjacent **replication bubbles** meet, they fuse to form the completely replicated DNA (Figure 2b). The sequences of eukaryotic chromosomal origins are much simpler than those of single replicon DNA molecules since the control of eukaryotic DNA replication operates primarily at the onset of S-phase (Section N1) rather than at each individual origin. All origins contain AT-rich sequences where the strands initially separate. AT-rich regions are more easily opened than GC-rich regions (Section B2).

Semi-discontinuous replication

In semi-conservative replication, both new strands of DNA are synthesized simultaneously at each replication fork. However, the catalytic mechanism of **DNA polymerases** allows only for synthesis in a 5′→3′ direction. Since the two strands of DNA are **antiparallel**, how is the parental strand that runs 5′→3′ past the replication fork copied? It appears to be made 3′→5′ but this cannot happen. If *E. coli* cells are labeled for just a few seconds with [³H]thymidine, a radioactive precursor of DNA, then a large fraction of the newly synthesized (**nascent**) DNA is found in small fragments 1000–2000 nt long (100–200 nt in eukaryotes) when the DNA is analyzed on **alkaline sucrose gradients**, which separates single strands of denatured DNA according to size. If the cells are incubated further with unlabeled thymidine, these **Okazaki fragments** rapidly join into high molecular weight DNA. These results can be explained by a **semi-discontinuous replication** model (Figure 3). One strand, the **leading strand**, is made continuously in a 5′→3′ direction from the origin. The second parental strand is not copied immediately but is displaced as a single strand for a distance of 1000–2000 nt (100–200 nt in eukaryotes). Synthesis of the second, **lagging**, strand is then initiated at the replication fork and proceeds 5′→3′ back towards the origin to form the first Okazaki fragment. As the replication fork progresses, the leading strand continues to be made as one long strand while further lagging strand fragments are made in a discontinuous fashion in the 'reverse' direction as the parental

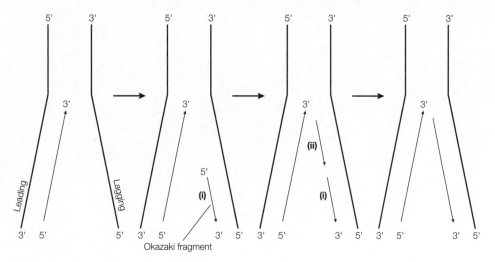

Figure 3. Semi-discontinuous replication. The first (i) and second (ii) Okazaki, or nascent, fragments to be made on the lagging strand are indicated. A single replication fork is shown.

lagging strand is displaced. Despite this difference in polarity, the *physical* direction of both leading and lagging strand synthesis is the same since the lagging strand is looped at the replication fork (Section D2). Soon after synthesis, the fragments are joined to make one continuous piece of DNA by the enzyme **DNA ligase** (Sections D2 and D3).

RNA priming

Close examination of Okazaki fragments has shown that the first few nucleotides at their 5′-ends are in fact ribonucleotides, as are the first few nucleotides of the leading strand. Hence, DNA synthesis is **primed** by RNA (Section D2). These primers are removed and the resulting gaps filled with DNA before the fragments are joined. The reason for initiating each piece of DNA with RNA appears to relate to the need for DNA replication to be of high **fidelity** (Section E2).

D2 Bacterial DNA replication

Key Notes

Experimental systems

Genetically simple bacteriophages and plasmids are useful model systems for studying the *in vitro* replication of the large and fragile bacterial chromosome. The simplest rely nearly exclusively on host cell replication proteins. Larger phages encode many of their own replication factors.

Initiation

Replication is regulated by the rate of initiation. Initiation at the *E. coli* origin, *oriC,* involves wrapping of the DNA around an initiator protein complex (DnaA-ATP) and separation of the strands at AT-rich sequences. The helicase DnaB then binds and extends the single-stranded region for copying. The level of DnaA is linked to growth rate.

Unwinding

As the parental DNA is unwound by DnaB helicase and covered by single-stranded binding protein, the resulting positive supercoiling has to be relieved by the topoisomerase DNA gyrase. Gyrase is a target for antibiotics.

Elongation

Multiple primers are synthesized on the lagging strand. A dimer of DNA polymerase III holoenzyme elongates both leading and lagging strands. The α subunits polymerize the DNA and the ε subunits proofread it. The β subunits clamp the polymerase to the template DNA. The $5'{\rightarrow}3'$ exonuclease activity of DNA polymerase I removes the lagging strand RNA primers, and the polymerase function simultaneously fills the gaps with DNA. DNA ligase joins the lagging strand fragments together. Many of the enzymes at the replication fork are associated in a large replisome.

Termination and segregation

In *E. coli,* both replication forks meet at a terminus region about 180° opposite the origin. The interlocked daughter molecules are separated by DNA topoisomerase IV.

Related topics

(Section C) Prokaryotic and eukaryotic chromosome structure	(D1) DNA replication: an overview
	(D3) Eukaryotic DNA replication

Experimental systems

Much of what we know about DNA replication has come from the use of *in vitro* systems consisting of purified DNA and all the proteins and other factors required for its complete replication. However, the chromosome of a bacterium such as *E. coli* is too large and fragile to be studied in this way. So, smaller and simpler bacteriophage and plasmid DNA molecules have been used extensively as model systems. Of these, the genetically simple phages M13 and φX174 have provided good models for the replication of the *E. coli* chromosome as they rely almost exclusively on cellular replication factors for their own replication. Their replicative forms comprise a supercoiled circle of only about 5.5–6.5 kb (Section M2). Larger phages, for example T7 (40 kb) and T4 (166 kb), encode many of their own replication proteins and employ some unusual mechanisms to complete their replication. Whilst not good model systems, they have been most informative in their own right and show the variety of solutions available for the complex problem of DNA replication. The description below relates to DNA replication in *E. coli*.

Initiation

One aspect of bacterial DNA replication for which phages and plasmids are not good models is initiation at the origin. This is because most of these DNAs are designed to replicate many times within a single cell division cycle whereas replication of the bacterial chromosome is tightly coupled to the growth cycle. The *E. coli* origin is within the genetic locus *oriC* and is bound to the cell membrane. This region has been cloned into plasmids that have had their own origins deleted to produce more easily studied **mini-chromosomes**, which behave like the *E. coli* chromosome. *OriC* contains multiple 9-bp binding sites for the initiator protein DnaA. Synthesis of this protein is coupled to growth rate so that the initiation of replication is also coupled to growth rate. At high cellular growth rates, bacterial chromosomes can re-initiate a second round of replication at the two new origins before the first round is completed. In this case, the daughter cells receive chromosomes that are already partly replicated (Figure 1). Once it has attained a critical level, DnaA protein forms a complex of 30–40 molecules, each bound to an ATP molecule, around which the *oriC* DNA becomes wrapped (Figure 2). This process requires that the DNA be negatively supercoiled (Section B3). This facilitates **melting** of three 13-bp AT-rich repeat sequences, which open to create a 'bubble.' The single strands of the bubble are coated with **single-stranded binding protein** (**Ssb**) to protect them from breakage and to prevent the DNA renaturing. Next, the hexameric, ring-shaped **DnaB** protein (Figure 3) binds to both strands (assisted by DnaC protein), one at each end of the bubble, to form a **pre-priming complex** (Figure 2). DnaB is a **DNA helicase**. Helicases are enzymes that use the energy of ATP hydrolysis to move into and melt double-stranded DNA (or RNA). The enzyme **DNA primase** then attaches to each DnaB helicase and synthesizes a short **RNA primer** to initiate synthesis of the **leading strands** of what are now two replication forks. Bidirectional replication then follows as described in Section D1.

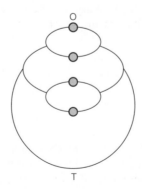

Figure 1. Re-initiation of bacterial replication at new origins (O) before completion of the first round of replication. T=terminus.

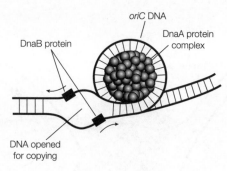

Figure 2. Opening of origin DNA for copying by wrapping of *oriC* around a DnaA protein complex.

Unwinding

For replication to proceed away from the origin, the DnaB helicases must travel along the template strands to open the double helix for copying. In a closed-circular DNA molecule, however, removal of helical turns at the replication fork leads to the introduction of additional turns in the rest of the molecule in the form of positive supercoiling (Section B3). Although the natural negative supercoiling of circular DNA partially compensates for this, it is insufficient to allow continued progression of the replication forks. This positive supercoiling must be relaxed continuously by the introduction of further negative supercoils by a type II **topoisomerase** called **DNA gyrase** (Section B3). Inhibitors of DNA gyrase, such as novobiocin and oxolinic acid, are effective inhibitors of bacterial replication and have antibiotic activity.

Elongation

At each replication fork, the DnaB helicase moves 5′→3′ on the lagging strand template displacing this strand while the leading strand is copied (Section D1). The displaced strand is continually covered and protected by **Ssb protein** (Figure 3). DNA primase then synthesizes RNA primers every 1000–2000 nt on the lagging strand. Both leading and lagging strand primers are elongated by **DNA polymerase III holoenzyme**. This multisubunit complex is a dimer, one half synthesizing the leading strand and the other the lagging strand. Having two polymerases in a single complex ensures that both strands are synthesized at the same rate. Despite the opposite polarity of the two DNA strands and the absolute need for 5′→3′ polymerization, the complete holoenzyme complex moves in the same physical direction (into the replication fork). This is made possible by looping of the lagging strand (Figure 3). Both halves of the dimer contain an α subunit, the actual polymerase, and an ε (**epsilon**) subunit, which is a **3′→5′ proofreading exonuclease**. Proofreading helps to maintain high replicational fidelity (Section E2). The dissociable β subunits (also known as sliding clamps) clamp the polymerases to the DNA. The τ (**tau**) subunits and the γ-**complex** (or **clamp loader**) hold the two polymerases together. Once the lagging strand primers have been elongated by DNA polymerase III, they are removed and the gaps filled by **DNA polymerase I**. This enzyme has 5′→3′ polymerase, 5′→3′ exonuclease and 3′→5′ proofreading exonuclease activities on a single polypeptide chain. The 5′→3′ exonuclease removes the primers while the polymerase function simultaneously fills the gaps with DNA by elongating the 3′-end of the adjacent Okazaki fragment (Figure 4). The final phosphodiester bond between the fragments is made by **DNA ligase**. The enzyme from *E. coli* uses the cofactor NAD⁺ as an unusual energy source.

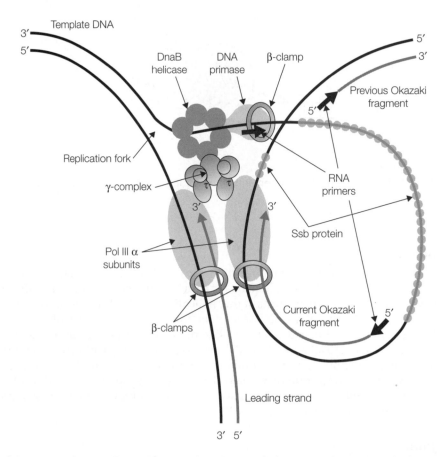

Figure 3. Polymerase cycling at the *E. coli* replication fork. DNA polymerase III holoenzyme comprises two catalytic α subunits, two β clamps, two τ subunits, two ε and θ subunits (not shown), and a multisubunit γ-complex that associates with the DnaB helicase. As the fork unwinds, the helicase-associated primase makes an RNA primer opposite the newly displaced lagging strand template, most of which is covered by Ssb protein. The γ-complex loads a β-clamp on to this primer. Once the current Okazaki fragment has been completed, the lagging strand polymerase dissociates from its β-clamp and swivels around the attached γ-complex to associate with the β-clamp on the new primer, releasing a loop of copied lagging strand DNA carrying the old β-clamp (which dissociates). The cycle then repeats.

The NAD^+ is cleaved to nicotinamide mononucleotide (NMN) and AMP, and the AMP is covalently attached to the 5'-end of one fragment via a 5'–5' linkage. The energy available from the hydrolysis of the pyrophosphate linkage of NAD^+ is thus maintained and subsequently used to link this 5'-end to the 3'-OH of the adjacent fragment with the release of AMP. *In vivo*, the DNA polymerase III holoenzyme dimer, helicase and primase are believed to be physically associated in a large, single complex called a **replisome**, which synthesizes DNA at a rate of 900 bp/s (Figure 3).

Termination and segregation

The two replication forks meet approximately 180° opposite *oriC*. Around this region are several terminator sites, which arrest the movement of the forks by binding the ***tus*** gene

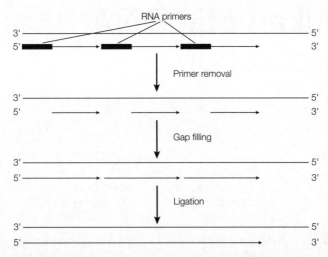

Figure 4. Removal of RNA primers from lagging strand fragments, and subsequent gap filling and ligation. Primer removal and gap filling by DNA polymerase I occur simultaneously.

product, an inhibitor of the DnaB helicase. Hence, if one fork is delayed for some reason, they will still meet within the terminus. Once replication is completed, the two daughter circles remain interlinked (**catenated**). They are unlinked by **topoisomerase IV**, a type II DNA topoisomerase (Section B3). They can then be segregated into the two daughter cells by movement apart of their membrane attachment sites (Section C1).

D3 Eukaryotic DNA replication

Key Notes

Experimental systems	Small animal viruses such as simian virus 40 are good models for elongation but not initiation. Yeast, with only 400 replicons, is much simpler than a typical mammalian cell, which may have 50 000–100 000 replicons. Extracts of *Xenopus* eggs replicate added DNA very efficiently.
Origins and initiation	Origins are activated at different times during S-phase in clusters of about 20–50. Individual yeast origins have been cloned, and all contain copies of a simple 11-bp consensus sequence that bind the Origin Recognition Complex. Licensing factors ensure that each origin is used only once per cell cycle.
Replication forks	Chromatin is disassembled ahead of the replication fork and reassembled after it has passed. Leading and lagging strands are initiated by the primase and polymerase activities of DNA polymerase α. DNA polymerase ε then takes over elongation of the leading strand while DNA polymerase δ elongates the lagging strand primers. DNA ligase I joins the fragments after primer removal. Archaeal DNA replication has many similarities to eukaryotic replication.
Nuclear matrix	Replication is spatially organized by a protein scaffold of insoluble protein fibers. Replication factories are immobilized on this matrix and the DNA moves through them.
Telomere replication	Telomeric DNA consists of multiple copies of a simple repeat sequence. The 3′-ends overhang the 5′-ends. Telomeric DNA is copied by telomerase, which carries the template for the repeat as a small RNA molecule. Telomerase is repressed in somatic cells but reactivated in many cancer cells.
Related topics	(Section C) Prokaryotic and eukaryotic chromosome structure (D1) DNA replication: an overview (D2) Bacterial DNA replication

Experimental systems

Because of the complexities of chromatin structure (Sections C2 and C3), eukaryotic replication forks move at around 50 bp/s. At this rate, it would take about 30 days to copy a typical mammalian chromosomal DNA molecule of 10^5 kb using two replication forks, hence the need for multiple replicons – typically 50 000–100 000 in a mammalian cell. To study all these simultaneously would be a daunting prospect. Fortunately, the yeast

Saccharomyces cerevisiae has a much smaller genome (14 Mb in 16 chromosomes) and only 400 replicons. The complete genome sequence is known. Simpler still are viruses such as simian virus 40 (SV40) (Section M3). SV40 DNA is a 5-kb double-stranded circle and forms nucleosomes when it enters the cell nucleus. It provides an excellent model for a mammalian replication fork. Another useful system is a cell-free extract prepared from the eggs of the African clawed frog, *Xenopus laevis*. Because of its high concentration of replication proteins (sufficient for several cell doublings *in vivo*), this extract can support the extensive replication of added DNA or even whole nuclei.

Origins and initiation

In eukaryotes, **clusters** (**tandem arrays**) of about 20–50 replicons initiate simultaneously at defined times throughout S-phase. Those that replicate early in S-phase comprise predominantly **euchromatin** (which includes transcriptionally active DNA) while those activated late in S-phase are mainly within **heterochromatin** (Section C3). Centromeric and telomeric DNA replicate last. This pattern reflects the accessibility of the different chromatin structures to initiation factors. Individual yeast replication origins have been cloned into bacterial plasmids. Since they allow these plasmids to replicate in yeast (a eukaryote) they are termed **autonomously replicating sequences** (**ARSs**). These yeast origins are about 200 bp in length and contain four 'subdomains' each of which includes an 11-bp AT-rich **consensus sequence** [A/T]TTTAT[A/G]TTT[A/T]. Two of these domains bind a set of six proteins, the **origin recognition complex** (**ORC**), which leads to opening of the DNA in one of the other subdomains. Defined origin sequences have not been isolated from mammalian cells, and it is believed that initiation of each replicon may occur at any one of several preferred but poorly defined sites within an **initiation zone,** which may be several kb in length. In contrast to bacteria, eukaryotic replicons can only initiate once per cell cycle. This is achieved through the involvement of proteins called **licensing factors**. Cdc6 and Cdt1 are synthesized and bind to the ORC only during G1 phase (Section N1). The **MCM helicase** (functionally equivalent to DnaB in *E. coli*) then binds to the ORC, but is inhibited by Cdc6 and Cdt1. At the start of S-phase, cyclin dependent kinase (CDK, Section N1) phosphorylates Cdc6 causing it to be degraded (in yeast), or exported from the nucleus (in metazoa). Cdt1 also becomes inhibited and further proteins bind to and activate the ORC, allowing MCM to unwind the DNA for copying. Since no new Cdc6 is available until the next G1 phase, any given origin cannot reinitiate until the next cell cycle. This system ensures that none of the complex eukaryotic genome is over- or under-replicated.

Replication forks

Before copying, the DNA must be unwound from the nucleosomes at the replication forks (Figure 1). This slows the movement of the forks to about 50 bp/s. After the fork has passed, new nucleosomes are assembled from a mixture of old and newly synthesized histones. As in bacteria, a DNA helicase (MCM) and a single-stranded binding protein, **replication protein A** (**RP-A**), are required to separate the strands. Three different DNA polymerases are involved in elongation. The leading strand and each lagging strand fragment are initiated with RNA by a primase activity that comprises two of the four subunits of **DNA polymerase α** — for that reason also known as **DNA polymerase-primase**. The polymerase function of this protein continues elongating the RNA primers with DNA, yielding a hybrid RNA–DNA primer which is further elongated by **DNA polymerase-ε** (**epsilon**) on the leading strand and the related but distinct **DNA polymerase-δ** (**delta**) on the lagging strand fragments. Both these enzymes have associated proofreading activity.

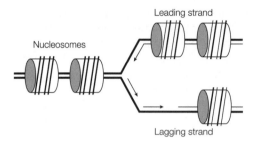

Figure 1. Nucleosomes at a eukaryotic replication fork.

These DNA polymerases are held on to the DNA by **proliferating cell nuclear antigen** (**PCNA**), the functional equivalent of the β subunit of *E. coli* DNA polymerase III holo-enzyme (Section D2). Primers are removed, not by an exonuclease as in *E. coli*, but by displacement by the growing 3′-end of the next fragment to be made, followed by cleav-age by the **Fen1** and **Dna2 endonucleases**. **DNA ligase I**, one of three eukaryotic DNA ligases, finally joins the fragments together. In addition to doubling the DNA, the histone content of the cell is also doubled during S-phase.

The DNA replication machinery of archaea is a simplified version of that found in eukary-otes and is evolutionarily distinct from bacterial systems. Although archaeal chromo-somes are small and circular like those of bacteria, some are complexed with histone-like proteins and some have multiple replication origins, which are bound by ORC-related proteins. The processing of Okazaki fragments occurs in a manner similar to that in eukaryotes.

Nuclear matrix

The nuclear matrix is a scaffold of insoluble protein fibers, which acts as an organizational framework for nuclear processes, including DNA replication (Sections C2 and C3). Huge **replication factories** containing all the enzymes and DNA associated with the replication forks of all replicons within a cluster are immobilized on the matrix, and the DNA moves through these sites as it replicates. These factories can be visualized in the microscope by pulse-labeling the replicating DNA with the thymidine analog, **bromodeoxyuridine** (**BUdR**), and visualizing the labeled DNA by immunofluorescence using an antibody that recognizes BUdR (Figure 2).

Telomere replication

The ends of linear eukaryotic chromosomes cannot be fully replicated by semi-discontinuous replication as there is no DNA to elongate to displace the RNA primers from the very 5′-ends of the lagging strands at each end of the chromosomal DNA molecule. Instead, these terminal primers are degraded by an **RNAse H**, which specifically hydrolyzes the RNA of an RNA–DNA hybrid duplex, leaving a short region of DNA uncopied. Thus, after multiple cell cycles, genetic information could be lost from the chromosome. To overcome this, the ends of eukaryotic chromosomes (**telomeres**, Section C3) consist of hundreds of copies of a simple, highly conserved, noninformational repeat sequence (e.g. 5′-TTAGGG-3′ in vertebrates) with the 3′-end overhanging the 5′-end (Figure 3). The enzyme **telomerase**, a type of **reverse transcriptase** (Section M1) is associated with a short RNA molecule, part of whose sequence is complementary to this repeat (5′--NNCCCUAANN--3′). This RNA acts as a template for the addition of these

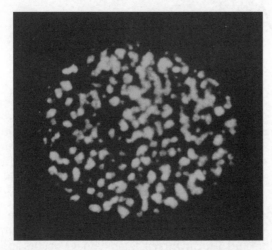

Figure 2. Replication factories in a eukaryotic nucleus. Reproduced from Siegel and Berezhny, *Journal of Cell Biology* (1999) **146**, with permission from Rockefeller University Press.

$$3'-\text{A A T C C C A A T C C C}-5'$$
$$5'-\text{T T A G G G T T A G G G (T T A G G G)}_n\ \text{T T A G G G}-3'$$

Figure 3. Sequence of human telomeric DNA (n=several hundred).

DNA repeats to the 3′-overhang by repeated cycles of elongation (polymerization) and translocation (movement of the template RNA to the next repeat). The complementary strand is then synthesized by normal lagging strand synthesis. Telomerase is always expressed in germ cells but its activity is repressed in the somatic cells of multicellular organisms, resulting in a gradual shortening of the chromosomes with each cell generation. As this shortening reaches informational DNA, the cells senesce and die. This phenomenon may be important in aging. The unlimited proliferative capacity and immortality of many cancer cells is associated with reactivation of telomerase activity.

E1 DNA damage

Key Notes

DNA lesions

The reactivity of DNA with exogenous chemicals or radiation can give rise to changes in its chemical or physical structure. These may block replication or transcription and so be lethal, or they may generate mutations through direct or indirect mutagenesis. The chemical instability of DNA can generate spontaneous lesions such as deamination and depurination.

Oxidative damage

Reactive oxygen species such as superoxide and hydroxyl radicals produce a variety of lesions including 8-oxoguanine and 5-formyluracil. Such damage occurs spontaneously but is increased by some exogenous agents including γ-rays, which also cause strand breaks.

Alkylation

Electrophilic alkylating agents such as methylmethane sulfonate and ethylnitrosourea can modify nucleotides in a variety of positions. Most lesions are indirectly mutagenic, but O^6-alkylguanine is directly mutagenic. Some alkylating agents crosslink the DNA.

Bulky adducts

Bulky lesions such as pyrimidine dimers and arylating agent adducts distort the double helix and cause localized denaturation. This disrupts the normal functioning of the DNA and can be both lethal and mutagenic.

Related topics

(A2) Nucleic acid structure and function	(E4) Recombination and transposition
(E2) Mutagenesis	(N4) Apoptosis
(E3) DNA repair	

DNA lesions

DNA is constantly exposed to internal and external chemical and physical agents that can modify the reactive nitrogen and carbon atoms in the heterocyclic ring systems of the bases and some of the exocyclic functional groups (i.e. the keto and amino groups of the bases), creating **lesions** – alterations to the normal chemical or physical structure of the DNA. The modified chemistry of the bases may lead to loss of base pairing or alternative base pairing (e.g. a modified A may base-pair with C instead of T). If such a lesion were allowed to remain in the DNA, a **mutation** could arise and become fixed in the DNA by **direct** or **indirect mutagenesis** (Section E2). Mutations in the germ line can be inherited while mutations in somatic cells can lead to altered cell function, including **carcinogenesis**. Alternatively, the chemical change may produce a physical distortion in the DNA that blocks replication and/or transcription, causing cell death. Thus, DNA lesions may be **mutagenic** and/or **lethal**. Some lesions are **spontaneous** and occur because of the inherent chemical reactivity of the DNA and the presence of normal, reactive chemical species within the cell. For example, the base C undergoes spontaneous hydrolytic **deamination** to give U (Figure 1). If left unrepaired, the resulting U would form a base pair with A during subsequent replication, giving rise to a point mutation (Section E2). In fact, the generation of U in DNA in this way is the probable reason why DNA contains T instead of U. Any U found in DNA is removed by an enzyme called uracil DNA-glycosylase and is replaced by C (Section E3). 5-Methylcytosine, a modified base found in small amounts in DNA (Sections A2 and C3), deaminates to T, a normal base, which can consequently escape repair.

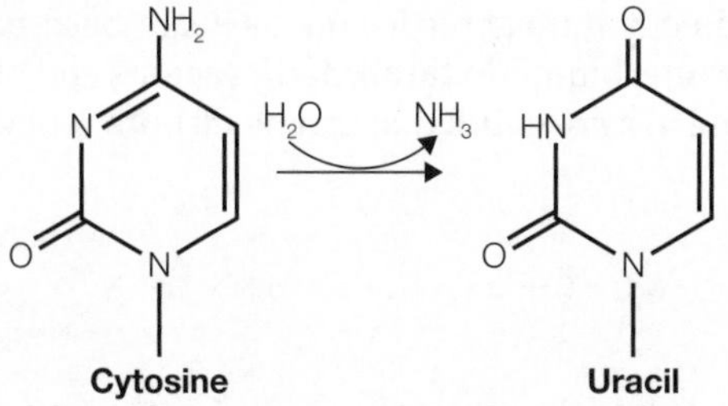

Figure 1. Spontaneous hydrolytic deamination of cytosine to uracil.

Depurination is another spontaneous hydrolytic reaction that involves cleavage of the **N-glycosylic (N-glycosidic)** bond between *N*-9 of the purine bases A and G and *C*-1′ of the deoxyribose sugar and hence loss of purine bases from the DNA. The sugar-phosphate backbone of the DNA remains intact. The resulting **apurinic (AP) site** is a **noncoding** lesion, as information encoded in the purine base is lost. Depurination in mammalian cells occurs at the rate of 10 000 purines lost per cell per day at 37°C. Though less frequent, **depyrimidination** can also occur. The term **abasic site** applies when either a purine or a pyrimidine is lost.

Oxidative damage

This occurs under normal conditions due to the presence of reactive oxygen species (ROS) in all aerobic cells, for example superoxide, hydrogen peroxide, and, most importantly, the hydroxyl radical (·OH). This radical can attack DNA at a number of points, producing a range of oxidation products with altered properties, for example 8-oxoguanine, 2-oxoadenine, and 5-formyluracil (Figure 2). The levels of these can be greatly increased by hydroxyl radicals generated from the radiolysis of water by ionizing radiation (X-rays and γ-rays). 8-oxo-G is directly mutagenic as it can base-pair with adenine during replication (Section E2). Oxidative damage also leads to single and double **strand breaks**, the latter being particularly dangerous (Sections E3 and E4).

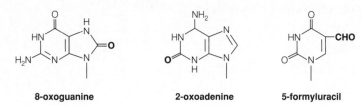

8-oxoguanine   2-oxoadenine   5-formyluracil

Figure 2. Examples of oxidized bases produced in DNA by reactive oxygen species.

Alkylation

Alkylating agents are electrophilic chemicals that readily add alkyl (e.g. methyl) groups to various positions on nucleic acids distinct from those methylated by normal methylating enzymes (Section O3). Common examples are **methylmethane sulfonate** (**MMS**) and **ethylnitrosourea** (**ENU**) (Figure 3a). Typical examples of methylated bases are 7-methylguanine, 3-methyladenine, 3-methylguanine and O^6-methylguanine (Figure 3b). Some of these lesions are potentially lethal as they can interfere with the unwinding of DNA during replication and transcription. Most are also indirectly mutagenic; however, O^6-methylguanine is a directly mutagenic lesion, as it can base-pair with thymine during replication (Section E2). Some bifunctional alkylating agents and other chemicals can react with both strands resulting in a **crosslink** that can block both replication and transcription.

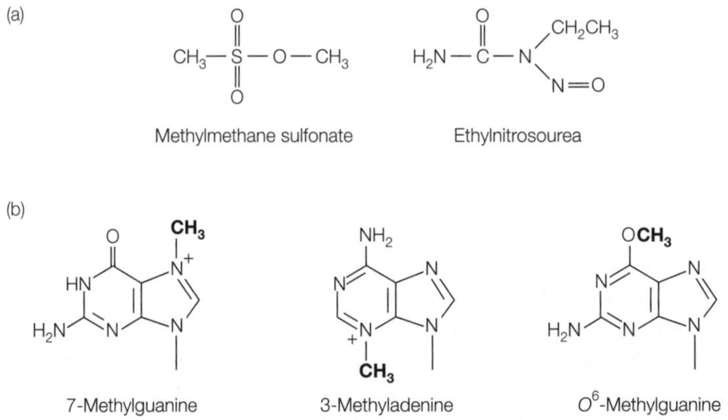

Figure 3. Examples of (a) alkylating agents; (b) alkylated bases.

Bulky adducts

Cyclobutane pyrimidine dimers are formed by ultraviolet light (a form of nonionizing radiation) from adjacent pyrimidines on one strand by cyclization of the double-bonded C5 and C6 carbon atoms of each base to give a cyclobutane ring (Figure 4a). The resulting loss of base pairing with the opposite strand causes localized denaturation of the DNA producing a **bulky lesion**, which can disrupt replication and transcription, and lead to indirect mutagenesis (Section E2). Another type of pyrimidine dimer, the **6,4-photoproduct**, results from the formation of a bond between C6 of one pyrimidine base and C4 of the adjacent base (see ring carbon numbers in Figure 4a). When the coal tar carcinogen **benzo[a]pyrene** is metabolized by cytochrome P-450 in the liver, one of its metabolites (a diol epoxide) can covalently attach to the 2-amino group of guanine residues (Figure 4b). Many other aromatic **arylating agents** form covalent **adducts** with DNA. The liver carcinogen **aflatoxin B1**, produced by *Aspergillus* fungi that sometimes contaminate crop plants, also covalently binds to DNA.

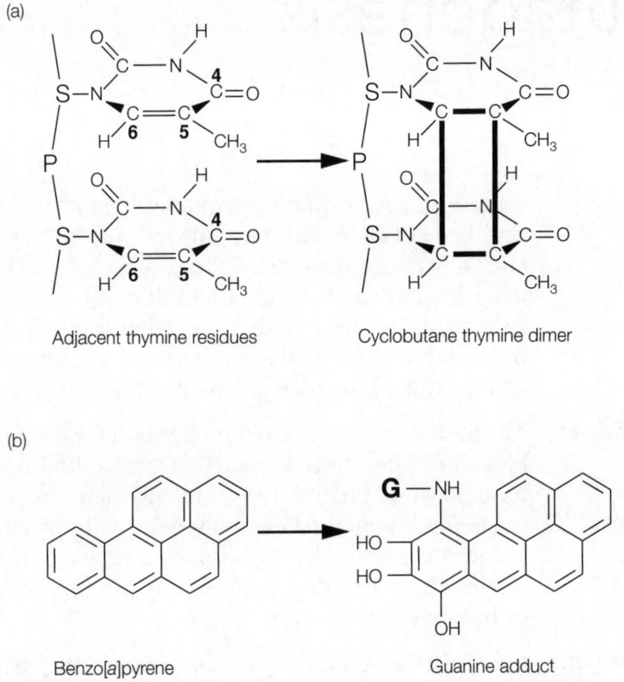

(a)

Adjacent thymine residues   Cyclobutane thymine dimer

(b)

Benzo[a]pyrene   Guanine adduct

Figure 4. Formation of (a) cyclobutane thymine dimer from adjacent thymine residues. S=sugar, P=phosphate; (b) guanine adduct of benzo[a]pyrene diol epoxide.

E2 Mutagenesis

Key Notes

Mutation

Mutations are heritable permanent changes in the base sequence of DNA. Point mutations may be transitions (e.g. G–C→A–T) or transversions (e.g. G–C→T–A). Deletions and insertions involve the loss or addition of bases and can cause frameshifts in the genetic code. Silent mutations have no phenotypic effect, while missense and nonsense mutations change the amino acid sequence of the encoded protein.

Replication fidelity

The high accuracy of DNA replication (one error per 10^{10} bases incorporated) depends on a combination of (i) proper base pairing of template strand and incoming nucleotide in the active site of the DNA polymerase, (ii) proofreading of the incorporated base by $3'{\rightarrow}5'$ exonuclease and (iii) mismatch repair. The use of RNA primers ensures high fidelity of initiation events.

Physical mutagens

Ionizing (e.g. X- and γ-rays) and nonionizing (e.g. UV) radiation produce a variety of DNA lesions. Pyrimidine dimers are the commonest product of UV irradiation.

Chemical mutagens

Base analogs can mispair during DNA replication to cause mutations. Nitrous acid deaminates cytosine and adenine. Alkylating and arylating agents generate a variety of adducts that can block transcription and replication and cause mutations by direct or, more commonly, indirect mutagenesis. Most chemical mutagens are carcinogenic.

Direct mutagenesis

If a base analog or modified base whose base pairing properties are different from the parent base is not removed by a DNA repair mechanism before passage of a replication fork, then an incorrect base will be incorporated. A second round of replication fixes the mutation permanently in the DNA.

Indirect mutagenesis and translesion DNA synthesis

Most lesions in DNA are repaired by error-free direct reversal or excision repair mechanisms before passage of a replication fork. If this is not possible, an error-prone form of translesion DNA synthesis may take place involving specialized DNA polymerases and one or more incorrect bases become incorporated opposite the lesion.

Related topics

(A2) Nucleic acid structure and function	(E3) DNA repair
(Section D) DNA replication	(E4) Recombination and transposition
(E1) DNA damage	

Mutation

Mutations are permanent, heritable alterations in the base sequence of the DNA. They arise either through spontaneous errors in DNA replication, or meiotic recombination (Section E4), or as a consequence of the damaging effects of physical or chemical agents on the DNA. The simplest mutation is a **point mutation** — a single base change. This can be either a **transition**, in which one purine (or pyrimidine) is replaced by the other, or a **transversion**, where a purine replaces a pyrimidine or vice versa. The phenotypic effects of such a mutation can be various. If it is in a noncoding or nonregulatory piece of DNA or in the third position of a codon, which often has no effect on the amino acid incorporated into a protein (Section K1), then it may be **silent**. If it results in an altered amino acid in a gene product then it is called a **missense** mutation, and its effect can vary from nothing to lethality, depending on the amino acid affected. Mutations that generate new stop codons are **nonsense** mutations and give rise to **truncated** (shortened) protein products. Insertions or deletions involve the addition or loss of one or more bases. These can produce **frameshift** mutations in genes, where the translated protein sequence to the C-terminal side of the mutation is completely changed. Mutations that affect the processes of cell growth and cell death can result in carcinogenesis (Sections N2 and N3). The accumulation of many silent and other nonlethal mutations in populations produces **genetic polymorphisms** – acceptable variations in the 'normal' DNA and protein sequences (Section C4).

Replication fidelity

The error rate of DNA replication is much lower than that of transcription because of the need to preserve the meaning of the genetic message from one generation to the next. For example, the **spontaneous mutation** rate in *E. coli* is about one error per 10^{10} bases incorporated during replication. One reason for replication errors is the presence of the minor tautomeric forms and conformational isomers of the bases, which have altered base pairing properties (Figure 1b, Section B1). For example, the (minor) *enol* form of G can form a base pair with the (major) *keto* form of T that has the same dimensions as normal 'Watson–Crick' A–T and G–C base pairs and so appears normal. Small additions of extra bases or deletions can also occur spontaneously during replication. The error rate is minimized by a variety of mechanisms. First, DNA polymerases will only efficiently incorporate an incoming nucleotide if it forms the correct base pair with the template nucleotide in its active site (base selection). Secondly, the occasional error (Figure 1a) is detected by the 3'→5' **proofreading exonuclease** associated with the polymerase (Section D2) because it has the wrong shape. This removes the incorrect nucleotide from the 3'-end before any further incorporation, allowing the polymerase then to insert the correct base (Figure 1b). In order for the proofreading exonuclease to work properly,

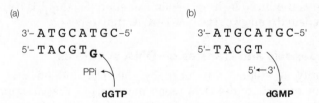

Figure 1. Proofreading of newly synthesized DNA. If, for example, a 'G' residue is wrongly incorporated opposite a template 'T' (a), it is removed by the 3' →5' proofreading exonuclease (b).

it must be able to distinguish a correct base pair from an incorrect one. The increased mobility of 'unanchored' base pairs at the very 5'-end of newly initiated lagging strand fragments of DNA (Section D1) means that they can never appear correct and so cannot be proofread. Hence, **ribo**nucleotides are used for the first few nucleotides (the RNA primer) so that they subsequently can be identified as low fidelity material and replaced later with DNA elongated (and proofread) from the adjacent fragment. Finally, errors that escape proofreading are corrected by **mismatch repair** (Section E3).

Physical mutagens

Absorption of high-energy ionizing radiation such as X-rays and γ-rays causes the target molecules to lose electrons. These electrons can cause extensive chemical alterations to DNA, including **strand breaks**, both single and double, and base and sugar destruction. Nonionizing radiation causes molecular vibrations or promotion of electrons to higher energy levels within the target molecules. This can lead to the formation of new chemical bonds. The most important form causing DNA damage is UV light, which produces **pyrimidine dimers** from adjacent pyrimidine bases (Section E1).

Chemical mutagens

Base analogs are derivatives of the normal bases with altered base pairing properties and can cause **direct mutagenesis**. A wide range of other natural and synthetic organic and inorganic chemicals can react with DNA and alter its properties. **Nitrous acid** promotes the deamination of C to U (Figure 1, Section E1), which base-pairs with A and causes G–C→A–T transitions upon subsequent replication. Deamination of A to the guanine analog hypoxanthine results in A–T→G–C transitions. **Alkylating agents**, such as **MMS** and **ENU**, and **arylating agents** (Section E1) produce **lesions** that usually have to be repaired to prevent serious disruption to the processes of transcription and replication. Processing of these lesions by the cell may give rise to mutations by **indirect mutagenesis**. Intercalators (Section B3) generate insertion and deletion mutations. Most chemical mutagens are **carcinogens** and cause cancer.

Direct mutagenesis

Direct mutagenesis results from the presence of a stable, modified base with altered base pairing properties in the DNA. All that is required for this lesion (nonpermanent) to be fixed as a mutation (permanent and heritable) is DNA replication. For example, the *keto* tautomer of the thymine analog 5-bromouracil base-pairs with adenine as expected, but the frequently occurring *enol* form (due to electronegativity of Br) pairs with guanine (Figure 2a). After one round of replication, a guanine may be inserted opposite the 5-bromouracil. After a second round of replication, a cytosine will be incorporated opposite this guanine and so the mutation is **fixed** (Figure 2b). The net effect is an A–T→G–C transition. 8-Oxo-G produced by intracellular oxidation (Section E1) can base-pair with A and, if left unrepaired, can give rise to A–T→C–G transversions.

Indirect mutagenesis and translesion DNA synthesis

The great majority of lesions introduced by chemical and physical mutagens are substrates for one or more of the DNA repair mechanisms that attempt to restore the original structure to the DNA in an error-free fashion before the damage is encountered by a replication fork (Section E3). Occasionally this is not possible, for example when a bulky or noncoding lesion is created immediately ahead of an advancing fork. This could lead to a potentially lethal stalling of the replication machinery so in such cases the cell

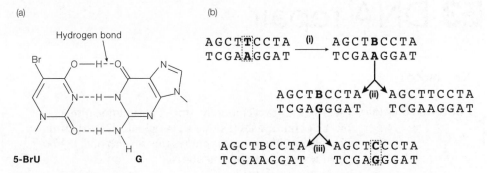

Figure 2. (a) Base-pairing of the enol form of 5-bromouracil with guanine; (b) direct mutagenesis by 5-bromouracil (B): (i) the 'T' analog 'B' can be readily incorporated in place of 'T'; (ii) after one round of replication 'G' may be incorporated opposite the 'B'; (iii) after a second round, a 'C' will be incorporated opposite the 'G'.

may resort to **translesion DNA synthesis** (**TLS**). This involves the temporary replacement of the normal replicative DNA polymerase by one of several specialized polymerases that are able to insert bases opposite the unrepaired lesion. After this, normal replication resumes. Thus, the lesion is not removed or repaired but tolerated and so the integrity of the replicated DNA is maintained. The lesion in the template strand can then be removed later, before the next round of replication.

All TLS polymerases lack proofreading activity and have relaxed active sites. Some are of relatively high fidelity and manage to incorporate the correct base opposite the lesion; however, others are of low fidelity and of necessity insert incorrect bases simply to ensure chromosomal integrity, leading to **indirect mutagenesis**. Indirect mutagenesis may be **targeted**, if the mutation occurs only at the site of the lesion, or **untargeted**, if mutations are also generated elsewhere in the genome. In *E. coli*, translesion DNA synthesis is part of the **SOS response** to DNA damage and is sometimes called **error-prone repair**, though it is not strictly a repair mechanism. The SOS response involves the induction of many genes whose functions are to increase survival after DNA damage. The damage-inducible **DNA polymerases IV** and **V** are essential for mutation induction by indirect mutagenesis. For example, pol V can insert a base opposite a noncoding lesion such as an **AP** site (Section E1) leading to targetted mutagenesis while, once induced, the error-prone pol IV can cause untargetted mutagenesis even where there is no lesion. In bacteria, untargetted mutagenesis may be a useful and deliberate response to a hostile, DNA-damaging environment in order to increase the mutation rate with the hope of producing a more resistant strain that could be selected for survival. Pol IV also creates many of the **adaptive mutations** that arise under the stressful conditions of stationary growth phase.

Mammalian cells have at least seven TLS polymerases, specialized for different types of lesion. **Pol-η** (**eta**) correctly inserts adenines opposite the linked thymines in a cyclobutane thymine dimer. **Pol-ι** (**iota**) is good at inserting the correct bases opposite a 6,4-photoproduct but inserts wrong bases opposite a cyclobutane dimer. These and **pol-κ** (**kappa**) can also bypass other noncoding lesions with varying degrees of fidelity. When some TLS polymerases insert a wrong base, they have difficulty extending the resulting mismatch at the 3'-end of the growing strand. In this case, **pol-ζ** (**zeta**) may continue polymerization before the normal replicative polymerase resumes. TLS is responsible for many of the point mutations occurring in cells, particularly the high level found in cancer cells.

E3 DNA repair

Key Notes

Photoreactivation
Cleavage of the cyclobutane ring of pyrimidine dimers by DNA photolyases restores the original DNA structure. Photolyases have chromophores that absorb blue light to provide energy for the reaction.

Alkyltransferase
An inducible protein specifically removes an alkyl group from the O^6 position of G and the O^4 position of T and transfers it to itself, causing inactivation of the protein.

Strand-break repair
After processing, most single-strand breaks are repaired in a manner similar to base excision repair. Breaks are detected by poly(ADP-ribose) polymerase, which leads to recruitment of repair factors. Double-strand breaks are repaired by homologous recombination or nonhomologous end joining.

Excision repair
In nucleotide excision repair, an endonuclease makes nicks on either side of the lesion, which is then removed to leave a gap. This gap is filled by a DNA polymerase, and a DNA ligase makes the final phosphodiester bond. In base excision repair, the lesion is removed by a specific DNA glycosylase. The resulting AP site is cleaved and expanded to a gap by an AP endonuclease plus exonuclease. Thereafter, the process is similar to nucleotide excision repair.

Mismatch repair
Replication errors that escape proofreading have a mismatch in the daughter strand. In *E. coli*, hemimethylation of the DNA after replication allows the daughter strand to be distinguished from the parental strand. The mismatched base is removed from the daughter strand by an excision repair mechanism.

Hereditary repair defects
Mutations in excision repair genes or a translesion DNA polymerase cause different forms of xeroderma pigmentosum, a sun-sensitive cancer-prone disorder. Other disorders include Cockayne syndrome, Fanconi anemia and ataxia telangiectasia.

Related topics

(A2) Nucleic acid structure and function
(Section D) DNA replication
(E1) DNA damage
(E2) Mutagenesis
(N4) Apoptosis

Photoreactivation

Cyclobutane pyrimidine dimers can be monomerized again by **DNA photolyases** (**photoreactivating enzymes**) in the presence of visible light. These enzymes have prosthetic groups (Section A3) that absorb blue light and transfer the energy to the cyclobutane ring, which is then cleaved. The *E. coli* photolyase has two chromophores,

N^5,N^{10}-methenyltetrahydrofolate, and reduced flavin adenine dinucleotide (FADH). Photoreactivation is specific for pyrimidine dimers. It is an example of **direct reversal** of a lesion and is error-free, but is not found in placental mammals.

Alkyltransferase

Another example of error-free direct reversal forms part of the **adaptive response** to alkylating agents. This response is induced in *E. coli* by low levels of alkylating agents and gives increased protection against the lethal and mutagenic effects of subsequent high doses. Mutagenic protection is afforded by an inducible **alkyltransferase** that removes the alkyl group from the directly mutagenic O^6-alkylguanines (which can base-pair with T, Section E1) and O^4-alkylthymines. Curiously, the alkyl group is transferred to a cysteine residue in the protein itself and inactivates it; thus, each alkyltransferase can only be used once. A similar protein specific for O^6-methylguanine is present in mammalian cells. Protection against lethality involves induction of a DNA glycosylase (see below) that removes other alkylated bases through **base excision repair**.

Strand-break repair

Single-strand breaks (**SSBs**) with 3'-OH and 5-P termini can simply be resealed by DNA ligase. However, strand breaks produced by ionizing radiation and other agents tend to have fragments of degraded sugar and bases attached that need to be removed.

In higher eukaryotes, SSBs are recognized by the enzyme **poly(ADP-ribose) polymerase-1** (**PARP1**), one of a family of ADP-ribosyltransferases. Upon binding to SSBs, PARP1 is activated and synthesizes long, branched-chain polymers of the unusual nucleic acid poly(ADP-ribose) (pADPR) using NAD$^+$ as substrate (Figure 1). PARP covalently attaches these chains to itself (**automodification**) and to other nuclear proteins including histones and repair proteins. This modification leads to changes in chromatin structure and the recruitment of repair proteins to the sites of damage by noncovalent

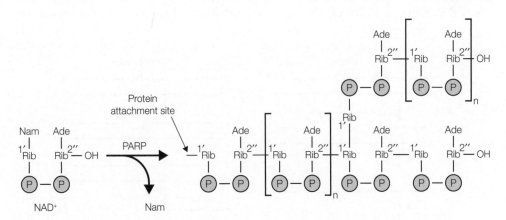

Figure 1. Synthesis of poly(ADP-ribose) from NAD$^+$. ADP-ribose units (shaded in NAD$^+$ structure) are linked between a 2' carbon of one unit and a 1' carbon of the next. See Section A2 for the identification of these carbon atoms. Branched chains are also linked via 1' and 2' carbons. Double prime marks are used to distinguish between these atoms in the two ribose moieties of an ADP-ribose unit. Nam=nicotinamide; Ade=adenine; Rib=ribose; PARP=poly(ADP-ribose) polymerase; $n \leq 60$.

binding to the pADPR chains and to PARP1 itself. One such protein is **XRCC1**, which acts as an organizing scaffold for many of the other proteins. Subsequent repair of the SSBs involves a process similar to short-patch BER (see below) and the pADPR is degraded.

Unrepaired SSBs are converted to **double-strand breaks** (**DSBs**) when encountered by a DNA replication fork, which can stall and collapse. DSBs created in this way or by ionizing radiation are potentially lethal as they can cause complete separation of a DNA molecule into two pieces. DSBs are repaired primarily by two mechanisms. **Homologous recombination** (**HR**) is largely error-free and is described in Section E4. **Nonhomologous end joining** (**NHEJ**) is not found in *E. coli* but is the main DSB repair pathway in mammalian cells, particularly during G1 phase and in nondividing cells when there is no sister copy of the DNA to participate in HR. It involves binding of the two broken ends by a protein, **Ku**, which then dimerizes to bring the two ends together for ligation by **DNA ligase IV**. However, if any bases have been lost from the ends or if there is more than one DSB, there is a risk of **deletions** and **chromosomal translocations** (joining together pieces from two different DNA molecules), as is often found in cancer cells.

Excision repair

These ubiquitous pathways operate on a wide variety of lesions and are essentially error-free. There are two forms, **nucleotide excision repair** (**NER**), operating mainly on bulky lesions, and **base excision repair** (**BER**), both with subdivisions. In NER, an endonuclease cleaves the DNA a precise number of bases on either side of the lesion (Figure 2a, step 1) and an oligonucleotide containing the lesion is removed, leaving a gap (Figure 2a, step 2). For example, in *E. coli*, the UvrABC endonuclease removes pyrimidine dimers and other bulky lesions by recognizing the distortion these produce in the double helix. In BER, individual modified bases are recognized by one of a group of relatively lesion-specific **DNA glycosylases** that cleave the *N*-glycosylic bond between the altered base and the sugar (Section A2), leaving an **apurinic** or **apyrimidinic** (**AP**) site (Figure 2b, step 1a). AP sites are also produced by spontaneous base loss (Section E1). An **AP endonuclease** then cleaves the DNA at this site and a gap may be created by further exonuclease activity (Figure 2b, steps 1b and 2). The gap is generally larger in NER and can be as small as one nucleotide in BER. From this point on, both forms of excision repair are essentially

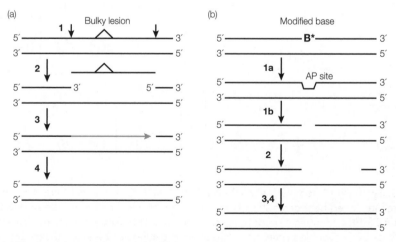

Figure 2. (a) Nucleotide excision repair; (b) base excision repair. Numbered stages are described in the text.

the same, though different enzymes are involved. In *E. coli*, the gap is filled by DNA polymerase I (Figures 2a and 2b, step 3) and the final phosphodiester bond made by DNA ligase (Figures 2a and 2b, step 4), much as in the final stages of processing of the lagging strand fragments during DNA replication (Section D1). In higher eukaryotes, the indirect strand breaks created by AP endonuclease recruit PARP1, which then recruits other repair factors (see above). Gap filling and ligation in **short-patch BER** (one nucleotide gap) involve **DNA polymerase β** and **DNA ligase III** whereas the longer gaps generated in **long-patch BER** and NER are filled by **DNA polymerases δ** or ε followed by **DNA ligase I** (as in replication). In eukaryotic NER, recognition and excision of DNA damage is a complex process involving at least 20 polypeptide factors. **Transcription-coupled NER** is a form of NER that favors lesions on the transcribed strands of transcriptionally active genes relative to the nontranscribed strands or transcriptionally inactive DNA. It is initiated mainly by stalling of **RNA polymerases** at lesions and uses protein factors such as the transcription factor **TFIIH** (Section H5) that are not used by **global NER**, which operates throughout the genome. Repair of damage to active genes that may encode essential proteins is clearly a priority for the cell.

Mismatch repair

This is a specialized form of excision repair that deals with any base mismatches produced during replication that have escaped proofreading (Section E2). In a replicational mispair, the wrong base is in the daughter strand. This system must, therefore, have a way of distinguishing the parental and daughter strands after the replication fork has passed to ensure that the mismatched base is removed only from the daughter strand. In *E. coli*, certain adenine residues are normally methylated in the sequence GATC on both strands (Section A2). Methylation of daughter strands lags several minutes behind replication. Thus, newly replicated DNA is **hemimethylated** – the parental strands are methylated but the daughter strands are not – so they can be readily distinguished. The mismatched base pair (e.g. GT or CA) is recognized and bound by a complex of the MutS and MutL proteins which then associates with the MutH endonuclease, which specifically nicks the daughter strand at a nearby GATC site. This nick initiates excision of a region containing the wrong base. In eukaryotes, strand discrimination is not methyl-directed but appears to rely on some form of strand asymmetry near the replication fork. Mismatch repair is clearly important in maintaining the overall error rate of DNA replication and, therefore, the spontaneous mutation rate: **hereditary nonpolyposis carcinoma of the colon** is caused by mutational loss of one of the human mismatch repair enzymes. Mismatch repair may also correct errors that arise from sequence misalignments during **meiotic recombination** in eukaryotes (Section E4).

Hereditary repair defects

Xeroderma pigmentosum (**XP**) is an autosomal recessive disorder characterized phenotypically by extreme sensitivity to sunlight and a high incidence of skin tumors. XP sufferers are defective in the NER of bulky DNA damage, including that caused by ultraviolet light. Defects in at least seven different genes can cause XP, indicating the complexity of excision repair in mammalian cells. **Xeroderma pigmentosum variant** (**XP-V**) is clinically very similar to classical XP but cells from XP-V individuals can carry out normal NER. In this case, the defect is in the gene encoding the TLS **pol-η**, which normally inserts A in an error-free fashion opposite the T residues in a cyclobutane thymine dimer (Sections E1 and E2). XP-V cells may, therefore, have to rely more heavily on alternative, error-prone TLS polymerases to maintain DNA integrity after radiation damage. Sufferers of **Cockayne**

syndrome are also sun-sensitive and defective in transcription-coupled excision repair, but are not cancer-prone. **Fanconi anemia** displays a defect in the repair of DNA strand crosslinks and some types of oxidative damage, while sufferers of **ataxia telangiectasia** have a mutation in **ATM kinase**, which controls and coordinates DSB repair.

E4 Recombination and transposition

Key Notes

Homologous recombination

The exchange of homologous regions between two DNA molecules occurs extensively in eukaryotes during meiosis. In bacteria, *recA*-dependent recombination involves a four-stranded Holliday intermediate that can resolve in two ways. The integrity of DNA containing double-strand breaks and unrepaired lesions can be maintained during replication by HR.

Site-specific recombination

The exchange of nonhomologous regions of DNA at specific sites is independent of *recA*. Integration of bacteriophage λ into the *E. coli* genome involves recombination at a 15-bp sequence present in both molecules and specific protein integration factors. Site-specific recombination also accounts for the generation of antibody diversity in animals and is useful in transgenic technology.

Transposition

Replicated copies of transposable DNA elements can insert themselves anywhere in the genome. All transposons encode a transposase that catalyzes the insertion. Retrotransposons replicate through an RNA intermediate and are related to retroviruses.

Related topics

(Section D) DNA replication   (E3) DNA repair
(E2) Mutagenesis   (M4) RNA viruses

Homologous recombination

Also known as **general recombination**, this process involves the exchange of homologous regions between two DNA molecules. In diploid eukaryotic germ cells, this occurs during **meiosis** when the homologous duplicated chromosomes line up in parallel in metaphase I and the nonsister chromatids exchange equivalent sections by **crossing over**. After meiosis is complete, the resulting haploid gametes contain information derived from both maternally and paternally inherited chromosomes, thus ensuring that an individual will inherit genes from all four grandparents. In somatic cells and in haploid micro-organisms, however, **HR** is used for the repair of DNA damage, including **DSBs** (Section E3). This requires the presence of two identical sections of DNA **duplex** (double helix) and so is restricted to growing cells that have at least partially replicated their genome but not yet divided. Bacteria may also use HR to incorporate features from DNA molecules taken up from the environment to introduce genetic variation.

In *E. coli* HR, the homologous DNA duplexes align with each other. One has a DSB, introduced either deliberately (as in meiosis in metazoa) or as a result of DNA damage (Figure 1a). **Strand resection** then takes place, whereby helicase and nuclease activities associated with the **RecBCD** protein complex trim back both strands at the break (Figure 1b). The single-stranded 3′ overhangs become coated with **RecA** protein and one of these protein-DNA filaments invades the other intact duplex by virtue of its complementarity, initially forming a **triplex** (three-stranded helix) in which the invading strand binds in the major groove of the intact duplex (Section A2). Displacement of the sister strand of the invaded duplex then creates a **heteroduplex** (a duplex made up of segments from the two different DNA molecules) and a **D-loop** (the displaced strand) (Figure 1c). The 3′-end of the invading strand is then extended by DNA polymerase I (Figure 1d) until the D-loop is large enough to capture and anneal to the 3′-end of the complementary strand of the broken DNA molecule (Figure 1e), whereupon this 3′-end is extended to copy the D-loop (Figure 1f). **Branch migration**, catalyzed by a combination of **RuvA** and **RuvB** proteins, then creates a second crossover (Figure 1g). The crossovers created in this way where four duplex DNA segments (two of which are heteroduplexes) are joined are called **Holliday junctions**. After ligation of the remaining breaks, these junctions can be **resolved** in one of two ways by the combined nicking and resealing action of **RuvC** endonuclease and DNA ligase. Crossover (Figure 1h) will occur if one junction is cut on the crossing strands and the other on the noncrossing strands while noncrossover (Figure 1i) results less frequently from cutting both junctions on the crossing strands.

In addition to repairing DSBs, HR can restore broken replication forks and allow replication forks to bypass bulky lesions. The lesions remain in the DNA but can be removed later by excision repair. In mammalian cells, the equivalent of RecA protein is **Rad51**. Rad51 associates with two proteins, **BRCA1** and **BRCA2**, which are also necessary for DSB repair. BRCA stands for BReast Cancer Associated and individuals with inherited mutations in either *BRCA* gene have a very high risk of developing breast and ovarian cancer as their cells have to rely on more error-prone modes of DNA repair than HR, including **NHEJ** (Section E3). PARP1 inhibitors have promise as anti-cancer drugs for individuals with *BRCA1* or *BRCA2* mutations as they cause accumulation of SSBs, which rapidly become DSBs in the rapidly replicating cancer cells (Section E3). As the *BRCA* cells are deficient in DSB repair, they are killed.

Errors in meiotic recombination can occur due the presence of repetitive sequences (Section C4) on the homologous chromosomes. **Unequal crossover** between misaligned chromosomes can lead to deletion or duplication of any gene between the repeats. Duplicated genes can evolve independently of the original and lead to gene/protein families of **paralogs** (Section S6). They can also become functionless **pseudogenes** by losing their original function through mutation without gaining a new one and so may be lost again. **Processed pseudogenes** arise by **retrotransposition** (see below).

Figure 1. HR between two DNA duplexes. (a) a double-strand break is introduced into one duplex; (b) strand resection trims back the 5′-ends of the break leaving 3′ overhangs; (c) one 3′-end invades the other duplex, displacing the sister strand as a D-loop; (d) DNA polymerase expands the D-loop by extending this 3′-end; (e) the D-loop captures and anneals the other 3′-end of the break; (f) DNA polymerase copies the D-loop by extending this 3′-end; (g) continued copying leads to 'branch migration' as the two crossover points (Holliday junctions) move further apart; (h) resolution of the right-hand junction by cutting the crossing strands (horizontal arrows in 1g) and the left-hand junction by cutting the noncrossing strands (vertical arrows in 1g) leads to crossover between the original two duplexes; (i) resolution of both junctions by cutting the crossing strands (horizontal arrows in 1g) results in noncrossover.

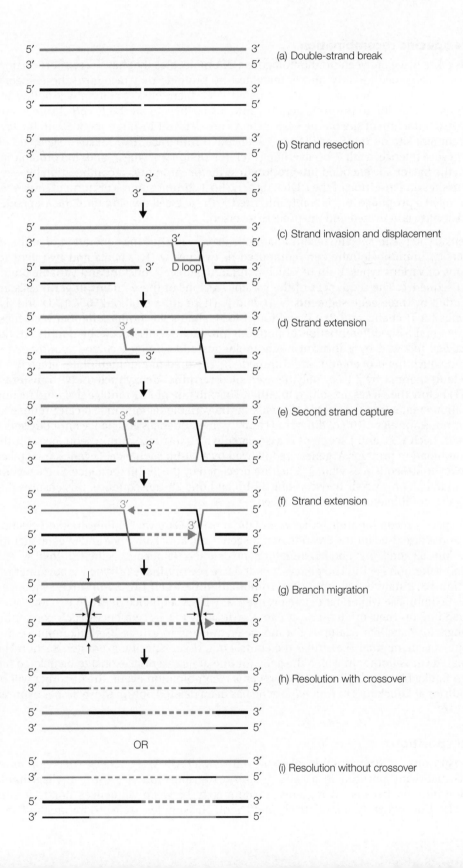

(a) Double-strand break

(b) Strand resection

(c) Strand invasion and displacement

D loop

(d) Strand extension

(e) Second strand capture

(f) Strand extension

(g) Branch migration

(h) Resolution with crossover

OR

(i) Resolution without crossover

Site-specific recombination

This takes place between nonhomologous DNA molecules that have very short sections of DNA sequence identity and is mediated by proteins that recognize these specific sequences. It does not require RecA-coated DNA filaments. Bacteriophage λ (Section M2) has the ability to insert its genome into a specific site on the *E. coli* chromosome. The **attP** attachment site on the λ DNA shares an identical 15-bp sequence with the **attB** attachment site on the *E. coli* DNA. A λ-encoded **integrase (Int)** makes staggered cuts in these sequences with 7-bp overhangs (cf. restriction enzymes), and, in combination with the bacterially encoded **integration host factor**, promotes recombination between these sites and insertion of the λ DNA into the host chromosome. This form of integrated λ is called a **prophage** and is stably inherited with each cell division until the λ-encoded **excisionase** is activated and the process reversed.

In eukaryotes, site-specific recombination is responsible for the generation of antibody diversity. Immunoglobulins are composed of two heavy (H) chains and two light (L) chains of various types, both of which contain regions of constant and variable amino acid sequence. The sequences of the variable regions of these chains in germ cells are encoded by three gene segments: V, D, and J. There are a total of 250 V, 15 D and five J genes for H chains and 250 V and four J for L chains. Recombination between these segments during differentiation of antibody-producing cells occurs by double strand-breakage followed by a form of nonhomologous end joining (see above and Section E3) in which the broken ends are trimmed and repaired in a nontemplate directed (i.e. random) manner by a DNA polymerase called **terminal deoxynucleotidyl transferase (TdT)** before the pieces are joined together. This mixture of TdT mutagenesis and recombination creates great sequence diversity in the variable region, and so can produce an enormous number ($>10^8$) of different H and L gene sequences and hence antibody specificities. Each V, D, and J segment is associated with short recognition sequences for the recombination proteins. V genes are followed by a 39-bp sequence, D genes are flanked by two 28-bp sequences while J genes are preceded by the 39-bp sequence. Since recombination can only occur between one 39-bp and one 28-bp sequence, this ensures that only VDJ combinations can be produced.

Site-specific recombination is the basis of the versatile Gateway® cloning method (Section P1) and is also used for the creation of transgenic and knockout mice and in gene therapy (Sections S2 and S5). A good system for gene deletion (knockout) is the **Cre-loxP** system from bacteriophage P1. The phage-encoded **Cre recombinase** (a Type I topoisomerase, Section B3) catalyzes site-specific recombination between two 34-bp **loxP** sequences. The recombinase will either delete or invert any DNA sequence between two loxP sites, depending on their orientation. It can also fuse two DNA sequences each containing a single loxP site. For example, if a mouse engineered to express Cre recombinase in a tissue-specifc manner (i.e. under the control of a tissue-specific promoter, Section H4) is mated with another in which the gene (or one of its essential exons) to be deleted has been flanked by two loxP sites by a previous recombination event, the offspring will be **conditional knockouts** in which the gene has been deleted only in those tissues expressing Cre.

Transposition

Transposons or **transposable elements** are small DNA sequences that can move to virtually any position in a cell's genome. Transposition is a form of **illegitimate recombination** because it requires no homology between sequences nor is it site-specific. Consequently, it is relatively inefficient. The simplest transposons are the *E. coli*

IS elements or **insertion sequences**, of which there may be several copies in the genome. These are 1–2 kb in length and comprise a **transposase** gene flanked by short (~20-bp) **inverted terminal repeats** (identical sequences but with opposite orientation). The transposase makes a staggered cut in the chromosomal DNA and, in a replicative process, a copy of the transposon inserts at the target site (Figure 2). The gaps are filled and sealed by DNA polymerase I and DNA ligase, resulting in a duplication of the target site and formation of a new **direct repeat** sequence. The gene into which the transposon inserts is usually inactivated, and genes between two copies of a transposon can be deleted by recombination between them. Inversions and other rearrangements of host DNA sequences can also occur. **Transposon mutagenesis** is a useful way of creating mutants. In addition to a transposase, the **Tn** transposon series carry other genes, including one for a β-**lactamase,** which confers penicillin resistance on the organism. The spread of antibiotic resistance among bacterial populations is a consequence of the transposition of resistance genes into plasmids, which replicate readily within the bacteria, and the ability of transposons to cross species barriers.

Many eukaryotic transposons have a structure similar to retroviral genomes (Section M4). The 6-kb yeast **Ty element** (with about 30 full-length copies per genome) encodes proteins similar to retroviral reverse transcriptases and integrases and is flanked by **long terminal repeat** (**LTR**) sequences. It replicates and moves to other genomic sites by transcription into RNA and subsequent reverse transcription of the RNA into a DNA duplex copy which then inserts elsewhere. The *copia* element from *Drosophila* (50 copies of a 5-kb sequence) is similar in structure. Such elements are called **retrotransposons** and it is believed that retroviruses are retrotransposons that have acquired the ability to exist outside the cell and pass to other cells. The dispersed repetitive sequences found in higher eukaryotic DNA (e.g. LINES and SINES) probably spread through the genome by transposition (Section C4). LINES are similar to retroviral genomes while the *Alu* element, a SINE, bears a strong resemblance to the **7SL RNA** component of the **signal recognition particle**, a complex involved in the secretion of newly synthesized polypeptides through the endoplasmic reticulum (Section L4), and probably originated as a reverse transcript of this RNA. **Processed pseudogenes** are produced by retrotransposition and, if they are copied from mRNAs, have no introns or promoters, unlike true pseudogenes (see above).

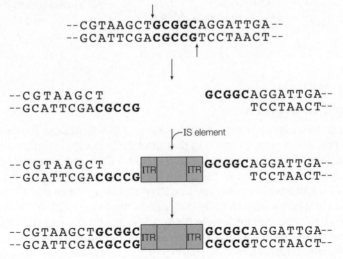

Figure 2. Transposition of an insertion sequence (IS) element into a host DNA with duplication of the target site (shown in bold). ITR, inverted terminal repeat sequence.

F1 Basic principles of transcription

Key Notes	
Transcription: an overview	Transcription is the synthesis of a single-stranded RNA from a double-stranded DNA template. RNA synthesis occurs in the 5′→3′ direction and its sequence corresponds to that of the sense strand of the DNA.
Initiation	RNA polymerase is the enzyme responsible for transcription. It binds to specific DNA sequences called promoters to initiate RNA synthesis. These sequences are upstream (to the 5′-end) of the region that codes for protein, and they contain short, conserved DNA sequences that are common to different promoters. The RNA polymerase binds to the dsDNA at a promoter sequence, resulting in local DNA unwinding. The position of the first synthesized base of the RNA is called the start site and is designated as position +1.
Elongation	RNA polymerase moves along the DNA and sequentially synthesizes the RNA chain. DNA is unwound ahead of the moving polymerase, and the helix is reformed behind it.
Termination	RNA polymerase recognizes the terminator, which ensures that no further ribonucleotides are incorporated. This sequence is commonly a hairpin structure. Some terminators require an accessory factor called rho for termination.
Related topics	(A2) Nucleic acid structure and function (F3) The *E. coli* σ⁷⁰ promoter (F2) *Escherichia coli* RNA polymerase (F4) Transcription initiation, elongation, and termination (G2) The *trp* operon

Transcription: an overview

Transcription is the enzymic synthesis of RNA on a DNA template. This is the first stage in the overall process of gene expression that ultimately leads to synthesis of the protein encoded by a gene. Transcription is catalyzed by an **RNA polymerase**, which requires a dsDNA template as well as the precursor ribonucleotides ATP, GTP, CTP, and UTP (Figure 1). RNA synthesis always occurs in a fixed direction, from the 5′- to the 3′-end of the RNA molecule (Section A2). Usually, only one of the two strands of DNA is transcribed into functional RNA (but see **antisense RNA** in Section I2). The strand whose sequence is copied is called the **template** (or **antisense** or **noncoding**) **strand** since it is used as the template to which ribonucleotides base-pair for the synthesis of the complementary RNA. The sequence of the RNA is, therefore, identical to that of the other **nontemplate** (or **sense**, or **coding**) **strand**, with U replacing T.

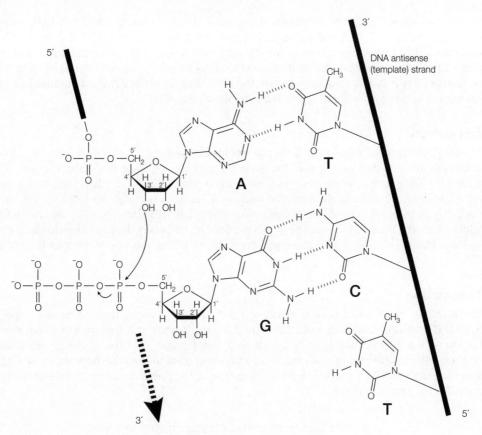

Figure 1. Formation of the phosphodiester bond in transcription.

Initiation

Initiation of transcription involves the binding of an RNA polymerase to the dsDNA. RNA polymerases are usually multisubunit enzymes. They bind to the dsDNA and initiate transcription at sites called **promoters**. Promoters are sequences of DNA at the start of genes, **upstream** (i.e. to the 5′-side with respect to the sense strand) of the coding region (Figure 2). Sequence elements of promoters are often conserved among different genes. Differences between the promoters of different genes give rise to differing efficiencies of transcription initiation and are involved in their regulation (Section G). The short conserved sequences within promoters are the sites at which the polymerase or other DNA-binding proteins bind to initiate or regulate transcription.

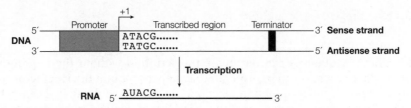

Figure 2. Structure of a typical transcription unit showing promoter and terminator sequences, and the RNA product.

In order to allow the template strand to be used for base pairing, the DNA helix must be locally unwound. Unwinding begins at the promoter site, to which the RNA polymerase binds. The polymerase then initiates the synthesis of the RNA strand at a specific nucleotide called the **start site** (**initiation site**). This is defined as position +1 of the gene sequence (Figure 2). When assembled on the DNA template, the RNA polymerase and its cofactors are often referred to as the **transcription complex**.

Elongation

The RNA polymerase covalently adds ribonucleotides to the 3′-end of the growing RNA chain (Figure 1). The polymerase therefore extends the growing RNA chain in a 5′→3′ direction. This occurs while the enzyme itself moves in a 3′→5′ direction along the template DNA strand. As the enzyme moves, it locally unwinds the DNA, separating the DNA strands, to expose the template for ribonucleotide base pairing and covalent addition to the 3′-end of the growing RNA chain. The helix is re-formed behind the polymerase. The *E. coli* RNA polymerase performs this reaction at a rate of around 40 nt/s at 37°C.

Termination

The termination of transcription involves the dissociation of the transcription complex from the DNA and the ending of RNA synthesis, and occurs at a specific DNA sequence known as the **terminator** (Figure 2 and Sections F2 and F3). These sequences often contain self-complementary regions that can form a **stem–loop** or **hairpin** secondary structure in the RNA product (Figure 3). These cause the polymerase to pause and subsequently cease transcription.

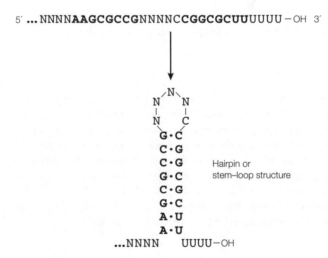

Figure 3. RNA hairpin (stem–loop) structure.

Some terminator sequences can terminate transcription without the requirement for accessory factors, whereas other terminator sequences require the **rho factor** (ρ) as an accessory protein. In the termination reaction, the strands of the RNA–DNA hybrid are separated, allowing the re-formation of the dsDNA, and the RNA polymerase and newly synthesized RNA (**nascent transcript**) are released from the DNA.

F2 *Escherichia coli* RNA polymerase

Key Notes

Escherichia coli RNA polymerase	RNA polymerase is responsible for RNA synthesis (transcription). The core enzyme, consisting of two alpha (α), one beta (β), one beta prime (β') and one omega (ω) subunits, is responsible for transcription elongation. The sigma factor (σ) is also required for correct transcription initiation. The complete enzyme, consisting of the core enzyme plus the σ factor, is called the holoenzyme.
α Subunit	Two α subunits are present in the RNA polymerase. They may be involved in promoter binding.
β Subunit	One β subunit is present in the RNA polymerase. The antibiotics rifampicin and the streptolydigins bind to the β subunit. The β subunit may be involved in both transcription initiation and elongation.
β' Subunit	One β' subunit is present in the RNA polymerase. It may be involved in template DNA binding. Heparin binds to the β' subunit.
σ factor	The σ factor is a separate component from the core enzyme. E. coli encodes several σ factors, the most common being σ^{70}. A σ factor is required for initiation at the correct promoter site. It does this by decreasing binding of the core enzyme to nonspecific DNA sequences and increasing specific promoter binding. The σ factor is released from the core enzyme when the transcript reaches 8–9 nt in length.
Related topics	(F1) Basic principles of transcription (F3) The *E. coli* σ^{70} promoter (F4) Transcription initiation, elongation, and termination	(G3) Transcriptional regulation by alternative σ factors and RNA (H1) The three RNA polymerases: characterization and function

Escherichia coli RNA polymerase

The *E. coli* RNA polymerase is a relatively large enzyme. The enzyme consists of at least five subunits; these are the alpha (α), beta (β), beta prime (β'), omega (ω), and sigma (σ) subunits. In the complete polymerase (called the **holoenzyme**) there are two α subunits and one each of the other four subunits (i.e. $\alpha_2\beta\beta'\sigma$). The complete enzyme is required for transcription initiation. However, the σ factor is not required for transcription elongation and is released from the transcription complex after initiation. The remaining enzyme, which translocates along the DNA, is known as the **core enzyme** and has the structure

$\alpha_2\beta\beta'$. The *E. coli* RNA polymerase can synthesize RNA at a rate of around 40 nt/s at 37°C and requires Mg^{2+} for its activity. The enzyme has a nonspherical structure with a projection flanking a cylindrical channel. The size of the channel suggests that it can bind directly to 16 bp of DNA, although the whole polymerase binds over a region of DNA covering around 60 bp.

Although most RNA polymerases like the *E. coli* polymerase have a multi-subunit structure, it is important to note that this is not an absolute requirement. The RNA polymerases encoded by bacteriophages T3 and T7 (Section P1) are single polypeptide chains that are much smaller than the bacterial multi-subunit enzymes. They synthesize RNA rapidly (200 nt/s at 37°C) and recognize their own specific DNA-binding sequences.

α Subunit

Two identical α subunits encoded by the *rpoA* gene are present in the core RNA polymerase enzyme. The α subunit is required for core protein assembly, but has had no clear transcriptional role assigned to it. When phage T4 infects *E. coli* the α subunit is modified on an arginine residue by ADP-ribosylation (addition of an ADP-ribose unit derived from NAD^+). This is associated with a reduced affinity for binding to promoters, suggesting that the α subunit may play a role in promoter recognition.

β Subunit

One β subunit is present in the core enzyme. This subunit is thought to contain the catalytic center of the RNA polymerase. Strong evidence for this has come from studies with antibiotics that inhibit transcription catalyzed by bacterial RNA polymerases. The important antibiotic **rifampicin** is a potent RNA polymerase inhibitor that blocks initiation but not elongation by binding to the β subunit. Mutations that give rise to resistance to rifampicin map to *rpoB,* the gene encoding the β subunit. This class of antibiotic does not inhibit eukaryotic polymerases and has, therefore, been used for treatment of bacterial infections such as tuberculosis. Another class of antibiotic, the **streptolydigins**, inhibits transcription elongation, and mutations that confer resistance to these antibiotics also map to *rpoB*. These studies suggest that the β subunit may contain two domains responsible for transcription initiation and elongation.

β′ Subunit

One β′ subunit encoded by the *rpoC* gene is present in the core enzyme. This subunit binds two Zn^{2+} ions, which are thought to participate in the catalytic function of the polymerase. The polyanion **heparin** also binds to the β′ subunit. It inhibits transcription *in vitro* and also competes with DNA for binding to the polymerase. This suggests that the β′ subunit may be responsible for binding to the template DNA.

σ Factor

The most common sigma factor in *E. coli* is σ^{70} (since it has a molecular mass of 70 kDa). Binding of the σ factor converts the core RNA polymerase enzyme into the **holoenzyme**. The σ factor has a critical role in promoter recognition, but is not required for transcription elongation. The σ factor contributes to promoter recognition by decreasing the affinity of the core enzyme for nonspecific DNA sites by a factor of 10^4 and increasing affinity for the promoter. Many bacteria (including *E. coli)* have multiple σ factors. They are involved in the recognition of specific classes of promoter sequences upstream of different genes (Section G3). The σ factor is released from the RNA polymerase when the RNA chain reaches 8–9 nt in length. The core enzyme then moves along the DNA synthesizing the growing RNA strand. The σ factor can then complex with a further core enzyme and re-initiate transcription. There is only 30% of the amount of σ factor present in the cell compared with core enzyme complexes. Therefore, only one-third of the polymerase complexes can exist as holoenzyme at any one time.

F3 The *E. coli* σ⁷⁰ promoter

Key Notes

Promoter sequences	Promoters contain conserved sequences that are required for specific binding of RNA polymerase and transcription initiation.
Promoter size	The promoter region extends for around 40 bp. Within this sequence, there are short regions of extensive conservation that are critical for promoter function.
–10 sequence	The –10 sequence is a 6-bp region present in almost all promoters. This hexamer is generally 10 bp upstream from the start site. The consensus –10 sequence is TATAAT.
–35 sequence	The –35 sequence is a further 6-bp region recognizable in most promoters. This hexamer is typically 35 bp upstream from the start site. The consensus –35 sequence is TTGACA.
Promoter efficiency	There is considerable variation between different promoter sequences and in the rates at which different genes are transcribed. Regulated promoters (e.g. the *lac* promoter) are activated by the binding of accessory activation factors such as catabolite activator protein (CAP). Alternative classes of consensus promoter sequence (e.g. heat-shock promoters) are recognized only by an RNA polymerase enzyme containing an alternative σ factor.
Related topics	(F1) Basic principles of transcription (F2) *Escherichia coli* RNA polymerase (F4) Transcription initiation, elongation, and termination (G1) The *lac* operon (G3) Transcriptional regulation by alternative σ factors and RNA (R4) Analysis of cloned genes

Promoter sequences

RNA polymerase binds to specific initiation sites upstream from transcribed sequences. These are called promoters. Although different promoters are recognized by different σ factors that interact with the RNA polymerase core enzyme, the most common σ factor in *E. coli* is σ⁷⁰. Promoters were first characterized through mutations that enhance or diminish the rate of transcription of genes, such as those in the *lac* operon (Section G1). The promoter lies upstream of the start site of transcription, generally assigned as position +1 (Section F1). In accordance with this, promoter sequences are assigned a negative number reflecting the distance upstream from the start of transcription.

Promoter size

The σ⁷⁰ promoter consists of a sequence of between 40 and 60 bp. The region from around –55 to +20 has been shown to be bound by the polymerase, and the region from –20 to +20 is strongly protected from nuclease digestion by DNase I (Section R4). This suggests that this region is tightly associated with the polymerase, which blocks access of the nuclease to the DNA. Mutagenesis of promoter sequences (Section R5) has shown that sequences up to around position −40 are critical for promoter function. Two 6-bp sequences at around positions −10 and −35 have been shown to be particularly important for promoter function in *E. coli*.

−10 sequence

The most conserved sequence in σ⁷⁰ promoters is a 6-bp sequence found in the promoters of many different *E. coli* genes. This sequence is centered around the −10 position with respect to the transcription start site (Figure 1). This is sometimes referred to as the **Pribnow box**, having first been recognized by David Pribnow in 1975. It has the **consensus sequence TATAAT**, where the consensus sequence is made up of the most frequently occurring nucleotides at each position when many sequences are compared. The first two bases (TA) and the final T are the most highly conserved. This hexamer is separated from the transcription start site by between 5 and 8 bp. This intervening sequence is not conserved, although the distance is critical. The −10 sequence appears to be the sequence at which DNA unwinding is initiated by the polymerase (Section F4).

Figure 1. Consensus sequences of *E. coli* promoters (the most conserved sequences are shown in bold).

−35 sequence

Upstream regions around position −35 also have a conserved hexamer sequence (Figure 1). This has the consensus sequence TTGACA, which is most highly conserved in efficient promoters. The first three positions of this hexamer are the most conserved. This sequence is separated from the −10 site by 16–18 bp in 90% of all promoters. The intervening sequence between these conserved elements is not important.

Promoter efficiency

The sequences described above are consensus sequences typical of strong promoters. However, there is considerable variation in sequence among different promoters, and they can vary in transcriptional efficiency by up to 1000-fold. Overall, the functions of different promoter regions can be defined as follows:

- The −35 sequence constitutes a recognition region which enhances recognition and interaction with the polymerase σ factor.

- The −10 region is important for DNA unwinding.

- The sequence around the start site influences initiation.

The sequence of the first 30 bases to be transcribed also influences transcription. Like all macromolecular synthetic reactions, the rate of transcription is controlled at the level of initiation. This sequence controls the rate at which the RNA polymerase clears the promoter, allowing re-initiation by another polymerase complex, thus influencing the rate of transcription and hence the overall promoter strength. The importance of strand separation in the initiation reaction is shown by the effect of negative supercoiling of the DNA template, which generally enhances transcription initiation, presumably because the supercoiled structure requires less energy to unwind the DNA (Section B3). Some promoter sequences are not sufficiently similar to the consensus sequence to be strongly transcribed under normal conditions. An example is the *lac* **promoter** P_{lac}, which requires an accessory activating factor called **catabolite activator protein** (**CAP**) to bind to a site on the DNA close to the promoter sequence in order to enhance polymerase binding and transcription initiation (Section G1). Other promoters, such as those of genes associated with heat shock, contain different consensus promoter sequences that can only be recognized by an RNA polymerase that is bound to a σ factor different from the general factor σ^{70} (Section G3).

F4 Transcription initiation, elongation, and termination

Key Notes

Promoter binding The σ factor enhances the specificity of the core $\alpha_2\beta\beta'$ RNA polymerase for promoter binding. The polymerase finds the promoter –35 and –10 sequences by sliding along the DNA and forming a closed complex with the promoter DNA.

DNA unwinding Around 17 bp of the DNA is unwound by the polymerase, forming an open complex. DNA unwinding at many promoters is enhanced by negative DNA supercoiling. However, the promoters of the genes for DNA gyrase subunits are repressed by negative supercoiling.

RNA chain initiation No primer is needed for RNA synthesis. The first 9 nt are incorporated without polymerase movement along the DNA or σ factor release. The RNA polymerase goes through multiple abortive chain initiations. Following successful initiation, the σ factor is released to form a ternary complex, which is responsible for RNA chain elongation.

RNA chain elongation The RNA polymerase moves along the DNA maintaining a constant region of unwound DNA called the transcription bubble. Ten to 12 nucleotides at the 5′-end of the RNA are constantly base-paired with the DNA template strand. The polymerase unwinds DNA at the front of the transcription bubble and rewinds it at the rear.

RNA chain termination Self-complementary sequences at the 3′-end of genes cause hairpin structures in the RNA, which act as terminators. The stem of the hairpin often has a high content of G–C base pairs giving it high stability, causing the polymerase to pause. The hairpin is often followed by four or more Us, which result in weak RNA–antisense DNA strand binding. This favors dissociation of the RNA strand, causing transcription termination.

Rho-dependent termination Some genes contain terminator sequences that require an additional protein factor, rho (ρ), for efficient transcription termination. Rho binds to specific sites in single-stranded RNA. It hydrolyzes ATP and moves along the RNA towards the transcription complex, where it enables the polymerase to terminate transcription.

Related topics	(B3) DNA supercoiling (F1) Basic principles of transcription (F2) *Escherichia coli* RNA polymerase	(F3) The *E. coli* σ⁷⁰ promoter (Section G) Regulation of transcription in bacteria (Section H) Transcription in eukaryotes

Promoter binding

The RNA polymerase core enzyme, $\alpha_2\beta\beta'$, has a general nonspecific affinity for DNA. This is referred to as **loose binding** and it is fairly stable. When σ factor is added to the core enzyme to form the holoenzyme, it markedly reduces the affinity for nonspecific sites on DNA by 20 000-fold. In addition, σ factor enhances holoenzyme binding to correct promoter-binding sites 100-fold. Overall, this dramatically increases the specificity of the holoenzyme for correct promoter-binding sites. The holoenzyme searches out and binds to promoters in the *E. coli* genome extremely rapidly. This process is too fast to be achieved by repeated binding and dissociation from DNA, and is believed to occur by the polymerase sliding along the DNA until it reaches the promoter sequence. At the promoter, the polymerase recognizes the double-stranded –35 and –10 DNA sequences. The initial complex of the polymerase with the base-paired promoter DNA is referred to as a **closed complex**.

DNA unwinding

In order for the antisense strand to become accessible for base pairing, the DNA duplex must be unwound by the polymerase. Negative supercoiling enhances the transcription of many genes, since this facilitates unwinding by the polymerase. However, some promoters are not activated by negative supercoiling, implying that differences in the natural DNA topology may affect transcription, perhaps due to differences in the steric relationship of the –35 and –10 sequences in the double helix. For example, the promoters for the enzyme subunits of **DNA gyrase** are inhibited by negative supercoiling. DNA gyrase is responsible for negative supercoiling of the *E. coli* genome (Section B3) and so this may serve as an elegant feedback loop for DNA gyrase protein expression. The initial unwinding of the DNA results in formation of an **open complex** with the polymerase and this process is referred to as **tight binding**.

RNA chain initiation

The transcription start site on the sense strand is a purine, most commonly G, in 90% of all *E. coli* genes (Section F3, Figure 1). Often, C or T is found on either side of the start site nucleotide (i.e. CGT or CAT). Unlike DNA synthesis (Section D), RNA synthesis does not require a primer (Figure 1). Thus, the chain is started with a GTP or ATP, from which synthesis of the rest of the chain is initiated. The polymerase initially incorporates the first two nucleotides and forms a phosphodiester bond between them, with the 5′-triphosphate group of the initiating GTP or ATP remaining at the 5′-end of the RNA. The first 9 nt are added without enzyme movement along the DNA. After each one of these 9 nt is added to the chain, there is a significant probability that the chain will be aborted. This process of **abortive initiation** is important for the overall rate of transcription since it has a major role in determining how long the polymerase takes to leave the promoter and allow another polymerase to initiate a further round of

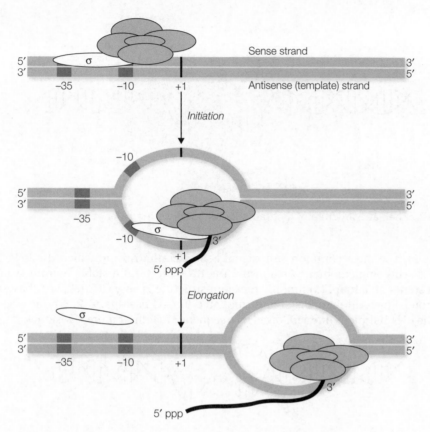

Figure 1. Formation of the transcription complex: initiation and elongation.

transcription. The minimum time for promoter clearance is 1–2 s, which is a long event relative to other stages of transcription.

RNA chain elongation

When initiation succeeds, the enzyme releases the σ factor and forms a **ternary complex** (three components) of polymerase–DNA–nascent RNA, causing the polymerase to progress along the DNA (**promoter clearance**) and allowing re-initiation of transcription from the promoter by a second RNA polymerase holoenzyme. The region of unwound DNA (the **transcription bubble**) appears to move along the DNA with the polymerase. The size of this region of unwound DNA remains constant at around 17 bp (Figure 2), and the 5′-end of the RNA forms a hybrid helix of about 12 bp with the antisense DNA strand. This corresponds to just less than one turn of the RNA–DNA helix. The *E. coli* polymerase moves at an average rate of 40 nt/s, but the rate can vary depending on local DNA sequence. Maintenance of the short region of unwound DNA indicates that the polymerase unwinds DNA in front of the transcription bubble and rewinds DNA at its rear. The RNA–DNA helix must rotate each time a nucleotide is added to the RNA.

RNA chain termination

The RNA polymerase remains bound to the DNA and continues transcription until it reaches a **terminator sequence** (**stop signal**) at the end of the transcription unit

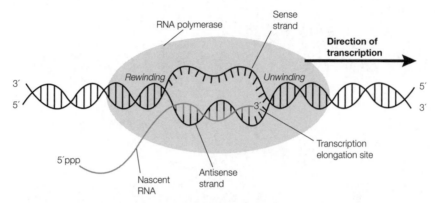

Figure 2. Schematic structure of the transcription bubble during elongation.

(Figure 3). The most common stop signal is an **RNA hairpin** in which the RNA transcript is self-complementary. As a result, the RNA can form a stable hairpin structure with a **stem** and a **loop**. Commonly the stem structure is very GC-rich; G–C base-pairs (3 H-bonds) have additional stability compared to A–U base-pairs (2 H-bonds, Section B2). The RNA hairpin is often followed by a sequence of four or more U residues. It seems

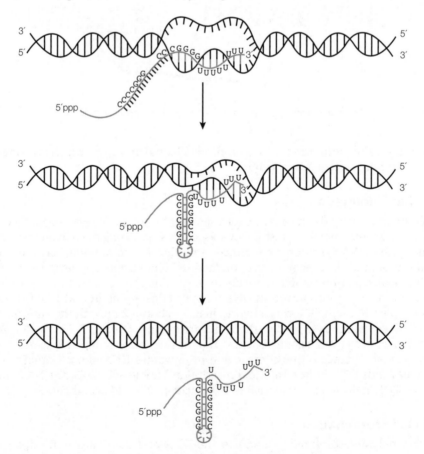

Figure 3. Schematic diagram of rho-independent transcription termination.

that the polymerase pauses immediately after it has synthesized the hairpin RNA. The subsequent stretch of U residues in the RNA base-pairs only weakly with the corresponding A residues in the antisense DNA strand. This favors dissociation of the RNA from the complex with the template strand of the DNA. The RNA is therefore released from the transcription complex. The nonbase-paired antisense strand of the DNA then re-anneals with the sense DNA strand, and the core enzyme dissociates from the DNA.

Rho-dependent termination

While the RNA polymerase can self-terminate at a hairpin structure followed by a stretch of U residues, other known terminator sites may not form strong hairpins. They need an accessory protein, the **rho factor** (ρ), to mediate transcription termination. Rho is a hexameric helicase. Helicases are enzymes that use the energy of ATP hydrolysis to move into and melt double-stranded nucleic acids (Section D2). Rho appears to bind to a stretch of 72 nucleotides near the 3′-end of the nascent RNA, probably through recognition of a specific structural feature rather than a consensus sequence. Rho moves 5′→3′ along the RNA into the RNA–DNA hybrid in the transcription complex and unwinds the RNA from the DNA template, releasing it. Thus, like rho-independent terminators, rho-dependent signals are recognized in the nascent RNA rather than in the template DNA. Sometimes, the rho-dependent terminators are hairpin structures that lack the subsequent stretch of U residues required for rho-independent termination.

G1 The *lac* operon

Key Notes

The operon	The concept of the operon was first proposed in 1961 by Jacob and Monod. An operon is a unit of bacterial and archaeal gene expression that includes coordinately regulated structural genes and control elements, which are recognized by regulatory gene products.
The lactose (*lac*) operon	The *lacZ*, *lacY* and *lacA* genes are transcribed from a *lacZYA* transcription unit under the control of a single promoter *Plac*. They encode enzymes required for the use of lactose as a carbon source within a polycistronic mRNA. The *lacI* gene product, the lac repressor, is expressed from a separate transcription unit upstream from *Plac*.
The lac repressor	The lac repressor is made up of four identical protein subunits. It has a symmetrical structure and binds to a palindromic (symmetrical) 28-bp operator DNA sequence *Olac* that overlaps the *lacZYA* RNA start site. DNA-bound repressor blocks transcription from *Plac*.
Induction	When lac repressor binds to the inducer (whose presence is dependent on lactose), it changes conformation and cannot bind to the *Olac* operator sequence. This allows rapid induction of *lacZYA* transcription.
Catabolite activator protein	The CAP is a transcription factor that is activated by binding to cAMP. cAMP levels rise when glucose is lacking. This complex binds to a site upstream from *Plac* and induces a 90° bend in the DNA. This induces RNA polymerase binding to the promoter and transcription initiation. The CAP activator mediates the global regulation of gene expression from catabolic operons in response to glucose levels.
Related topics	(F1) Basic principles of transcription (F2) *Escherichia coli* RNA polymerase (F3) The *E. coli* σ⁷⁰ promoter (F4) Transcription initiation, elongation, and termination (G2) The *trp* operon

The operon

François Jacob and Jacques Monod proposed the operon model for the coordinate regulation of transcription of genes involved in specific metabolic pathways in 1961, and shared a 1965 Nobel prize for this groundbreaking work. The **operon** is a unit of gene expression and regulation, and typically includes:

- The **structural genes** (any gene other than a regulator) for enzymes involved in a specific biosynthetic pathway and whose expression is coordinately controlled

- Control elements such as an **operator sequence**, which is a DNA sequence that regulates transcription of the structural genes

- **Regulator gene**(s) whose products recognize the control elements, for example a repressor that binds to and regulates an operator sequence

The lactose (*lac*) operon

E. coli can use lactose as a source of carbon. The enzymes required for the use of lactose as a carbon source are only synthesized when lactose is available as the sole carbon source. The ***lac* operon** (Figure 1) consists of three structural genes, or **cistrons** (another name for a gene based on inheritance properties). These are: ***lacZ,*** which codes for β-**galacto-sidase**, an enzyme responsible for hydrolysis of lactose to galactose and glucose; ***lacY,*** which codes for a **galactoside permease**, which is responsible for lactose transport across the bacterial cell wall; and ***lacA***, which codes for **galactoside *O*-acetyltransferase**, which may detoxify nonmetabolizable galactoside sugars and help exclude them from the cell. The three structural genes are encoded in a single transcription unit, ***lacZYA***, which has a single promoter, P_{lac}. This organization means that the three structural proteins are expressed together as a single **polycistronic mRNA** containing more than one coding region under the same regulatory control. The *lacZYA* transcription unit contains an **operator site** O_{lac}, which is positioned between bases –5 and +21 at the 5′-end of the P_{lac} promoter region. This site binds a protein called the **lac repressor**, which is a potent inhibitor of *lac* transcription when bound to the operator. The lac repressor is encoded by a separate regulatory gene ***lacI***, which is also a part of the *lac* operon; *lacI* is situated just upstream from P_{lac}.

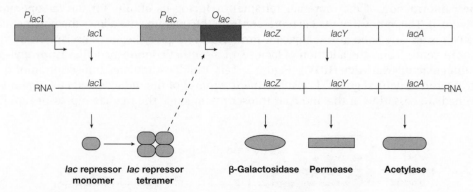

Figure 1. Structure of the lactose operon.

The lac repressor

The *lacI* gene encodes the lac repressor, which is active as a tetramer of identical subunits. It has a very strong affinity for the *lac* operator-binding site, O_{lac}, and also has a generally high affinity for DNA. The *lac* operator site consists of a 28-bp **palindrome** (a palindrome has the same DNA sequence when one strand is read left to right and the complementary strand is read right to left, each in a 5′ to 3′ direction, see Section O3). This inverted repeat symmetry of the operator matches the inherent symmetry of the lac repressor, which is made up of four identical subunits. In the absence of lactose, the repressor occupies the operator-binding site. It seems that both the lac repressor and the RNA polymerase can bind simultaneously to the *lac* promoter and operator sites. The lac repressor actually increases the binding of the RNA polymerase to the *lac* promoter by two orders of magnitude. This means that when the lac repressor is bound to the O_{lac} operator DNA sequence, the polymerase is also likely to bind to the adjacent P_{lac} promoter sequence.

Induction

In the absence of an inducer, the lac repressor blocks all but a very low level of transcription of *lacZYA*. When lactose is added to cells, the low basal level of the permease allows its uptake, while the equally limited amount of intracellular β-galactosidase catalyzes the conversion of some of this lactose to its isomer **allolactose** (Figure 2) in an alternative reaction to the usual hydrolysis.

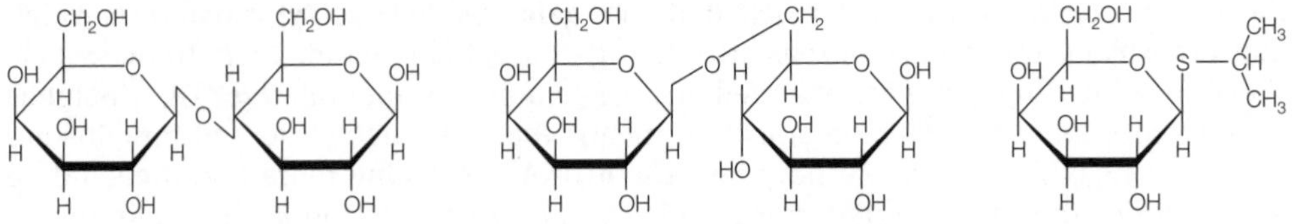

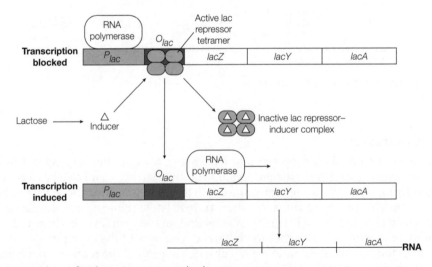

Figure 2. Structures of lactose, allolactose and IPTG.

Allolactose acts as an inducer and binds to the lac repressor. This causes a change in the conformation of the repressor tetramer, reducing its affinity for the *lac* operator (Figure 3). The removal of the lac repressor from the operator site allows the polymerase (which is already sited at the adjacent promoter) to rapidly begin transcription of the *lacZYA* genes. Thus, the addition of lactose, or a synthetic inducer such as **isopropyl-β-D-thiogalactopyranoside** (**IPTG**) (Figure 2) very rapidly stimulates transcription of the *lac* operon structural genes. The subsequent removal of the inducer leads to an almost immediate cessation of this induced transcription, since the free lac repressor rapidly

Figure 3. Binding of inducer inactivates the lac repressor.

re-occupies the operator site and the existing *lacZYA* RNA transcript is extremely unstable (Section J3). IPTG is commonly used in the laboratory to activate the *lac* promoter in cloning vectors (Section P1).

Catabolite activator protein

The P_{lac} promoter is not a strong promoter. P_{lac} and related promoters do not have strong −35 sequences and some even have weak −10 consensus sequences. For high-level transcription, they require the assistance of another protein called **Catabolite Activator Protein (CAP)**, also known as **cAMP receptor protein (CRP)**. As long as glucose is present, *E. coli* does not require alternative carbon sources such as lactose. Therefore, catabolic operons, such as the lactose operon, are not normally activated when glucose is available. This regulation is mediated by CAP, which exists as a dimer that cannot bind to DNA on its own, nor regulate transcription. Glucose reduces the level of the regulatory nucleotide **cyclic AMP (cAMP)** in the cell. When glucose is absent, the levels of cAMP in *E. coli* increase and CAP binds to cAMP. The CAP–cAMP complex binds to the P_{lac} promoter just upstream from the site for RNA polymerase. CAP binding induces a 90° bend in DNA, and this is believed to enhance RNA polymerase binding to the promoter, increasing transcription 50-fold.

The CAP-binding site is an inverted repeat and can be adjacent to the promoter (as in the *lac* operon), can lie within the promoter itself, or can be much further upstream from the promoter. Differences in the CAP-binding sites of the promoters of different catabolic operons may mediate different levels of response of these operons to cAMP *in vivo*.

G2 The *trp* operon

Key Notes

The tryptophan (*trp*) operon	The *trp* operon encodes five structural genes involved in tryptophan biosynthesis. One transcript encoding all five enzymes is synthesized using single promoter (*Ptrp*) and operator (*Otrp*) sites.
The trp repressor	The trp repressor is the product of a separate operon, the *trpR* operon. The repressor is a dimer that interacts with the *trp* operator only when it is complexed with tryptophan. Repressor binding reduces transcription 70-fold.
The attenuator	A terminator sequence is present in the 162 bp *trp* leader before the start of the *trpE*-coding sequence. It is a rho-independent terminator that terminates transcription at base +140, which is in a run of eight Us just after a hairpin structure. This structure is called the attenuator, because it can cause premature termination of *trp* RNA synthesis.
Leader RNA structure	The *trp* leader RNA contains four regions of complementary sequence that are able to form alternative hairpin structures. One of these structures is the attenuator hairpin.
The leader peptide	The leader RNA contains an efficient ribosome-binding site and encodes a 14-amino acid leader peptide. Codons 10 and 11 of this peptide encode tryptophan. When tryptophan is low the ribosome will pause at these codons.
Attenuation	The RNA polymerase pauses on the DNA template at a site at the end of the leader peptide-encoding sequence. When a ribosome initiates translation of the leader peptide, the polymerase continues to transcribe the RNA. If the ribosome pauses at the tryptophan codons (i.e. tryptophan levels are low), it changes the availability of the complementary leader sequences for base pairing so that an alternative RNA hairpin forms instead of the attenuator hairpin. As a result, transcription does not terminate. If the ribosome is not stalled at the tryptophan residues (i.e. tryptophan levels are high), then the attenuator hairpin is able to form and transcription is terminated prematurely.
Importance of attenuation	Attenuation gives rise to 10-fold regulation of transcription by tryptophan. Transcription attenuation occurs in at least six operons involved in amino acid biosynthesis. In some operons (e.g. histidine), it is the only mechanism for feedback regulation of amino acid synthesis.

Related topics	(F1) Basic principles of transcription	(F3) The *E. coli* σ⁷⁰ promoter
	(F2) *Escherichia coli* RNA polymerase	(F4) Transcription initiation, elongation, and termination
		(G1) The *lac* operon

The tryptophan (*trp*) operon

The *trp* operon encodes five structural genes whose activities are required for tryptophan synthesis (Figure 1). The operon encodes a single transcription unit producing a 7-kb transcript, which is synthesized downstream from the *trp* promoter (P_{trp}) and *trp* operator (O_{trp}) sites. Like many of the operons involved in amino acid biosynthesis, the *trp* operon has evolved systems for coordinated expression of these genes when the product of the biosynthetic pathway, tryptophan, is in short supply in the cell. As with the *lac* operon, the RNA product of this transcription unit is very unstable, enabling bacteria to respond rapidly to changing needs for tryptophan.

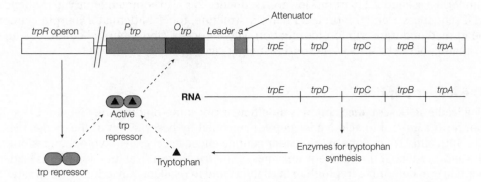

Figure 1. Structure of the *trp* operon and function of the trp repressor.

The trp repressor

One gene product of the separate *trpR* operon, the trp repressor, specifically interacts with the operator site of the *trp* operon. The symmetrical operator sequence forming the trp repressor-binding site overlaps with the *trp* promoter sequence between bases –21 and +3. The core binding site is a palindrome of 18 bp. The trp repressor binds tryptophan and can only bind to the operator when it is complexed with tryptophan. The repressor is a dimer of two subunits that have structural similarity to the CAP protein and lac repressor (Section G1). The repressor dimer has a structure with a central core and two flexible DNA-reading heads, each formed from the C-terminal half of one subunit. Only when tryptophan is bound to the repressor are the reading heads the correct distance apart, and the side chains in the correct conformation, to interact with successive major grooves (Section A2) of the DNA at the *trp* operator sequence. Tryptophan, the end-product of the enzymes encoded by the *trp* operon, therefore acts as a **corepressor** and inhibits its own synthesis through **end-product inhibition**. The repressor reduces transcription initiation by around 70-fold. This is a much smaller transcriptional effect than that mediated by the binding of the lac repressor.

The attenuator

At first, it was thought that the repressor was responsible for all of the transcriptional regulation of the *trp* operon. However, it was observed that the deletion of a sequence between the operator and the coding region of the *trpE* gene resulted in an increase in both the basal and the activated (derepressed) levels of transcription. This site is termed the **attenuator** and lies towards the end of the transcribed leader sequence of 162 nt that precedes the *trpE* initiator codon (Section K1). The attenuator is a rho-independent terminator site with a short GC-rich palindrome followed by eight successive U residues (Section F4). If this sequence is able to form a hairpin structure in the RNA transcript, then it acts as a highly efficient transcription terminator and only a functionless 140-bp transcript is synthesized.

Leader RNA structure

The leader sequence of the *trp* operon RNA contains four regions of complementary sequence that can form different base-paired RNA structures (Figure 2). These are termed sequences 1, 2, 3, and 4. The attenuator hairpin is the product of the base pairing of sequences 3 and 4 (3:4 structure). Sequences 1 and 2 are also complementary and can form a second 1:2 hairpin. However, sequence 2 is also complementary to sequence 3. If sequences 2 and 3 form a 2:3 hairpin structure, the 3:4 attenuator hairpin cannot be formed and transcription termination will not occur. Under normal conditions, the formation of the 1:2 and 3:4 hairpins is energetically favorable (Figure 2a).

The leader peptide

The leader RNA sequence contains an efficient ribosome-binding site (Section L1) and can form a 14-amino acid leader peptide encoded by bases 27–68 of the leader RNA. The 10th and 11th codons of this leader peptide encode successive tryptophan residues, the end-product of the synthetic enzymes of the *trp* operon. The function of this leader peptide is to determine tryptophan availability and to regulate transcription termination (see below). Tryptophan is a relatively rare amino acid (about 1% of the residues in *E. coli* proteins), therefore the appearance of two successive Trp codons is unusual. Thus, under conditions of low tryptophan availability, the ribosome would be expected to pause at this site.

Attenuation

Attenuation depends on the fact that transcription and translation are tightly coupled in *E. coli;* translation can occur as an mRNA is being transcribed (Section J3). The 3′-end of the *trp* leader peptide coding sequence overlaps complementary sequence 1 (Figure 2); the two Trp codons are within sequence 1 and the stop codon is between sequences 1 and 2. The availability of tryptophan is sensed by the translation apparatus, and determines whether the transcription-terminating (3:4) hairpin forms in the mRNA.

As transcription of the *trp* operon proceeds, the RNA polymerase pauses at the end of sequence 2 until a ribosome begins to translate the leader peptide. Under conditions of high tryptophan availability, the ribosome rapidly incorporates tryptophan at the two Trp codons in the leader peptide (see above) and thus translates to the end of the leader message. The ribosome is then occluding sequence 2 and, as the RNA polymerase reaches the terminator sequence, the 3:4 hairpin can form, and transcription may be terminated (Figure 2b). This is the process of **attenuation**.

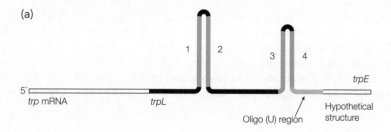

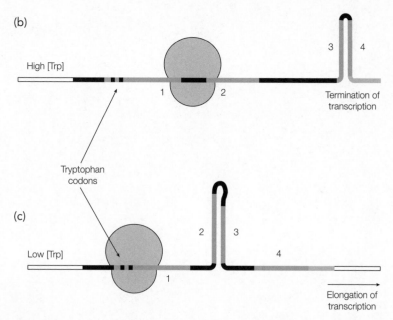

Figure 2. Transcriptional attenuation in the *trp* operon: (a) Trp RNA structure under normal conditions; (b) attenuation of transcription under conditions of high [Trp]; (c) formation of the anti-terminator under conditions of low [Trp].

Alternatively, if tryptophan is in scarce supply, it will not be available as an aminoacyl tRNA for translation (Section K2), and the ribosome will tend to pause at the two Trp codons, occluding sequence 1. This leaves sequence 2 free to form a hairpin with sequence 3 (Figure 2c), known as the **anti-terminator**. The terminator (3:4) hairpin cannot form, and transcription continues into *trpE* and beyond. Thus the level of the end-product, tryptophan, determines the probability that transcription will terminate early (attenuation), rather than proceeding through the whole operon.

Importance of attenuation

The presence of tryptophan gives rise to a 10-fold repression of *trp* operon transcription through the process of attenuation alone. Combined with control by the trp repressor

(70-fold), this means that tryptophan levels exert a 700-fold regulatory effect on expression of the *trp* operon. Attenuation occurs in at least six operons that encode enzymes concerned with amino acid biosynthesis. For example, the *his* operon has a leader that encodes a peptide with seven successive His codons. Not all of these other operons have the same combination of regulatory controls found in the *trp* operon. The *his* operon has no repressor–operator regulation, and attenuation forms the only mechanism of feedback control.

G3 Transcriptional regulation by alternative σ factors and RNA

Key Notes	
Sigma factors	The sigma (σ) factor is responsible for recognition of consensus promoter sequences and is only required for transcription initiation. Many bacteria produce alternative sets of σ factors.
Promoter recognition	In *E. coli*, σ^{70} is responsible for recognition of the -10 and -35 consensus sequences. Differing consensus sequences are found in sets of genes that are regulated by the use of alternative σ factors.
Heat shock	Around 17 proteins are specifically expressed in *E. coli* when the temperature is increased above 37°C. The genes for these proteins are transcribed by RNA polymerase using an alternative sigma factor σ^{32}. σ^{32} recognizes its own specific promoter consensus sequences.
Sporulation in *Bacillus subtilis*	Under nonoptimal environmental conditions, *B. subtilis* cells form endospores through a cell differentiation process involving cell partition into a mother cell and a forespore. This process is closely regulated by a set of σ factors that regulate each step in this process.
Bacteriophage σ factors	Many bacteriophages synthesize their own σ factors in order to 'take over' the host cell's own transcription machinery by substituting the normal cellular σ factor and altering the promoter specificity of the RNA polymerase. *Bacillus subtilis* SPO1 phage expresses a cascade of σ factors, which allow a defined sequence of expression of early, middle and late phage genes.
Regulation by noncoding RNA	The untranslated regions (UTRs) of many bacterial mRNAs contain *cis*-acting sequences called riboswitches that bind small metabolites and change their structure to become transcriptional or translational terminators. Other *trans*-acting small RNAs bind directly to proteins or to target mRNAs.
Related topics	(F1) Basic principles of transcription (F2) *Escherichia coli* RNA polymerase (F3) The *E. coli* σ^{70} promoter (F4) Transcription initiation, elongation, and termination (I2) Examples of transcriptional regulation

Sigma factors

The $\alpha_2\beta\beta'$ core enzyme of RNA polymerase is unable to start transcription at promoter sites (Section F). In order to specifically recognize the consensus –35 and –10 elements of general promoters, it requires the σ factor subunit. This subunit is only required for transcription initiation, being released from the core enzyme after initiation and before RNA elongation takes place (Section F4). Thus, σ factors bind both to core RNA polymerase and to specific promoter sequences in DNA. Many bacteria, including *E. coli,* produce a set of σ factors that recognize different sets of promoters. Transcription initiation from single promoters or small groups of promoters is commonly regulated by single transcriptional repressors, such as the lac repressor, or transcriptional activators, such as the CAP (Section G1). However, some environmental conditions require a massive change in the overall pattern of gene expression in the cell. Under such circumstances, bacteria may use a different set of σ factors to direct RNA polymerase binding to different promoter sequences. This process allows the diversion of the cell's basic transcription machinery to the specific transcription of different classes of genes.

Promoter recognition

The σ^{70} factor is the most common σ factor in *E. coli* and is responsible for recognition of general promoters with consensus –35 and –10 elements (Section F3). However, the binding of an alternative σ factor to RNA polymerase can confer different promoter specificity on this enzyme. Comparisons of promoters activated by the polymerase when complexed to specific σ factors show that each σ factor recognizes a different combination of sequences centered approximately around the –35 and –10 sites. It seems likely that the factors themselves contact both of these regions, with the –10 region being most important.

Heat shock

The response to heat shock is one example in *E. coli* where gene expression is altered significantly by the use of different σ factors. When *E. coli* is subjected to an increase in temperature, the synthesis of a set of around 17 proteins, called **heat-shock proteins (HSPs)**, is induced. If the cells are transferred from 37 to 42°C, this burst of HSP synthesis is transient. However if the increase in temperature is more extreme, such as to 50°C, where growth of *E. coli* is not possible, then the HSPs are the only proteins synthesized. This requires the synthesis of all other proteins to be switched off and involves both transcriptional and translational control. As far as transcription is concerned, the promoters for *E. coli* HSP genes are recognized by a unique form of RNA polymerase holoenzyme containing a variant σ factor, σ^{32}, encoded by the *rpoH* gene. σ^{32} is a minor protein and is much less abundant than σ^{70}. Holoenzyme containing σ^{32} acts exclusively on the promoters of HSP genes and does not recognize the general consensus promoters of most other

Consensus promoter	–35 sequence	–10 sequence
Standard (σ^{70})	- - - - - - - - T T G A C A ⋯⋯16–18 bp⋯⋯	T A T A A T
Heat shock (σ^{32})	T - - C - C - C T T G A A ⋯⋯ 13–15 bp ⋯⋯C C C	C A T - T

Figure 1. Comparison of the heat-shock (σ^{32}) and general (σ^{70}) responsive promoters.

genes (Figure 1). Accordingly, heat-shock promoters have different sequences to other general promoters that bind to σ^{70}.

Sporulation in *Bacillus subtilis*

Vegetatively (i.e. 'normally') growing *B. subtilis* cells form bacterial **endospores** in response to a sub-optimal environment. An endospore is a tough, stress-resistant, non-reproductive cell. The formation of an endospore (**sporulation**) requires drastic changes in gene expression, including the cessation of the synthesis of almost all of the proteins required for vegetative existence as well as the production of proteins necessary for the resumption of protein synthesis when the endospore germinates (resumes growth) under more optimal conditions. The process of endospore formation involves the asymmetrical division of the bacterial cell into two compartments, the **forespore**, which forms the endospore, and the **mother cell**, which is eventually discarded. This system is considered one of the most fundamental examples of **cell differentiation**. The RNA polymerase in *B. subtilis* is functionally identical to that in *E. coli* and the vegetatively growing *B. subtilis* contains a diverse set of σ factors. Sporulation is regulated by a further set of σ factors in addition to those of the vegetative cell. Different σ factors are specifically active before cell partition occurs, in the forespore and in the mother cell. Cross-regulation of this compartmentalization permits the forespore and mother cell to tightly coordinate the differentiation process.

Bacteriophage σ factors

Some bacteriophages encode novel σ factors that endow the host RNA polymerase with a new promoter specificity that causes them to selectively transcribe the phage genes (e.g. phage T4 in *E. coli* and SPO1 in *B. subtilis*). This strategy is an effective alternative to the need for the phage to encode its own complete polymerase (cf. bacteriophage T7, Section F2). The *B. subtilis* bacteriophage SPO1 expresses a '**cascade**' of σ factors in sequence to allow its own genes to be transcribed at specific stages during virus infection. Initially, **early genes** are expressed by the normal bacterial holoenzyme. Among these early genes is the gene encoding σ^{28}, which then displaces the bacterial σ factor from the RNA polymerase. The σ^{28}-containing holoenzyme is then responsible for expression of the **middle genes**. The phage middle genes include genes 33 and 34, which specify a further σ factor that is responsible for the specific transcription of **late genes**. In this way, the bacteriophage can use the host's RNA polymerase machinery and expresses its genes in a defined sequential order.

Regulation by noncoding RNA

The attenuator in the *trp* operon (Section G2) is an example of a *cis*-acting regulatory noncoding RNA, i.e. a sequence that regulates the activity of the protein-coding region of the *same* molecule. **Riboswitches** are another example. Riboswitches are *cis*-acting sequences found mainly in bacteria that alter their secondary structure in response to binding a regulatory metabolite. This can lead to the formation or disruption of a transcription terminator. Alternatively, riboswitches may regulate translation by binding to the Shine–Dalgarno sequence of bacterial mRNAs (Section L2). Examples found in the UTRs (Section L2) of various mRNAs bind glycine, lysine, purines, and the cofactor thiamin pyrophosphate (the only eukaryotic example) and regulate the metabolism and transport of these metabolites. The existence of riboswitches and ribozymes (Section J2) that recognize and bind small molecules in a manner similar to proteins supports the importance of RNA in early evolution.

Bacteria have several hundred genes encoding **small RNAs (sRNAs)** of about 50–200 nt that regulate the expression of *other* genes, i.e. they act in ***trans***. They either bind directly to a protein to influence its activity, e.g. binding of 6S RNA to RNA polymerase, or the RNA of RNase P (Section J2), or they base-pair to a target RNA whose expression they regulate, often assisted by the binding of **Hfq** protein to the sRNA.

H1 The three RNA polymerases: characterization and function

Key Notes

Eukaryotic RNA polymerases	Three nuclear eukaryotic polymerases transcribe different sets of genes. Their activities are distinguished by their different sensitivities to the fungal toxin α-amanitin.

- RNA polymerase I is located in the nucleolus. It is responsible for the synthesis of the precursors of most rRNAs.

- RNA polymerase II is located in the nucleoplasm and is responsible for the synthesis of pre-mRNA and most snRNAs and miRNAs.

- RNA polymerase III is located in the nucleoplasm. It is responsible for the synthesis of the precursors of 5S rRNA, tRNAs and some other small nuclear and cytosolic RNAs.

RNA polymerase subunits	Each RNA polymerase has 12 or more different subunits. The two largest subunits are similar to each other and to the β and β′ subunits of *E. coli* RNA polymerase. Other subunits in each enzyme have homology to the α subunit of the *E. coli* enzyme. Five additional subunits are common to all three polymerases, while others are polymerase specific.
Eukaryotic RNA polymerase activities	Like bacterial RNA polymerases, the eukaryotic enzymes do not require a primer and synthesize RNA in a 5′→3′ direction. Unlike bacterial polymerases, they require accessory factors for DNA binding.
The CTD of RNA Pol II	The largest subunit of RNA polymerase II has a seven amino acid repeat at the C-terminus called the C-terminal domain (CTD). This sequence, Tyr-Ser-Pro-Thr-Ser-Pro-Ser, is repeated 52 times in mouse RNA polymerase II and is subject to phosphorylation.
Related topics	(F2) *Escherichia coli* RNA polymerase (H2) RNA Pol I genes: the ribosomal repeat (H3) RNA Pol III genes: 5S and tRNA transcription (H4) RNA Pol II genes: promoters and enhancers (H5) General transcription factors and RNA Pol II initiation (I2) Examples of transcriptional regulation

Eukaryotic RNA polymerases

The mechanism of eukaryotic transcription is similar to that in prokaryotes. However, the large number of polypeptides associated with the eukaryotic transcription machinery makes it far more complex. Three different nuclear RNA polymerase complexes are responsible for the transcription of different types of eukaryotic genes. The different RNA polymerases were identified by chromatographic purification of the enzymes and elution at different salt concentrations. Each RNA polymerase has a different sensitivity to the fungal toxin **α-amanitin** and this can be used to distinguish their activities.

- **RNA polymerase I** (RNA Pol I) synthesizes the major 45S pre-rRNA transcript (Sections H2 and J1). It is located in the nucleolus and is insensitive to α-amanitin.

- **RNA polymerase II** (RNA Pol II) transcribes all protein-coding genes and most snRNA and miRNA genes (Section J2). It is located in the nucleoplasm and is very sensitive to α-amanitin.

- **RNA polymerase III** (RNA Pol III) transcribes the genes for tRNA, 5S rRNA, U6 snRNA and certain other small RNAs. It is located in the nucleoplasm and is moderately sensitive to α-amanitin.

In addition to these nuclear enzymes, eukaryotic cells contain additional polymerases in mitochondria and chloroplasts while plants have yet further polymerases involved in small RNA synthesis. Like bacteria, archaea contain a single RNA polymerase but it is structurally and functionally homologous to eukaryotic RNA Pol II.

RNA polymerase subunits

All three polymerases are large enzymes containing 12 or more subunits. The genes encoding the two largest subunits of each RNA polymerase have homology (related DNA coding sequences) to each other. All of the three eukaryotic polymerases contain subunits with homology to subunits within the *E. coli* core RNA polymerase $\alpha_2\beta\beta'$ (Section F2). The largest subunit of each eukaryotic RNA polymerase is similar to the β' subunit of the *E. coli* polymerase, and the second largest subunit is similar to the β subunit, which contains the active site of the *E. coli* enzyme. The functional significance of this homology is supported by the observation that the second largest subunits of the eukaryotic RNA polymerases also contain the active sites. Two subunits common to RNA Pol I and RNA Pol III, and a further subunit specific to RNA Pol II, have homology to the *E. coli* RNA polymerase α subunit. At least five other smaller subunits are common to the three different polymerases. Each polymerase also contains an additional four to seven subunits that are only present in one type.

Eukaryotic RNA polymerase activities

Like bacterial RNA polymerases, each of the eukaryotic enzymes catalyzes transcription in a 5′→3′ direction and synthesizes RNA complementary to the antisense template strand. The reaction requires the precursor nucleotides ATP, GTP, CTP, and UTP and does not require a primer for transcription initiation. Unlike the purified bacterial enzymes, the purified eukaryotic RNA polymerases require the presence of additional initiation proteins (**transcription factors**) before they are able to bind to promoters and initiate transcription.

The CTD of RNA Pol II

The C-terminal end of RNA Pol II contains a stretch of seven amino acids that is repeated 52 times in the mouse enzyme and 26 times in yeast. This heptapeptide has the sequence **Tyr-Ser-Pro-Thr-Ser-Pro-Ser** and is known as the **carboxy-terminal domain** or **CTD**. These repeats are essential for viability. The CTD sequence may be phosphorylated at the serines and some tyrosines. *In vitro* studies have shown that the CTD is unphosphory-lated at transcription initiation, but phosphorylation occurs during transcription elonga-tion as the RNA polymerase leaves the promoter. Since RNA Pol II transcribes all of the eukaryotic protein-coding genes, it is the most important RNA polymerase for the study of differential gene expression. The CTD has been shown to be an important target for differential activation of transcription elongation and it recruits protein factors necessary for RNA processing (Sections H5, I2, and J3).

H2 RNA Pol I genes: the ribosomal repeat

Key Notes

Ribosomal RNA genes	The pre-rRNA transcription units contain three sequences that encode the 18S, 5.8S and 28S rRNAs. Pre-rRNA transcription units are arranged in clusters in the genome as long tandem arrays separated by non-transcribed spacer sequences.
Role of the nucleolus	Pre-rRNA is synthesized by RNA polymerase I (RNA Pol I) in the nucleolus. The arrays of rRNA genes loop together to form the nucleolus and are known as nucleolar organizer regions.
RNA Pol I promoters	The pre-rRNA promoters consist of two transcription control regions. The core element includes the transcription start site. The upstream control element (UCE) is approximately 50 bp long and begins at around position –100.
Upstream binding factor	Upstream binding factor (UBF) binds to the UCE. It also binds to a different site in the upstream part of the core element, causing the DNA to loop between the two sites.
Selectivity factor 1	Selectivity factor 1 (SL1) binds to and stabilizes the UBF–DNA complex. SL1 then allows binding of RNA Pol I and initiation of transcription.
TBP and TAF$_i$s	SL1 is made up of four subunits. These include the TATA-binding protein (TBP), which is required for transcription initiation by all three RNA polymerases. The other factors are RNA Pol I-specific TBP-associated factors called TAF$_i$s.
Related topics	(C4) Genome complexity (F2) *Escherichia coli* RNA polymerase (F3) The *E. coli* σ^{70} promoter (F4) Transcription initiation, elongation, and termination (H1) The three RNA polymerases: characterization and function (J1) rRNA processing and ribosomes

Ribosomal RNA genes

RNA polymerase I (RNA Pol I) is responsible for the continuous synthesis of rRNAs during interphase. Human cells contain five clusters each of around 40 copies of the rRNA gene situated on different chromosomes (Figure 1 and Section C4). Each rRNA gene produces a 45S rRNA transcript which is about 13 000-nt long. This transcript is cleaved to give one copy each of the 28S (5000 nt), 18S (2000 nt), and 5.8S (160 nt) rRNAs (Section J1). The

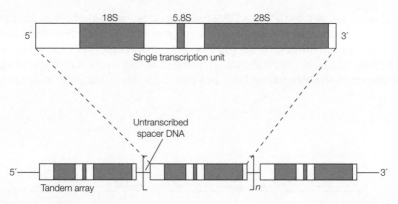

Figure 1. Ribosomal RNA transcription units.

continuous transcription of multiple copies of these rRNA genes is essential for sufficient production of the abundant processed rRNAs that are packaged into ribosomes.

Role of the nucleolus

Each rRNA gene cluster is known as a **nucleolar organizer region**, since the nucleolus contains large loops of DNA corresponding to the gene clusters. After a cell emerges from mitosis, rRNA synthesis restarts and tiny nucleoli appear at the chromosomal locations of the rRNA genes. During active rRNA synthesis, the pre-rRNA transcripts appear along the length of the rRNA genes and may be visualized in the electron microscope as '**Christmas tree structures**.' In these structures, the RNA transcripts are densely packed along the DNA and stick out perpendicularly. Short transcripts can be seen at the start of the gene, and these get longer towards the end of the transcription unit, which is indicated by the disappearance of the transcripts.

RNA Pol I promoters

Mammalian pre-rRNA gene promoters have a bipartite transcription control region (Figure 2). The core element includes the transcription start site and encompasses bases –31 to +6. This sequence is essential for transcription. An additional element of around 50–80 bp named the **upstream control element** (**UCE**) begins about 100 bp upstream from the start site (–100). The UCE is responsible for an increase in transcription of around 10- to 100-fold compared with that from the core element alone.

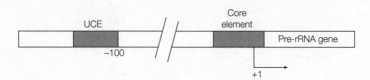

Figure 2. Structure of a mammalian pre-rRNA promoter.

Upstream binding factor

A specific DNA-binding protein, called **upstream binding factor** (**UBF**) binds to the UCE. As well as binding to the UCE, UBF binds to a sequence in the upstream part of the core element. The sequences of the two UBF-binding sites have no obvious similarity. One molecule of UBF is thought to bind to each sequence element. The two molecules of UBF may then bind to each other through protein–protein interactions, causing the intervening DNA to form a loop between the two binding sites (Figure 3). A low rate of basal transcription is seen in the absence of UBF, but this is greatly stimulated in the presence of UBF.

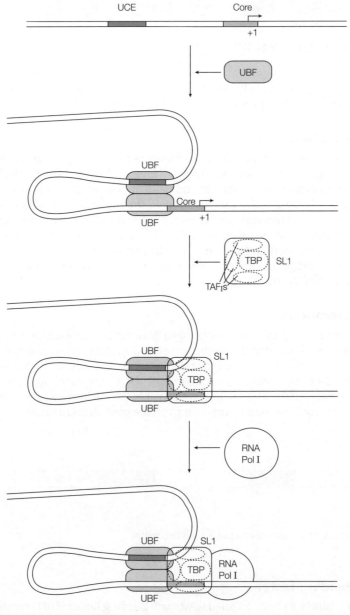

Figure 3. Schematic model for rRNA transcription initiation.

Selectivity factor 1

An additional factor, called **selectivity factor** (**SL1**) is essential for RNA Pol I transcription. SL1 binds to and stabilizes the UBF-DNA complex and interacts with the free downstream part of the core element. SL1 binding allows RNA Pol I to bind to the complex and initiate transcription and is essential for rRNA transcription.

TBP and TAF$_I$s

SL1 contains several subunits, including a protein called **TBP** (**TATA-binding protein**). TBP is required for initiation by all three eukaryotic RNA polymerases (Sections H1, H3 and H5), and seems to be a critical factor in eukaryotic transcription. The other three subunits of SL1 are referred to as **TBP-associated factors** or **TAFs**, and those subunits required for RNA Pol I transcription are referred to as **TAFIs**.

H3 RNA Pol III genes: 5S and tRNA transcription

Key Notes

RNA polymerase III
RNA polymerase III (RNA Pol III) has 16 or more subunits. The enzyme is located in the nucleoplasm and it synthesizes the precursors of 5S rRNA, the tRNAs and other small nuclear and cytosolic RNAs.

tRNA genes
Two transcription control regions, called the A box and the B box, lie downstream from the transcription start site. These sequences are therefore both conserved sequences in the tRNAs as well as promoter sequences in the DNA. TFIIIC binds to the A and B boxes in the tRNA promoter while TFIIIB binds to the TFIIIC-DNA complex and interacts with DNA upstream from the TFIIIC-binding site. TFIIIB contains three subunits, TBP, BRF and B″, and is responsible for RNA Pol III recruitment and hence transcription initiation.

5S rRNA genes
The genes for 5S rRNA are organized in a tandem cluster. The 5S rRNA promoter contains a conserved C box 81–99 bases downstream from the start site, and a conserved A box at around 50–65 bases downstream. TFIIIA binds strongly to the C box promoter sequence. TFIIIC then binds to the TFIIIA DNA complex, interacting also with the A box sequence. This complex allows TFIIIB to bind, recruit the polymerase, and initiate transcription.

Alternative RNA Pol III promoters
A number of RNA Pol III promoters are regulated by upstream as well as downstream promoter sequences. Other promoters require only upstream sequences, including the TATA box and other sequences found in RNA Pol II promoters.

RNA Pol III termination
The RNA polymerase can terminate transcription without accessory factors. A cluster of A residues is often sufficient for termination.

Related topics
(F2) *Escherichia coli* RNA polymerase
(H1) The three RNA polymerases: characterization and function
(H2) RNA Pol I genes: the ribosomal repeat
(H4) RNA Pol II genes: promoters and enhancers

(H5) General transcription factors and RNA Pol II initiation
(J1) rRNA processing and ribosomes
(J2) tRNA and other small RNA processing
(K2) tRNA structure and function

RNA polymerase III

RNA polymerase III (RNA Pol III) is a complex of at least 16 different subunits. Like RNA Pol II, it is located in the nucleoplasm. RNA Pol III synthesizes the precursors of 5S rRNA and the tRNAs as well as some snRNAs, cytosolic RNAs and long non-coding RNAs (Section I2).

tRNA genes

The initial transcripts produced from tRNA genes are precursor molecules that are processed into mature tRNAs (Section J2). Unusually, the transcription control regions of tRNA genes lie *after* the transcription start site within the transcription unit. There are two highly conserved sequences within the DNA encoding the tRNA, called the A box (5′-TGGCNNAGTGG-3′) and the B box (5′-GGTTC-GANNCC-3′). These sequences also encode important sequences in the tRNA itself, called the D-loop and the TψC loop (Section K2). This means that these highly conserved sequences within the tRNAs also act as promoter DNA sequences.

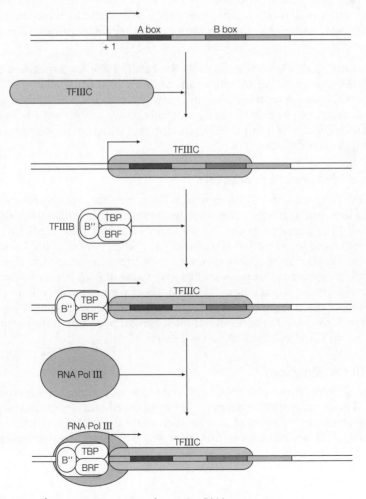

Figure 1. Initiation of transcription at a eukaryotic tRNA promoter.

Two complex DNA-binding factors have been identified that are required for tRNA transcription initiation by RNA Pol III (Figure 1). **TFIIIC** binds to both the A and B boxes in the tRNA promoter. **TFIIIB** binds 50 bp upstream from the A box. TFIIIB consists of three subunits, one of which is **TBP**, the general initiation factor required by all three RNA polymerases (see Sections H2 and H5). The second is called **BRF** (TFIIB-related factor, since it has homology to TFIIB, the RNA Pol II initiation factor, Section H5). The third subunit is called **B″**. TFIIIB has no sequence specificity and therefore its binding site appears to be determined by the position of TFIIIC binding to the DNA. TFIIIB allows RNA Pol III to bind and initiate transcription. Once TFIIIB has bound, TFIIIC can be removed without affecting transcription. TFIIIC is therefore an assembly factor for the positioning of the initiation factor TFIIIB.

5S rRNA genes

RNA Pol III synthesizes the 5S rRNA component of the large ribosomal subunit. This is the only rRNA to be transcribed separately. Like the other rRNA genes that are transcribed by RNA Pol I, the 5S rRNA genes are tandemly arranged in a gene cluster. In humans, there is a single cluster of around 2000 genes. The promoters of 5S rRNA genes contain an internal control region called the C box, which is located 81–99 bp downstream from the transcription start site. A second sequence termed the A box around bases +50 to +65 is also important.

The C box of the 5S rRNA promoter acts as the binding site for a specific DNA-binding protein, **TFIIIA** (Figure 2). TFIIIA acts as an assembly factor that allows TFIIIC to interact with the 5S rRNA promoter. The A box may also stabilize TFIIIC binding. TFIIIC is then bound to the DNA at an equivalent position relative to the start site as in the tRNA promoter. Once TFIIIC has bound, TFIIIB can interact with the complex and recruit RNA Pol III to initiate transcription.

Alternative RNA Pol III promoters

Many RNA Pol III genes also rely on upstream sequences for the regulation of their transcription. Some promoters such as those of the U6 small nuclear RNA (**U6 snRNA**, Section J3) and small RNA genes from the Epstein–Barr virus use only regulatory sequences upstream from their transcription start sites. The coding region of the U6 snRNA has a characteristic A box. However, this sequence is not required for transcription. The U6 snRNA upstream sequence contains sequences typical of RNA Pol II promoters, including a TATA box (Section H4) at bases –30 to –23. These promoters also share several other upstream transcription factor binding sequences with many U snRNA genes that are transcribed by RNA Pol II. These observations suggest that common transcription factors can regulate both RNA Pol II and RNA Pol III genes.

RNA Pol III termination

Termination of transcription by RNA Pol III appears only to require polymerase recognition of a simple nucleotide sequence. This consists of clusters of dA residues whose termination efficiency is affected by the surrounding sequence. Thus the sequence 5′-GCAAAAGC-5′ is an efficient termination signal in the *Xenopus borealis* somatic 5S rRNA gene.

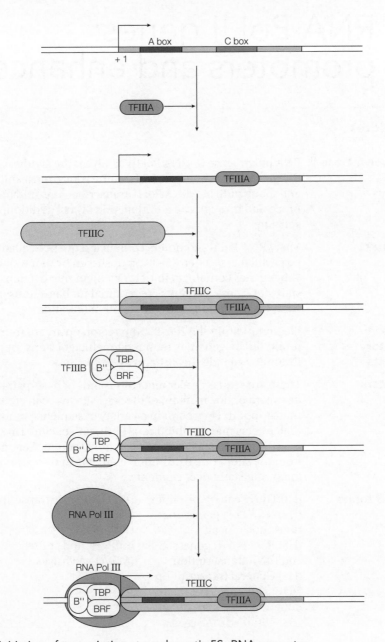

Figure 2. Initiation of transcription at a eukaryotic 5S rRNA promoter.

H4 RNA Pol II genes: promoters and enhancers

Key Notes

RNA polymerase II	RNA polymerase II (RNA Pol II) catalyzes the synthesis of the pre-mRNAs for all protein-coding genes, most snRNAs and many miRNAs. RNA Pol II-transcribed pre-mRNAs are processed through cap addition, poly(A) tail addition, and splicing.
Promoters	Many RNA Pol II promoters contain a sequence called a TATA box, which is situated 25–30 bp upstream from the start site. Other genes contain an initiator element that overlaps the start site. These elements are required for basal transcription complex formation, and transcription initiation.
Upstream regulatory elements	Elements within the 100–200-bp region upstream from the promoter are generally required for efficient transcription. Examples include the SP1 and CCAAT boxes.
Enhancers	These are sequence elements thousands of base-pairs upstream or downstream of the start site that can activate transcription. They contain a variety of sequence motifs and their effects may be ubiquitous or tissue-specific. Thus, there is a continuous spectrum of regulatory sequence elements, from the short-range promoter elements to the extreme long-range enhancer elements.
Related topics	(F3) The *E. coli* σ^{70} promoter (H2) RNA Pol I genes: the ribosomal repeat (H3) RNA Pol III genes: 5S and tRNA transcription (H5) General transcription factors and RNA Pol II initiation	(I1) Eukaryotic transcription factors (J3) mRNA processing, hnRNPs, and snRNPs (M3) DNA viruses

---

RNA polymerase II

RNA polymerase II (RNA Pol II) is located in the nucleoplasm. It is responsible for the transcription of all protein-coding genes, most snRNA genes and sequences encoding miRNAs, siRNAs and lncRNAs (Sections I2, J2 and L4). The pre-mRNAs must be processed after synthesis by 5′-capping, 3′-polyadenylation and removal of introns by splicing (Section J3).

Promoters

Many eukaryotic promoters contain a sequence called the **TATA box** around 25–35 bp upstream from the start site of transcription (Figure 1). It has the 7-bp consensus sequence 5′-TATA(A/T)A(A/T)-3′ although it is now known that the protein that binds to the TATA box, TBP, binds to an 8-bp sequence that includes an additional downstream base-pair, whose identity is not important (Section H5). The TATA box acts in a similar way to an *E. coli* promoter –10 sequence to position the RNA Pol II for correct transcription initiation (Section F3). While the sequence around the TATA box is critical, the sequence between the TATA box and the transcription start site is not critical. However, the spacing between the TATA box and the start site is important. Around 50% of the time, the start site of transcription is an A (on the sense strand). Some eukaryotic genes contain an **initiator element** instead of a TATA box. The initiator element is located around the transcription start site. Many initiator elements have a C at position –1 and an A at +1. Other promoters have neither a TATA box nor an initiator element. These genes are generally transcribed at low rates, and initiation of transcription may occur at different start sites over a length of up to 200 bp. These genes often contain a GC-rich 20–50-bp region within the first 100–200 bp upstream from the start site.

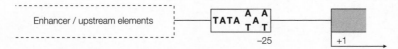

Figure 1. RNA Pol II promoter containing TATA box.

Upstream regulatory elements

The low activity of short basal promoters may be greatly increased by the presence of other elements located upstream of the promoter. These elements are found in many genes that vary widely in their levels of expression in different tissues. Two common examples are the **SP1 box**, which is found upstream of many genes both with and without TATA boxes, and the **CCAAT box**. Promoters may have one, both, or multiple copies of these sequences. These sequences, which are often located within 100–200-bp upstream of the promoter, are referred to as **upstream regulatory elements** (**UREs**) and play an important role in ensuring efficient transcription from the promoter.

Enhancers

Transcription from many eukaryotic promoters can be stimulated by control elements that are located many thousands of base-pairs away from the transcription start site. This was first observed in the genome of the DNA virus SV40 (Section M3). A sequence of around 100 bp from SV40 DNA can significantly increase transcription from a basal promoter even when it is placed far upstream. Enhancer sequences are characteristically 100–200-bp long and contain multiple sequence elements that contribute to the total activity of the enhancer. They may be ubiquitous or cell type-specific. Classically, enhancers have been described as having the following general characteristics:

• They exert strong activation of transcription of a linked gene from the correct start site.

• They activate transcription when placed in either orientation with respect to linked genes.

- They are able to function over long distances of more than 1 kb whether from an upstream or downstream position relative to the start site.

- They exert preferential stimulation of the closest of two tandem promoters.

However, as more enhancers and promoters have been identified, it has been shown that the upstream promoter and enhancer motifs overlap physically and functionally. There seems to be a continuum between classic enhancer elements and those promoter elements that are orientation specific and which must be placed close to the promoter to have an effect on transcriptional activity.

H5 General transcription factors and RNA Pol II initiation

Key Notes

RNA Pol II basal transcription factors
A complex series of basal transcription factors have been characterized that bind to RNA Pol II promoters and together initiate transcription. They are named TFIIA, B, C, etc.

TFIID
TFIID binds to the TATA box. It is a multiprotein complex of the TATA-binding protein (TBP) and multiple accessory factors called TBP-associated factors, or TAF$_{II}$s.

TBP
TBP is a transcription factor required for transcription initiation by all three RNA polymerases. It has a saddle structure that binds to the minor groove of the DNA at the TATA box, unwinding the DNA and introducing a 45° bend.

TFIIA
TFIID binding to the TATA box is enhanced by TFIIA. TFIIA appears to stop inhibitory factors binding to TFIID. These inhibitory factors would otherwise block further assembly of the transcription complex.

TFIIB and RNA polymerase binding
TFIIB binds to TFIID and acts as a bridge factor for RNA polymerase binding. The RNA polymerase binds to the complex associated with TFIIF.

Factors binding after RNA polymerase
After RNA polymerase binding, TFIIE, TFIIH, and TFIIJ associate with the transcription complex in a defined binding sequence. Each of these proteins is required for transcription *in vitro*.

CTD phosphorylation by TFIIH
TFIIH phosphorylates the CTD of RNA Pol II. This results in formation of a processive polymerase complex.

The initiator transcription complex
TBP is recruited to initiator-containing promoters by a further DNA-binding protein. The TBP is then able to initiate transcription initiation by a mechanism similar to that in TATA box-containing promoters.

Related topics	(F2) *Escherichia coli* RNA polymerase	(H3) RNA Pol III genes: 5S and tRNA transcription
	(F4) Transcription initiation, elongation, and termination	(H4) RNA Pol II genes: promoters and enhancers
	(H1) The three RNA polymerases: characterization and function	(I1) Eukaryotic transcription factors
	(H2) RNA Pol I genes: the ribosomal repeat	(I2) Examples of transcriptional regulation

RNA Pol II basal transcription factors

A series of nuclear transcription factors has been identified, cloned, and purified. These are required for basal transcription initiation from RNA Pol II promoter sequences *in vitro*. These multisubunit factors are named transcription factor IIA, IIB, etc. (TFIIA, etc.). They have been shown to assemble on basal promoters in a specific order (Figure 1) and they may be subject to multiple levels of regulation (Section I1). Basal transcription factors homologous to TBP and TFIIB are also important in archaea.

TFIID

In promoters containing a TATA box (Section H4), the RNA Pol II transcription factor **TFIID** is responsible for binding to this key promoter element. The binding of TFIID to the TATA box is the earliest stage in the formation of the RNA Pol II transcription initiation complex. TFIID is a multiprotein complex in which only one polypeptide, the **TATA-binding protein** (TBP), binds to the TATA box. The complex also contains other polypeptides known as **TBP-associated factors** ($TAF_{II}s$). In mammalian cells, TBP binds to the TATA box and is then joined by at least eight $TAF_{II}s$ to form TFIID.

TBP

TBP is present in all three eukaryotic transcription complexes (in SL1, TFIIIB, and TFIID) and clearly plays a major role in transcription initiation (Sections H2 and H3). TBP is a monomeric protein. All eukaryotic TBPs analyzed have very highly conserved CTDs of 180 residues and this conserved domain functions as well as the full-length protein in transcription *in vivo*. TBP has been shown to have a saddle structure with an overall dyad symmetry, but the two halves of the molecule are not identical. TBP interacts with DNA in the minor groove so that the inside of the saddle binds to DNA at the TATA box and the outside surface of the protein is available for interactions with other protein factors. Binding of TBP deforms the DNA so that it is bent into the inside of the saddle and unwound. This results in a kink of about 45° between the first two and last two base pairs of the 8-bp TATA element. A TBP with a mutation in its TATA box-binding domain retains its function for transcription by RNA Pol I and Pol III (Sections H2 and H3), but it inhibits transcription initiation by RNA Pol II. This indicates that the other two polymerases use TBP to initiate transcription, but the precise role of TBP in these complexes remains unclear.

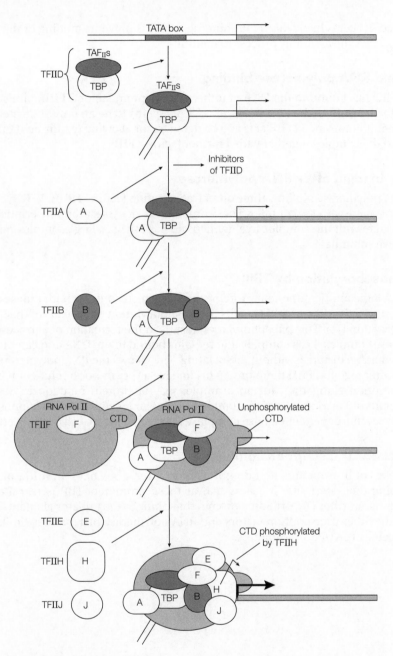

Figure 1. A schematic diagram of the assembly of the RNA Pol II transcription initiation complex at a TATA box promoter.

TFIIA

TFIIA binds to TFIID and enhances TFIID binding to the TATA box, stabilizing the TFIID-DNA complex. TFIIA is made up of at least three subunits. In transcription studies *in vitro*, the requirement for TFIIA is lost as TFIID is purified. *In vivo*, TFIIA appears to counteract the effects of inhibitory factors such as DR1 and DR2 with which TFIID is associated. It seems likely that TFIIA binding to TFIID prevents binding of these inhibitors and allows the assembly process to continue.

TFIIB and RNA polymerase binding

Once TFIID has bound to the DNA, another transcription factor, **TFIIB**, binds to TFIID. TFIIB can also bind to the RNA polymerase. This seems to be an important step in transcription initiation since TFIIB acts as a bridging factor allowing recruitment of the polymerase to the complex together with a further factor, **TFIIF**.

Factors binding after RNA polymerase

After RNA polymerase binding, three other transcription factors, **TFIIE**, **TFIIH**, and **TFIIJ**, rapidly associate with the complex. These proteins are necessary for transcription *in vitro* and associate with the complex in a defined order. TFIIH is a large complex made up of at least five subunits.

CTD phosphorylation by TFIIH

TFIIH is a large multicomponent protein complex which contains both **kinase** and **helicase** activity. Activation of TFIIH results in **phosphorylation** of the CTD of the RNA polymerase (Section H1). This phosphorylation results in the formation of a **processive** RNA polymerase complex, i.e. a complex that remains bound to the DNA and can move along it a considerable distance without dissociating. This allows the RNA polymerase to leave the promoter region. TFIIH therefore seems to have a very important function in control of transcription elongation (for an example, see Tat protein function in Section I2). Components of TFIIH are also important in DNA repair (Section E3) and in phosphorylation of the cyclin-dependent kinase complexes that regulate the cell cycle (Section N1).

The initiator transcription complex

Many RNA Pol II promoters that do not contain a TATA box have an **initiator element** overlapping their start site. It seems that at these promoters TBP is recruited to the promoter by a further DNA-binding protein that binds to the initiator element. TBP then recruits the other transcription factors and RNA polymerase in a manner similar to that which occurs at TATA box promoters.

I1 Eukaryotic transcription factors

Key Notes

Transcription factor domain structure
Transcription factors have a modular structure consisting of DNA-binding and transcription activation domains. Some transcription factors have dimerization domains.

DNA-binding domains

- Helix–turn–helix domains are found in DNA-binding proteins such as the lac repressor and the 60-amino-acid domain encoded by the homeobox sequence. A recognition α-helix interacts with the DNA and is separated from another α-helix by a characteristic right-angle β-turn.

- Zinc finger domains include the C_2H_2 zinc fingers, which bind Zn^{2+} through two Cys and two His residues, and the C_4 fingers, which bind Zn^{2+} through four Cys residues. Proteins with C_2H_2 zinc fingers bind to DNA using three or more fingers, while C_4 fingers occur in pairs, and the proteins bind to DNA as dimers.

- Basic domains are associated with leucine zipper and helix–loop–helix (HLH) dimerization domains. Dimerization is generally necessary for basic domain binding to DNA.

Dimerization domains

- Leucine zippers have a hydrophobic leucine residue at every seventh position in an α-helical region, which results in a leucine at every second turn on one side of the α-helix. Two monomeric proteins dimerize through interactions between their leucine zippers.

- HLH proteins have two α-helices separated by a nonhelical peptide loop. Hydrophobic amino acids on one side of the C-terminal α-helix allow dimerization. As with the leucine zipper, the HLH motif is often adjacent to an N-terminal basic domain that requires dimerization for DNA binding.

Transcription activation domains

- Acidic activation domains contain a high proportion of acidic amino acids and are present in many transcription activators.

- Glutamine-rich domains contain a high proportion of glutamine residues, and are present in the activation domains of, for example, the transcription factor SP1.

	• Proline-rich domains contain a continuous run of proline residues and can activate transcription. A proline-rich activation domain is present in the product of the proto-oncogene c-Jun.	
Repressor domains	Repressors may block transcription factor activity indirectly by masking DNA binding or transcriptional activation. Alternatively, they may contain a specific direct repressor domain.	
Targets of transcriptional regulators	Different activation domains may have many different targets within the basal transcription complex. Proposed targets include chromatin, TAF$_{II}$s in TFIID, TFIIB, and the CTD of RNA Pol II, which may undergo differential phosphorylation by TFIIH.	
Chromatin modification	Histone acetyltransferases (HATs) and ATP-dependent remodeling enzymes can facilitate transcription through epigenetic changes to chromatin structure. Acetylation of histones by HATs tends to activate gene expression. This is opposed by the action of histone deacetylases (HDACs) and methyltransferases. Chromatin remodeling is the enzyme-assisted movement of nucleosomes on DNA, which can allow transcription factors to gain access to DNA. It may also involve the action of long noncoding RNAs.	
Related topics	(A3) Protein structure and function (C2) Chromatin structure (C3) Eukaryotic chromosome structure (Section G) Regulation of transcription in bacteria	(H4) RNA Pol II genes: promoters and enhancers (H5) General transcription factors and RNA Pol II initiation (I2) Examples of transcriptional regulation

Transcription factor domain structure

Transcription factors other than the general transcription factors of the basal transcription complex (Section H5) were first identified through their affinity for specific motifs in promoters, UREs, or enhancer regions (Section H4). These factors have two distinct activities. First, they bind specifically to their DNA-binding site and, secondly, they activate transcription. Mutagenesis of the yeast transcription factors **Gal4** and **Gcn4** showed that their DNA-binding and transcription activation activities were in separate parts of the proteins – the **activation domain** and **DNA-binding domain** (Section A3). In **domain swap experiments**, the activation domains of these proteins were fused to the bacterial **LexA repressor** (see Section G1 for a description of bacterial repressors and operators). These hybrid fusion proteins (Section A5) activated transcription from a promoter containing the *lexA* **operator** sequence, indicating that the transcriptional activation function of the yeast proteins was separable from their DNA-binding activity.

In addition, many transcription factors occur as homo- or heterodimers, held together by **dimerization domains**. A few transcription factors have **ligand-binding domains** that allow regulation of transcription factor activity by binding of an accessory small molecule. The steroid and thyroid hormone receptors (Section I2) are an example containing all four of these types of domain.

DNA-binding domains

The helix–turn–helix domain

This domain is characteristic of DNA-binding proteins containing a 60-amino-acid **homeodomain**, which is encoded by a sequence called the **homeobox** (Section I2). In the **Antennapedia** transcription factor of *Drosophila*, this domain consists of four α-helices in which helices II and III are at right angles to each other and are separated by a characteristic β-turn. The characteristic helix–turn–helix structure (Figure 1) is also found in bacteriophage DNA-binding proteins such as the phage λ cro repressor (Section M2), lac and trp repressors (Sections G1 and G2), and catabolite activator protein, CAP (Section G1). The domain binds so that one helix, known as the recognition helix, lies partly in and interacts with the major groove of the DNA (Section A2). The recognition helices of two homeodomain factors Bicoid and Antennapedia can be exchanged, and this swaps their DNA-binding specificities. Indeed, the specificity of this interaction is demonstrated by the observation that the exchange of only one amino acid residue swaps the DNA-binding specificities.

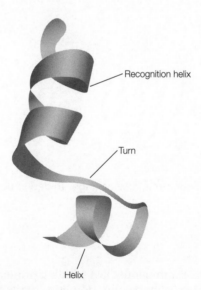

Figure 1. The helix–turn–helix core structure.

The zinc finger domain

This domain exists in several forms. For example, the C_2H_2 zinc finger has a loop of 12 amino acids anchored by two cysteine and two histidine residues that tetrahedrally coordinate a zinc ion (Figure 2a). This **motif** (Section A3) folds into a compact structure comprising two β-strands and one α-helix, the latter binding in the major groove of the DNA (Figure 2b). The α-helical region contains conserved basic amino acids that are responsible for interacting with the DNA. This structure is repeated nine times in TFIIIA, the RNA Pol III transcription factor (Section H3). It is also present in three copies in transcription factor SP1 (Section I2). Usually, three or more C_2H_2 zinc fingers are required for DNA binding. A related motif, in which the zinc ion is coordinated by four cysteine residues, occurs in steroid hormone receptor transcription factors (Section I2). These factors consist of homo- or hetero-dimers, in which each monomer contains two C_4 zinc finger motifs (Figure 2c). The two motifs fold together into a more complex conformation stabilized by zinc, which binds to DNA by the insertion of one α-helix from each monomer into successive major grooves, in a manner reminiscent of the helix–turn–helix proteins.

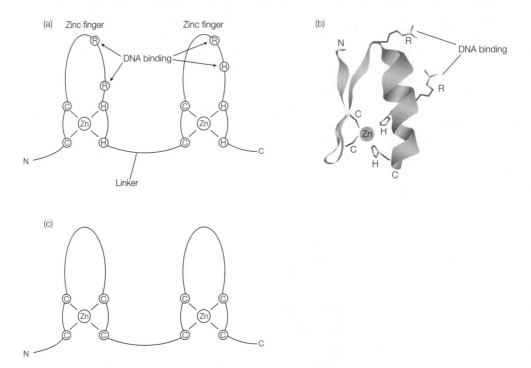

Figure 2. (a) The C_2H_2 zinc finger motif; (b) zinc finger folded structure; (c) the C4 zinc finger motif.

The basic domain

A **basic domain** is found in a number of DNA-binding proteins and is generally associated with one or other of two dimerization domains, the **leucine zipper** or the **helix–loop–helix** (**HLH**) motif (see below). These are referred to as basic leucine zipper (**bZIP**) or basic HLH (**bHLH**) proteins. Dimerization of the proteins brings together two basic domains, which can then interact with the DNA.

Dimerization domains

Leucine zippers

Leucine zipper proteins contain a hydrophobic leucine residue at every seventh position in a region that is often in the C-terminal part of the DNA-binding domain. These leucines lie in an α-helical region and the regular repeat of these residues forms a hydrophobic surface on one side of the α-helix, with a leucine every second turn of the helix. These leucines are responsible for dimerization through interactions between the hydrophobic faces of an α-helix on each monomer (Figure 3). This interaction forms a **coiled-coil structure**. bZIP transcription factors contain a basic DNA-binding domain N-terminal to the leucine zipper. This is present on an α-helix which is a continuation of the leucine zipper α-helical C-terminal domain. The N-terminal basic domains of each helix form a symmetrical structure in which each basic domain lies along the DNA in opposite directions, interacting with a symmetrical DNA recognition site so that the protein in effect forms a clamp around the DNA. The leucine zipper is also used as a dimerization domain in proteins that use DNA-binding domains other than the basic domain, including some homeodomain proteins.

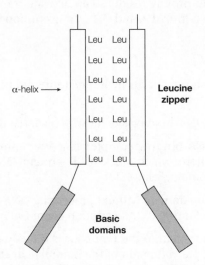

Figure 3. The leucine zipper and basic domain dimer of a bZIP protein.

The helix–loop–helix domain

The overall structure of this domain is similar to the leucine zipper, except that a non-helical loop of polypeptide chain separates two α-helices in each monomeric protein. Hydrophobic residues on one side of the C-terminal α-helix allow dimerization. This structure is found in the **MyoD** family of proteins (Section I2). As with the leucine zipper, the HLH motif is often found adjacent to a basic domain that requires dimerization for DNA binding. With both bHLH and bZIP proteins the formation of heterodimers allows much greater diversity and complexity in the transcription factor repertoire.

Transcription activation domains

Acidic activation domains

Comparison of the transcription activation domains of yeast Gcn4 and Gal4, the mammalian glucocorticoid receptor, and herpes virus activator VP16 shows that they all have a very high proportion of acidic amino acids. These have been called **acidic activation domains** or 'acid blobs' or 'negative noodles' and are characteristic of many transcription activation domains. It is still uncertain what other features are required for these regions to function as efficient transcription activation domains.

Glutamine-rich domains

Glutamine-rich domains were first identified in two activation regions of the transcription factor SP1. As with acidic domains, the proportion of glutamine residues seems to be more important than overall structure. Domain-swap experiments between glutamine-rich activation regions from the diverse transcription factors SP1 and the *Drosophila* protein Antennapedia showed that these domains could substitute for each other.

Proline-rich domains

Proline-rich domains have been identified in several transcription factors. As with glutamine, a continuous run of proline residues can activate transcription. This domain is found, for example, in the c-Jun, AP2, and Oct-2 transcription factors.

Repressor domains

Repression of transcription may occur by indirect interference with the function of an activator, for example by:

- **Blocking** the activator DNA-binding site (as with bacterial repressors; Section G)

- Formation of a **non-DNA-binding complex** (e.g. the repressors of steroid hormone receptors, or the Id protein, which blocks HLH protein–DNA interactions because it lacks a DNA-binding domain; Section I2)

- **Masking** of the activation domain without preventing DNA binding (e.g. Gal80 masks the activation domain of the yeast transcription factor Gal4)

In other cases, a specific domain of the repressor is directly responsible for inhibition of transcription. For example, a domain of the mammalian **thyroid hormone receptor** can repress transcription in the absence of thyroid hormone and activates transcription when bound to its ligand (Section I2). The product of the **Wilms tumor gene**, *WT1*, is a tumor-suppressor protein having a specific proline-rich repressor domain that lacks charged residues.

Targets of transcriptional regulators

The presence of diverse activation domains raises the question of whether they each have the same or different targets in the basal transcription complex and elsewhere for the activation of transcription. They are distinguishable from each other, since the acidic activation domain can activate transcription from a downstream enhancer site, while the proline domain only activates weakly and the glutamine domain not at all. While proline and acidic domains are active in yeast, glutamine domains have no activity, implying that they have a different target that is not present in the yeast transcription complex. Proposed targets of different transcriptional activators include:

- Chromatin structure

- Interaction with TFIID through specific TAFIIs

- Interaction with TFIIB

- Interaction or modulation of the TFIIH complex activity leading to differential phosphorylation of the CTD of RNA Pol II (Sections H1 and H5)

It seems likely that different activation domains have different targets, and almost any component or stage in transcription initiation and elongation could be a target, resulting in multistage regulation.

Chromatin modification

One major target for transcriptional activation is chromatin (Section C2). Changes to chromatin structure can control the activity of genes and the 'islands' of unmethylated CpG sequences around active promoters provide one example of **epigenetic** regulation (Section C3). Several classes of enzymes can facilitate transcription through changes to chromatin, including the **histone acetyltransferases (HATs)**, **histone methyltransferases**, and the **ATP-dependent remodeling enzymes**.

Histone acetyltransferases acetylate conserved lysines in histones by transferring an acetyl group from acetyl-CoA. Acetylation of histones is associated with transcriptional activation and with **euchromatin**. **Histone deacetylases** have the opposite activity to HATs. It was first thought that lysine acetylation of histones neutralized their positive charge, reducing the binding of the histone to negatively charged DNA, rendering DNA more accessible to transcription factors. More recently, it has become clear that histone acetylation allows specific histone-protein interactions. Histone acetyltransferases can also acetylate nonhistone proteins such as transcription factors, to facilitate gene expression. Histones can also be **methylated** on lysine and arginine residues, and this is usually associated with transcriptional repression, although some cases of activation have been recorded.

Chromatin remodeling involves the enzyme-catalyzed movement of nucleosomes along DNA to allow transcription factors access to bind to the DNA. This is believed to involve disassembly and reassembly of the nucleosome core (Section C3). Remodeling is promoted by enzyme complexes containing ATPase and other subunits that are responsible for the specificity and control of the complex. RNA polymerase II-mediated transcription of lncRNAs (Section I2) may also control chromatin remodeling and activation of protein-coding genes. Chromatin may be progressively converted to a more open configuration as several species of lncRNAs are transcribed. This underlines a growing theme suggesting that, as in prokaryotes (Section F4), eukaryotic transcription is a surprisingly dynamic process involving several cycles of abortive synthesis of short RNAs from the **5′-untranslated region (5′-UTR)** of genes before a full length protein-coding transcript is synthesized.

12 Examples of transcriptional regulation

Key Notes	
Constitutive transcription factors: SP1	SP1 is a ubiquitous transcription factor that contains three zinc finger motifs and two glutamine-rich transactivation domains.
Hormonal regulation: steroid hormone receptors	Steroid hormones enter the cell and bind to a steroid hormone receptor. The receptor dissociates from a bound inhibitor protein, dimerizes and translocates to the nucleus. The DNA-binding domain of the steroid hormone receptor binds to response elements, giving rise to activation of target genes. Thyroid hormone receptors act as transcription repressors until they are converted to activators by hormone binding. Heterodimers of the glucocorticoid receptor with other transcription factors can repress gene expression.
Regulation by phosphorylation: STAT proteins	Interferon-γ activates JAK kinase, which phosphorylates STAT1α. STAT1α dimerizes and translocates to the nucleus, where it activates expression of target genes.
Transcription elongation: HIV Tat	Tat protein binds to an RNA sequence called TAR, present at the 5′-end of all human immunodeficiency virus (HIV) RNAs. In the absence of Tat, HIV transcription terminates prematurely. The Tat–TAR complex activates TFIIH in the transcription initiation complex at the promoter, leading to phosphorylation of the RNA Pol II CTD. This permits full-length transcription by the polymerase.
Cell determination: MyoD	The expression of *MyoD* and related genes *(Myf5, Mrf4,* and *myogenin)* can convert nonmuscle cells into muscle cells. Their expression activates muscle-specific gene expression and blocks cell division. Each gene encodes an HLH transcription factor. HLH heterodimer formation, and the non-DNA binding inhibitor, Id, give rise to diversity and regulation of these transcription factors.
Embryonic development: homeodomain proteins	The homeobox, which encodes a DNA-binding domain, was originally found in *Drosophila melanogaster* homeotic genes. These encode transcription factors that specify the development of body parts. The conservation of both function and organization of the homeotic (*Hox*) gene clusters between *Drosophila* and mammals suggests that these proteins have important common roles in development.

Regulation by noncoding RNAs	Long noncoding RNAs regulate transcription in a number of ways including binding to RNA polymerase and other transcription factors and altering chromatin structure. MicroRNAs can also affect transcription by regulating the translation of transcription factors and via chromatin modification.
Related topics	(H1) The three RNA polymerases: characterization and function
	(H5) General transcription factors and RNA Pol II initiation
	(I1) Eukaryotic transcription factors
	(N1) The cell cycle

Constitutive transcription factors: SP1

SP1 binds to a GC-rich sequence with the consensus sequence GGGCGG. It is a constitutive transcription factor whose binding site is found in the promoter of many **housekeeping** genes (Section C3). SP1 is present in all cell types. It contains three zinc finger motifs and two glutamine-rich transactivation domains. The glutamine-rich domains of SP1 interact specifically with TAF$_{II}$110, one of the TAF$_{II}$s that bind to the TATA-binding protein (TBP) to make up TFIID (Section H5). This represents one target through which SP1 may interact with and regulate the basal transcription complex.

Hormonal regulation: steroid hormone receptors

Many transcription factors are activated by **hormones** that are secreted by one cell type and act upon a different cell type to alter gene expression. For example, **steroid hormones** are lipid-soluble and can diffuse through the plasma membranes and bind to a cytoplasmic transcription factor called a **steroid hormone receptor**, a member of the **nuclear receptor family**. In the absence of the steroid, the receptor is bound to an inhibitor, the heat shock protein HSP90 (Figure 1). When the hormone binds to the **ligand-binding domain** of the receptor, the receptor is released from the inhibitor, a previously masked **nuclear localization signal** (Section L4) is revealed, and the receptor dimerizes and translocates to the nucleus. The DNA-binding domain of the steroid hormone receptor then interacts with its specific DNA-binding sequence, or **response element**, and this gives rise to activation of the target gene. This action of **Type I** nuclear receptors is not true of all. For example, the **Type II** thyroid hormone and retinoic acid receptors are always nuclear and act as DNA-bound repressors in the absence of hormone. In the presence of the hormone, the receptors are converted from transcriptional repressors to transcriptional activators by replacement of **corepressor** proteins by **coactivators**. Also, while the ligand-bound glucocorticoid receptor can form a **homodimer** that activates transcription of target genes, it can also form a **heterodimer** with other transcription factors that prevents them from working, thus repressing the expression of other genes.

Regulation by phosphorylation: STAT proteins

Many hormones and signaling factors do not diffuse into the cell. Instead, they bind to cell-surface receptors and pass a signal to proteins within the cell through a process called **signal transduction**. This process often involves reversible protein phosphorylation.

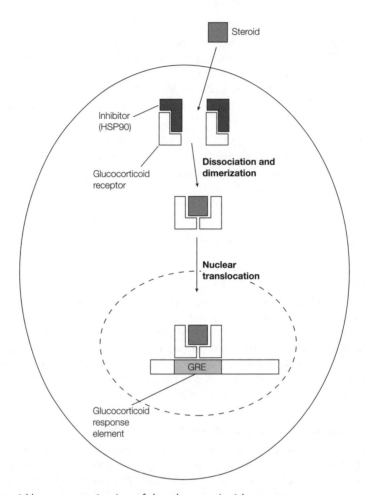

Figure 1. Steroid hormone activation of the glucocorticoid receptor.

Interferon-γ induces phosphorylation of a transcription factor called **STAT1α** through activation of an intracellular kinase (Section A3) called JAK. When STAT1α protein is unphosphorylated, it exists as a monomer in the cell cytoplasm and has no transcriptional activity. However, when STAT1α becomes phosphorylated on a specific tyrosine residue, it is able to form a homodimer, which moves from the cytoplasm into the nucleus. In the nucleus, STAT1α is able to activate the expression of target genes whose promoter regions contain a consensus DNA-binding motif (Figure 2).

Transcription elongation: HIV Tat

HIV encodes an activator protein called **Tat**, which is required for productive HIV gene expression. Tat binds to an RNA stem–loop structure called **TAR**, which is present in the 5′-UTR of all HIV RNAs, just after the HIV transcription start site. The predominant effect of Tat in mammalian cells is at the level of transcription elongation. In this respect it can be seen to be analogous to a *cis*-linked regulatory ncRNA (see below and Section L4). In the absence of Tat, the HIV transcripts terminate prematurely due to poor **processivity** of the RNA Pol II transcription complex (Section H5). Tat is thought to bind to TAR on one

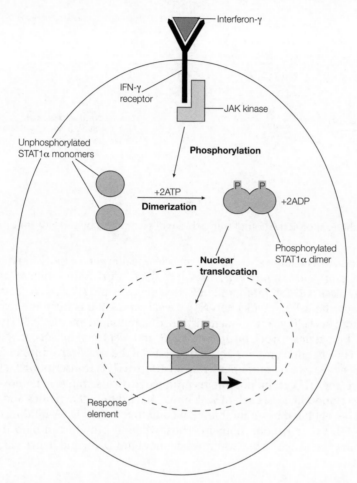

Figure 2. Interferon-γ-mediated transcription activation caused by phosphorylation and dimerization of the STAT1α transcription factor.

transcript in a complex together with cellular RNA-binding factors. This protein–RNA complex may loop backwards and interact with the new transcription initiation complex that is assembled at the promoter. This interaction is thought to result in the activation of the kinase activity of TFIIH. This leads to phosphorylation of the CTD of RNA Pol II, making the RNA polymerase a processive enzyme (Sections H1 and H5). As a result, the polymerase is able to read through the HIV transcription unit, leading to the productive synthesis of HIV proteins (Figure 3).

Cell determination: *MyoD*

Muscle cells arise from mesodermal embryonic cells called **somites**. Somites become committed to forming muscle cells (**myoblasts**) before there is an appreciable sign of cell differentiation to form skeletal muscle cells (called **myotomes**). This process is called **cell determination**. *MyoD* was identified originally as a gene expressed in undifferentiated cells that were committed to form muscle and had therefore undergone cell determination. Overexpression of *MyoD* turns fibroblasts (cells that lay down the basement matrix in many tissues) into muscle-like cells that express muscle-specific genes and resemble myotomes.

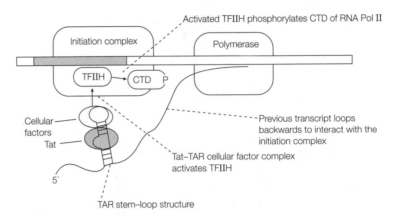

Figure 3. Mechanism of activation of transcriptional elongation by the HIV Tat protein.

MyoD protein has been shown to activate muscle-specific gene expression directly. MyoD also activates expression of *p21/Waf1/Cip1*. p21/Waf1/Cip1 is an inhibitor of the **cyclin-dependent kinases** (**CDKs**). Inhibition of CDK activity by p21/Waf1/Cip1 causes arrest in the G1-phase of the cell cycle (Section N1). *MyoD* expression is therefore responsible for withdrawal from the cell cycle, a characteristic of differentiated muscle cells. The expression of any one of four genes, *MyoD*, *myogenin*, *Myf5*, and *Mrf4*, can in fact convert fibroblasts into muscle. The encoded proteins are all members of the HLH transcription factor family. MyoD is most active as a heterodimer with constitutive HLH transcription factors E12 and E47. Therefore, the HLH group of proteins produces a diverse range of hetero- and homo-dimeric transcription factors, each of which may have different activities and roles. These proteins can be regulated by an inhibitor called **Id** that lacks a DNA-binding domain, but contains the HLH dimerization domain. Thus, while Id protein can bind to MyoD and related proteins, the resulting heterodimers cannot bind DNA, and hence cannot regulate transcription.

Embryonic development: homeodomain proteins

The **homeobox** is a conserved DNA sequence that encodes a helix–turn–helix DNA binding protein structure called the **homeodomain** (Section I1). The homeodomain was first discovered in the transcription factors encoded by **homeotic genes** of *Drosophila*. Homeotic genes are responsible for the correct specification of body parts. For example, mutation of one of these genes, *Antennapedia*, causes the fly to form a leg where the antenna should be. These genes are very important in spatial pattern formation in the embryo. The homeobox sequence has been conserved among a wide range of eukaryotes, and homeobox-containing genes have been shown to be important in mammalian development. In *Drosophila* and mammals, the homeobox genes are arranged in gene clusters called *Hox* **loci**, in which homologous genes are arranged in the same order. The gene homologs are also expressed in a similar order in the embryo on the anterior to posterior axis. This suggests that the conserved homeobox-encoded DNA-binding domain is a feature of a group of transcription factors with a conserved function in embryonic development.

Regulation by noncoding RNAs

Recently, high-throughput methods of global gene expression analysis (Section S2) have revealed the existence of an abundant class of ncRNAs within the **hnRNA** fraction (Section

J3) known as **long (intergenic) noncoding RNAs (lncRNAs or lincRNAs)**, which have been implicated as important transcriptional regulators. They are transcribed from a surprisingly large part of the genome, including repetitive elements such as SINES (Section C4), and may outnumber protein-coding mRNAs by a factor of 10 to 20. Many, though not all, lncRNAs are transcribed by RNA Pol II and are capped and polyadenylated (Sections H4 and J3). They are synthesized from both DNA strands and include **antisense** transcripts of protein-coding genes (where the sense transcript encodes the protein; Section F1).

LncRNAs can act either in *cis* or in *trans*. *Cis*-acting lncRNAs are encoded by the same gene or structure whose expression they regulate whereas *trans*-acting lncRNAs may regulate many genes at unlinked sites in the genome. One way in which lncRNAs can regulate transcription is by modifying the functions of protein transcription factors. For example, the stress-responsive transcripts of the dispersed *Alu* element (Section C4) are synthesized by RNA Pol III and bind directly to RNA Pol II to inhibit transcription of RNA Pol II-dependent genes as a way of inhibiting cell growth. Others have been found to play a major role in the action of the key tumor suppressor p53 (Section N3). p53, itself a transcription factor, activates the expression of dozens of *trans*-acting lncRNAs including one called *linc-p21*. This in turn associates with the protein hnRNP-K, which then binds to the promoters of many p53-dependent genes and represses their transcription, thus promoting apoptosis. On the other hand, a further p53-regulated lncRNA called *PANDA* interacts with a transcription factor NF-YA to inhibit apoptosis and cell division. Thus, lncRNAs appear to be able to regulate transcription in a complex and integrated way similar to protein transcription factors.

LncRNAs can also regulate transcription by modifying chromatin structure. In the females of placental mammals, one X chromosome is inactivated by packaging into a transcriptionally silent structure called a **Barr body** that consists entirely of heterochromatin (Section C3). As a result, XX females have only one active X chromosome and so have the same **gene dosage** of X-linked genes as XY males. X-inactivation is achieved through the action of **X-inactive specific transcript (*Xist*)**, a *cis*-acting lncRNA encoded on the X chromosome. *Xist* functions through the irreversible epigenetic repression of transcription by recruiting histone-modifying enzymes (Section I1). *Xist* expression is negatively regulated by the overlapping antisense RNA *Tsix*.

MicroRNAs (miRNAs) are another type of regulatory ncRNA that operate mainly at the level of translation (Section L4). However, they can influence transcription indirectly by regulating the translation of protein transcription factors. An important example is the control of *Hox* gene expression in animal development (see above). Several miRNA genes have been found in *Hox* gene clusters and their expression is colinear with the overall anatomical expression pattern. This implies that they use the same enhancers as the protein-coding genes to control their transcription. Translation of the Hox mRNAs is down-regulated by binding of the miRNAs to their 3′-UTR regions (Section L4). As part of the **RNA-induced transcriptional silencing (RITS)** complex, a structure analogous to the RISC complex of translational control (Section L4), small ncRNAs can also repress transcription directly in certain regions of the genome in some organisms. The RITS causes histone methylation and formation of heterochromatin (Sections C3 and I1), and the ncRNA may direct destruction of complementary nascent transcripts by analogy with RISC.

J1 rRNA processing and ribosomes

Key Notes

Types of RNA processing

In both bacteria and eukaryotes, primary RNA transcripts undergo various alterations or processing events to become mature RNAs. The three commonest types are: (i) nucleotide removal by nucleases, (ii) nucleotide addition to the 5′- or 3′-end, and (iii) nucleotide modification on the base or the sugar.

rRNA processing in bacteria

An initial 30S transcript is made in *E. coli* by RNA polymerase transcribing one of the seven rRNA operons. Each contains one copy of the 5S, 16S, and 23S rRNA coding regions, together with some tRNA sequences. This 6000-nt transcript folds and complexes with proteins, becomes methylated, and is then cleaved by specific nucleases (RNase III, M5, M16, and M23) to release the mature rRNAs.

rRNA processing in eukaryotes

In the nucleolus of eukaryotes, RNA Pol I transcribes the rRNA genes, which usually exist in tandem repeats, to yield a long, single pre-rRNA which contains one copy each of the 18S, 5.8S, and 28S sequences. Many specific ribose methylations, and pseudouridylations take place directed by small nucleolar ribonucleoprotein particles (snoRNPs), and the maturing rRNA molecules fold and complex with ribosomal proteins. Various spacer sequences are removed from the long pre-rRNA molecule by a series of specific cleavages. RNA Pol III synthesizes the 5S rRNA from unlinked genes. It undergoes little processing.

RNPs and their study

Cells contain a variety of RNA–protein complexes (RNPs). Their structures and functions can be studied by dissociation, re-assembly, electron microscopy, use of antibodies, RNase protection, RNA binding, cross-linking, and neutron and X-ray diffraction.

Bacterial ribosomes

Ribosomes are complexes of rRNA molecules and specific ribosomal proteins. These large RNPs are the machines the cell uses to carry out translation. The *E. coli* 70S ribosome is formed from a large 50S and a small 30S subunit. The large subunit contains 31 different proteins and one each of the 23S and 5S rRNAs. The small subunit contains a 16S rRNA molecule and 21 different proteins.

Eukaryotic ribosomes

Eukaryotic ribosomes are larger and more complex than their bacterial counterparts, but carry out the same role. The complete mammalian 80S ribosome is composed of one large 60S subunit and one small 40S subunit. The 40S subunit contains an 18S rRNA molecule and 33 distinct proteins. The 60S subunit contains one 5S rRNA, one 5.8S rRNA, one 28S rRNA, and 49 proteins.

Related topics	(F1) Basic principles of transcription	(H3) RNA Pol III genes: 5S and tRNA transcription
	(H2) RNA Pol I genes: the ribosomal repeat	(J2) tRNA and other small RNA processing

Types of RNA processing

Very few RNA molecules are transcribed directly into the final **mature RNA** product (Sections F and H). Most newly transcribed RNA molecules (**primary transcripts**) undergo various alterations to yield the mature product. **RNA processing** is the collective term used to describe these alterations. The commonest include:

(i)   The **removal of nucleotides** by both endonucleases and exonucleases

(ii)   The **addition of nucleotides** usually to the 5′- or 3′-ends of the primary transcripts or their cleavage products, though internal additions also occur (Section J4)

(iii) The **modification of certain nucleotides** on either the base or the sugar moiety

These processing events take place on the major classes of RNA in both bacteria and eukaryotes.

rRNA processing in bacteria

In *E. coli* there are seven different rRNA operons dispersed throughout the genome, named rrnH, rrnE, etc. Each operon contains one copy of each of the **5S, 16S,** and **23S rRNA** sequences (see Section A4 for a definition of the sedimentation coefficient, S). Between one and four sequences encoding tRNA molecules are also present in these rRNA operons, and so the primary transcripts are processed to give both rRNA and tRNA molecules. The initial transcript has a sedimentation coefficient of 30S (approximately 6000 nt) and is normally quite short-lived (Figure 1a). In *E. coli* mutants defective in **RNase III**, this 30S transcript accumulates, indicating that RNase III is involved in rRNA processing. Mutants defective in other RNases such as **M5, M16,** and **M23** have also revealed the involvement of these RNases in *E. coli* rRNA processing.

The post-transcriptional processing of *E. coli* rRNA takes place in a series of defined steps (Figure 1a). Both during and following transcription of the 6000-nt primary transcript, the RNA folds into a series of stem-loop structures by base pairing between complementary sequences in the transcript (Section A2). The formation of this secondary structure allows several proteins to bind to form a **ribonucleoprotein** (**RNP**) complex. Many of these proteins remain attached to the RNA and become part of the ribosome. After the binding of these proteins, base and sugar modifications take place, including 24 specific base methylations in which **S-adenosylmethionine** is the methyl group donor and adenine the most commonly methylated base. Primary cleavage events then follow, carried out mainly by RNase III, to release precursors of the 5S, 16S, and 23S molecules. Further cleavage reactions at the 5′- and 3′-ends of these precursors by RNases M5, M16, and M23 release the mature rRNA molecules (Figure 1a).

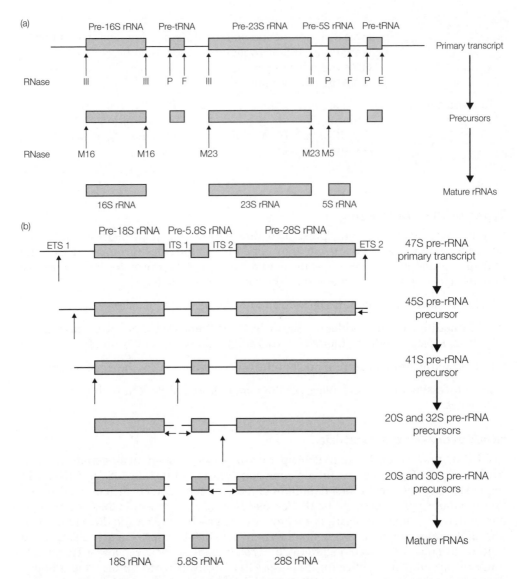

Figure 1. (a) Processing of the *E. coli* rRNA primary transcript; (b) mammalian pre-rRNA processing (↑=endonuclease; ← and →=exonuclease.

rRNA processing in eukaryotes

rRNA in eukaryotes is also generated from a single, long precursor molecule by specific modification and cleavage steps. In many eukaryotes, the rRNA genes are present in a tandemly repeated cluster in the nucleolus containing 100 or more copies of the transcription unit and are transcribed by **RNA Pol I** (Section H2). The precursor has a characteristic size in each organism, being about 7000 nt in yeast and 13 500 nt in mammals (Figure 1b). It contains one copy of the **18S** coding region and one copy each of the **5.8S** and **28S** coding regions, which together are the equivalent of the 23S rRNA in bacteria. The eukaryotic **5S** rRNA is transcribed by **RNA Pol III** from unlinked genes to give a 120-nt transcript that undergoes little or no processing.

The 13 500-nt (47S) precursor of mammalian pre-rRNA undergoes two types of chemical modification, methylation and **pseudouridylation** (see below), followed by a number of endonucleolytic cleavages, initially in the **external transcribed spacers (ETSs) 1 and 2** (Figure 1b). As with bacterial pre-rRNA, the precursor folds and complexes with proteins and maturation factors as it is being transcribed. This takes place in the nucleolus. A 90S precursor containing the U3 snoRNP (see below) bound to 18S pre-rRNA sequences can be detected as the earliest precursor in the complex pathway of ribosome maturation. **Methylation** takes place at over 100 conserved sites to give 2′-O-methylribose and this is carried out by a subset of small nuclear RNP particles that are abundant in the nucleolus, i.e. **small nucleolar RNPs (snoRNPs)**. This group of snoRNPs is known as the **C/D group** because each member contains a single snoRNA molecule with two short conserved sequences, the C box and the D box. In addition, each snoRNA contains a short stretch of complementarity to a specific part of the pre-rRNA and, by base pairing to it, defines where methylation takes place. Mammalian pre-rRNAs are also modified by **pseudou-ridylation** at approximately 100 conserved sites. Another group of snoRNPs, the **H/ACA group**, defines which uridine residues are modified to pseudouridine (Section K2). These snoRNPs each contain a single snoRNA that has an internal H box (5′-ANANNA-3′) and a 5′-ACA-3′ triplet near the 3′-end, and forms two hairpin structures. Unpaired bases in the stems of these hairpins form base pairs with complementary regions in the pre-rRNA and thus identify the specific uridine residue for modification. Cleavages in the **internal transcribed spacers (ITSs)** then release the 20S pre-rRNA from the 32S pre-rRNA. Both of these precursors must be trimmed further, mainly by exonucleases (some of which are thought to be components of the **exosome**, a complex of several exonucleases that are involved in mRNA degradation), and the 5.8S region must base-pair to the 28S rRNA before the mature molecules are produced.

At least one eukaryote, the protozoon *Tetrahymena thermophila*, makes a pre-rRNA that undergoes an unusual form of processing before it can function. It contains an intron (Section J3) in the precursor for the largest rRNA that must be removed during process-ing. Although this process occurs *in vivo* in the presence of protein, it has been shown that the intron can actually excise itself *in vitro* in the complete absence of protein. The RNA folds into an enzymatically active form, or **ribozyme** (Section J2), to perform self-cleavage and ligation.

RNPs and their study

Many RNA molecules exist in cells as **RNPs** (Section A4), with specific proteins attaching to specific RNAs. **Ribosomes** are the largest and most complex RNPs and are formed by the rRNA molecules complexing with specific ribosomal proteins during processing. Other RNPs are discussed in Sections J2 and J3. Several methods are used to study RNPs, including **dissociation**, where the RNP is purified and separated into its RNA and protein components, which are then characterized. **Re-assembly** is used to discover the order in which the components fit together and, if the components can be modified, it is possible to gain clues as to their individual functions. **Electron microscopy** can allow direct visualization if the RNPs are large enough, otherwise it can roughly indicate overall shape. **Antibodies to RNPs** or their individual components can be used for purification, inhibition of function, and, in combination with electron microscopy, can show the overall positions of the components in the overall structure. **RNA binding experiments** can show whether a particular protein binds to an RNA, and subsequent treatment of the RNA–protein complex with RNase (**RNase protection experiment**) can show which parts of the RNA are protected by bound protein (i.e. the site of binding). **Cross-linking experiments** using UV light with or without chemical cross-linking agents can show which regions of the RNA and protein molecules are in close contact in the complex. Physical methods such as **neutron** and **X-ray diffraction** can ultimately give the complete three-dimensional structure (Section A5). Collectively, these methods have provided much information on the structure of the RNPs described in this and subsequent sections.

Bacterial ribosomes

The importance of ribosomes to a cell is well illustrated by the fact that in *E. coli* ribosomes account for 25% of the dry weight (10% of total protein and 80% of total RNA). Figure 2 shows the components of the *E. coli* ribosome. The **70S ribosome** of molecular mass 2.5×10^6 Da is made up of a large **50S** subunit and a small **30S** subunit. The latter is composed of one copy of the 16S rRNA molecule and 21 different proteins denoted S_1 to S_{21}. The large subunit contains one 23S and one 5S rRNA molecule and 31 different proteins, named L_1 to L_{34} (the original L_7, L_8, and L_{26} were misassigned). The sizes of these ribosomal proteins vary widely, from L_{34}, which is only 46 amino acids, to S_1, which is 557. Most of these relatively small proteins are basic, which might be expected since they bind to RNA. It is possible to re-assemble functional *E. coli* ribosomes from the RNA and protein components, and there is a defined pathway of assembly. The various methods of studying RNPs have led to the structures for the *E. coli* ribosomal subunits shown in Figure 3.

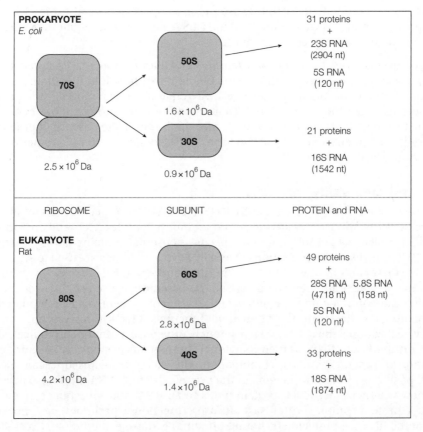

Figure 2. Composition of typical bacterial and eukaryotic ribosomes.

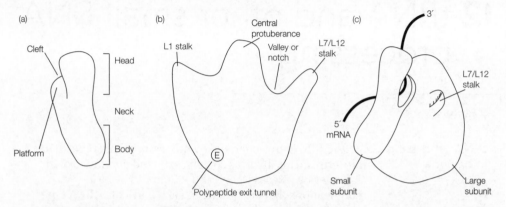

Figure 3. Features of the *E. coli* ribosome. (a) The 30S subunit; (b) the 50S subunit; (c) the complete 70S ribosome.

Eukaryotic ribosomes

The corresponding sizes and components of the 80S rat ribosome are shown in Figure 2. In the large 60S subunit, which contains 49 proteins and one 5S rRNA molecule, the 5.8S rRNA and the 28S rRNA molecules together are the equivalent of the bacterial 23S molecule. The small 40S subunit contains the 18S rRNA and 33 different proteins. Although all the rRNAs are larger in eukaryotes, there is a considerable degree of conservation of secondary structure in each of the corresponding molecules. The ribosomes in a typical eukaryotic cell can collectively make about a million peptide bonds per second, which equates to about two per ribosome per second. Different ribosomes are present in mitochondria and chloroplasts. Their sizes and structures are more similar to those of bacteria, from which they are derived.

J2 tRNA and other small RNA processing

Key Notes

tRNA processing in bacteria

Mature tRNAs are generated by processing longer pre-tRNA transcripts. This involves specific exo- and endonucleolytic cleavages by RNases D, E, F and P followed by base modifications that are unique to each particular tRNA type. Following an initial 3'-cleavage by RNase E or F, RNase D can trim the 3'-end to within 2 nt of mature length. RNase P then cuts to give the mature 5'-end. RNase D finally removes the two 3'-residues, and base modifications take place.

tRNA processing in eukaryotes

Many eukaryotic pre-tRNAs are synthesized with an intron as well as extra 5'- and 3'-nucleotides, all of which are removed during processing. In contrast to bacterial tRNA, the 3'-terminal CCA is added by the enzyme tRNA nucleotidyl transferase. Many base modifications also occur.

RNase P

E. coli RNase P is a simple RNP comprising a 377-nt RNA and a single 13.7-kDa protein. In both bacteria and eukaryotes, its function is to cleave 5'-leader sequences from pre-tRNAs. The RNA component alone can cleave pre-tRNAs *in vitro* and so it is a catalytic RNA, or ribozyme.

Ribozymes

Several biochemical reactions are carried out by RNA enzymes (ribozymes). These catalytic RNAs can cleave themselves or other RNA molecules, or perform ligation or self-splicing reactions. They can work alone but are often complexed *in vivo* with protein(s), which enhance their catalytic activity. Novel ribozymes can be designed in the laboratory as specific RNA-cutting tools.

Processing of other small RNAs

Many eukaryotes make microRNAs that act via an RNA-induced silencing complex (RISC) to inhibit the translation of mRNAs containing a complementary sequence. These small RNAs are processed from longer transcripts by the enzymes Dicer and Drosha.

Related topics

(H3) RNA Pol III genes: 5S and tRNA transcription
(J1) rRNA processing and ribosomes

(K2) tRNA structure and function
(L4) Translational control and post-translational events

tRNA processing in bacteria

In Section J1, Figure 1, it was seen that the rRNA operons of *E. coli* contain coding sequences for tRNAs. In addition, there are other operons in *E. coli* that contain up to seven tRNA genes separated by spacer sequences. Mature tRNA molecules are processed from precursor transcripts of both these types of operon by **RNases D, E, F, and P** in an ordered series of steps, illustrated for *E. coli* tRNA^{Tyr} in Figure 1. Once the primary transcript has folded and formed characteristic stems and loops (Section K2), an endonuclease (RNase E or F) cuts off a flanking sequence at the 3′-end at the base of a stem to leave a precursor with nine extra nucleotides. The exonuclease RNase D then removes seven of these 3′-nucleotides one at a time. RNase P can then make an endonucleolytic cut to produce the mature 5′-end of the tRNA. In turn, this allows RNase D to trim the remaining 2 nt from the 3′-end, giving the molecule its mature 3′-end. Finally, the tRNA undergoes a series of **base modifications**. Different pre-tRNAs are processed in a similar way, but the base modifications are unique to each particular tRNA type. The more common tRNA base modifications are shown in Section K2, Figure 1.

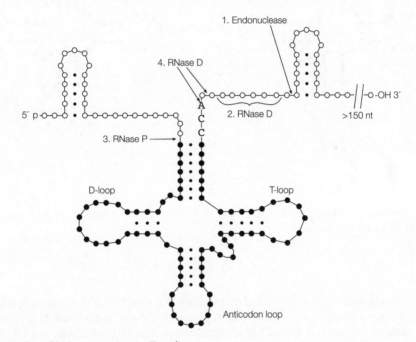

Figure 1. Pre-tRNA processing in *E. coli*.

tRNA processing in eukaryotes

For comparison, the processing of the eukaryotic yeast tRNA^{Tyr} is shown in Figure 2. In this case, the **pre-tRNA** is synthesized with a 16-nt 5′-leader, a 14-nt **intron** (intervening sequence), and two extra 3′-nucleotides. Again, the primary transcript forms a secondary structure with characteristic stems and loops that allow endonucleases to recognize and remove the 5′-leader and the two 3′-nucleotides. A major difference between bacteria and eukaryotes is that, in the former, the 5′-CCA-3′ at the 3′-end of the mature tRNAs is encoded in the genes. In eukaryotic nuclear-encoded tRNAs this is not the case. After the two 3′-nucleotides have been removed, the enzyme **tRNA nucleotidyl transferase** adds the sequence 5′-CCA-3′ to the 3′-end to generate the mature 3′-end of the tRNA. The next step is the removal of the intron, which occurs by endonucleolytic cleavage at each end of the intron followed by ligation of the half molecules of tRNA. The tRNA introns are all capable of base-pairing with their respective anticodon sequences, thus forming an extended anticodon arm structure. The endonucleolytic cleavages at each end of the intron leave the two tRNA half molecules needing modification to their unusual end structures before ligation can take place. The introns of yeast pre-tRNAs can be processed in vertebrate cells, thus the eukaryotic tRNA processing machinery seems to have been conserved during evolution.

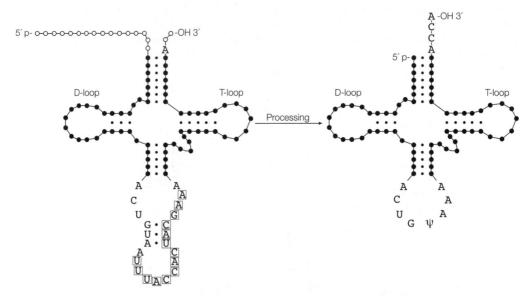

Figure 2. Processing of yeast pre-tRNA^Tyr. Intron nucleotides are boxed and some base-pair with the anticodon.

RNase P

RNase P is an **endonuclease** composed of one RNA and one protein molecule. It is therefore a very simple **RNP**. Its role in cells is to generate the mature 5′-end of tRNAs from their precursors. RNase P is found in both prokaryotes and in eukaryotic nuclei, where it is one of the small nuclear RNPs (snRNPs). In *E. coli*, the endonuclease is composed of a 377-nt RNA and a small basic protein of 13.7 kDa. The secondary structure of the RNA has been highly conserved during evolution. Surprisingly, the RNA component alone will work *in vitro* as an endonuclease on pre-tRNA. This RNA is therefore one example of a **catalytic RNA**, or **ribozyme**, capable of catalyzing a chemical reaction in the absence of protein. *In vitro*, the RNase P ribozyme reaction requires a higher Mg^{2+} concentration than exists *in vivo* in order to stabilize the catalytic conformation of the RNA. In cells, this function is provided by the protein component.

Ribozymes

Ribozymes are naturally occurring **catalytic RNA** molecules that can catalyze particular biochemical reactions. In addition to RNAse P (above), several other types are involved in RNA processing. An intron in the large subunit rRNA of *Tetrahymena* can remove itself from the transcript *in vitro* in the absence of protein (Section J1). This process of **self-splicing** requires guanosine, or a phosphorylated derivative, as cofactor. The *in vitro* reaction is about 50 times less efficient than the *in vivo* reaction, so it is probable that cellular proteins may assist the reaction *in vivo*. During the replication of some plant RNA viruses, **concatameric** (multiple length) molecules of the genomic RNA are produced. These are caused by the polymerase continuing to synthesize RNA after it has completed one circle of template. These molecules are able to fold up in such a way as to self-cleave themselves into monomeric, genome-sized lengths. Studies of these self-cleaving molecules (*cis* cleavage) have identified the minimum sequences needed, and researchers have managed to develop ribozymes that can cleave other target RNA molecules (*trans* cleavage) using *in vitro* selection techniques. In protein synthesis, 23S (and 28S) rRNA is also catalytically active and provides the peptidyl transferase activity of the large ribosomal subunit (Section L2).

Processing of other small RNAs

Eukaryotes contain many small, noncoding RNAs that regulate the expression of other genes (Sections A2 and L4). These closely related **miRNAs, siRNAs,** and **piRNAs** are processed from longer primary transcripts. For example, miRNA genes are transcribed by RNA Pol II or III to give a **primary miRNA** several hundred nucleotides long that can contain several different miRNA sequences, each on a 70–80-nt hairpin. In the nucleus, the RNase III-related endonuclease **Drosha** liberates individual **pre-miRNAs** from this long precursor. These are then transported to the cytoplasm where another RNAse III called **Dicer** processes them further, cutting the hairpin loop to produce the mature 22-nt double-stranded miRNA with 2 nt overhangs at the 3'-ends. In germ cells, piRNAs are produced from long ssRNAs to produce slightly longer mature forms of 24–31 nt. Usually just one strand of the miRNA becomes incorporated into an RNP complex called **RISC (RNA-induced silencing complex)**, which includes Dicer and other proteins and is responsible for inhibiting the translation of specific mRNAs to which the miRNA is at least partially complementary (Section L4).

J3 mRNA processing, hnRNPs, and snRNPs

Key Notes

Processing of pre-mRNA

There is essentially no processing of bacterial mRNA; it can start being translated before it has finished being transcribed. In eukaryotes, mRNAs are synthesized by RNA Pol II as longer precursors (pre-mRNAs), the population of different pre-mRNAs being called heterogeneous nuclear RNA (hnRNA). Specific proteins bind to hnRNA to form hnRNP and then small nuclear RNP (snRNP) particles interact with hnRNP to carry out some of the RNA processing events. Processing of eukaryotic hnRNA involves four events: 5'-capping, 3'-cleavage and polyadenylation, splicing, and methylation.

hnRNP

RNA Pol II transcripts (hnRNA) complex with the three most abundant hnRNP proteins, the A, B, and C proteins, to form hnRNP particles. These contain three copies of three tetramers and around 600–700 nucleotides of hnRNA. They assist RNA processing events.

snRNP particles

Several different uracil-rich small nuclear RNA (snRNA) molecules made by RNA Pol II complex with specific proteins to form snRNPs. The most abundant are involved in splicing, and many define methylation sites in pre-rRNA. Those containing the sequence 5'-RA(U)3–6GR-3' bind eight common proteins in the cytoplasm, become hypermethylated, and are imported back into the nucleus.

5' Capping

This is the addition of a 7-methylguanosine nucleotide (m^7G) to the 5'-end of an RNA Pol II transcript when it is about 25-nt long. The m^7G, or cap, is added in reverse polarity (5' to 5'), thus acting as a barrier to 5'-exonuclease attack, but it also promotes splicing, transport, and translation.

3' Cleavage and polyadenylation

Most eukaryotic pre-mRNAs bind specific factors at conserved sequences near the 3'-end that form a polyadenylation site. They are cleaved at this site and poly(A) polymerase (PAP) then adds a poly(A) tail of around 250 nt to generate the mature 3'-end. Histone pre-mRNA 3'-ends are formed by a different mechanism.

Splicing

In eukaryotic pre-mRNA processing, intervening sequences (introns) that interrupt the coding regions (exons) are removed and the two flanking exons are joined. This splicing reaction occurs in the nucleus and requires the intron to have a 5'-GU, an AG-3' and a branchpoint sequence. In a two-step reaction, the intron is removed as a lariat and is degraded. Splicing involves the binding of snRNPs to the conserved sequences in an ordered series of six steps to form a spliceosome in which the cleavage and ligation reactions take place. An exon junction complex binds to mark the point in the mRNA where an intron has been removed.

Role of Pol II CTD in processing   The CTD of RNA Pol II helps to recruit factors and coordinate the three main RNA processing events undergone by pre-mRNA.

Pre-mRNA methylation   A small percentage of A residues in pre-mRNA, those in the sequence 5′-RRACX-3′ where R=purine, become methylated at the N6 position.

mRNA degradation   Most eukaryotic mRNAs are degraded by first removing the poly(A) tail and then either decapping followed by 5′→3′ exonuclease digestion, or 3′→5′ digestion by the exosome. Other mRNAs are degraded by one of four other pathways.

Related topics

(H4) RNA Pol II genes: promoters and enhancers
(J1) rRNA processing and ribosomes
(L4) Translational control and post-translational events

Processing of pre-mRNA

There appears to be little or no processing of mRNA transcripts in bacteria, although a few cases of intron-containing genes have been described in bacteria, phages, and plasmids. In fact, because the directions of transcription and translation are both 5′→3′ and both processes take place in the same cellular compartment in bacteria, ribosomes can assemble on and begin to translate mRNA molecules before they have been completely synthesized. Eukaryotic RNA Pol II transcribes a wide variety of different genes, from the small snRNA genes of 60–300 bp to the huge dystrophin gene, whose transcript is nearly 2.5 Mbp in length (Section H4), as well and the long noncoding RNAs (Section I2). The collection of products made by this enzyme is therefore referred to as **heterogeneous nuclear RNA (hnRNA)**. Transcripts that will be processed to give mRNAs are called **pre-mRNAs**. Pre-mRNA molecules are processed to mature mRNA by 5′-capping, 3′-cleavage and polyadenylation, splicing, and methylation.

hnRNP

The pre-mRNAs within the hnRNA fraction rapidly become associated with proteins to form **heterogeneous nuclear ribonucleoprotein (hnRNP)**. The proteins involved have been classified as hnRNP proteins A–U. There are two forms of each of the three more abundant hnRNP proteins, the A, B, and C proteins. Purification of this material from nuclei gives a fairly homogeneous preparation of 30–40S particles called hnRNP particles. These particles are about 20 nm in diameter and each contains about 600–700 nt of RNA complexed with three copies of three different tetramers. These tetramers are $(A_1)_3B_2$, $(A_2)_3B_1$, and $(C_1)_3C_2$. The hnRNP proteins are thought to maintain the hnRNA in a single-stranded form and to assist in the various RNA processing reactions.

snRNP particles

RNA Pol II also transcribes most **snRNAs**, which complex with specific proteins to form **snRNPs**. These RNAs are rich in the base uracil and are thus denoted U1, U2, etc. The most abundant are those involved in **pre-mRNA splicing** (see below) — **U1, U2, U4, U5, and U6** (U6 is transcribed by RNA Pol III, Section H3). However, further snRNAs exist, the majority being involved in determining the sites of methylation and pseudouridylation of pre-rRNA; these are located in the nucleolus and so are called snoRNAs (Section J1). The major nucleoplasmic snRNPs are formed by the association of individual snRNAs with a common set of eight small, basic proteins and a variable number of snRNP-specific proteins. These core proteins, known as the **Sm proteins**, require the sequence $5'$-RA(U)$_{3-6}$GR-$3'$ to be present in a single-stranded region of the RNA. U6 does not have this sequence but it is usually base-paired to U4, which does. The snRNPs are formed as follows. They are synthesized in the nucleus by RNA Pol II and have a normal $5'$-cap (see below). They are exported to the cytoplasm where they associate with the common core proteins and with other specific proteins. Their $5'$-cap gains two methyl groups and they are then imported back into the nucleus where they function in splicing.

5′ Capping

Initially, the $5'$-end of a transcript retains the three phosphate groups from the first nucleoside triphosphate to be incorporated. Very soon after RNA Pol II starts making a transcript, when the RNA chain is around 30-nt long, it pauses (Section H4) while the $5'$-end is chemically modified. This involves removal of one of the three terminal phosphates and addition of the nucleotide GMP in a $5'$–$5'$ orientation, the reverse of the normal $3'$–$5'$ linkage, by the enzyme **mRNA guanylyltransferase**, using GTP as the GMP donor. The G residue is then methylated to form **7-methylguanosine** (m^{7}G) (Figure 1). Subsequent methylations of the sugars on the first and second transcribed nucleotides then take place, particularly in vertebrates. This $5'$ modification is called a **cap** and it forms a barrier to $5'$-exonucleases and thus stabilizes the transcript; however, the cap is also important in other reactions undergone by pre-mRNA and mRNA, such as splicing, nuclear transport and translation.

3′ Cleavage and polyadenylation

The $3'$-end of most mature eukaryotic mRNAs is generated by cleavage of the $3'$ terminal noncoding sequence followed by the addition of a run, or tail, of A residues called the **poly(A) tail**. This specific feature of mRNAs allowed their separation from the other types of cellular RNAs, permitting the construction of cDNA libraries as described in Section Q2. The cleavage and polyadenylation reaction requires that specific sequences be present in the DNA and its pre-mRNA transcript. These consist of a $5'$-AAUAAA-$3'$, the **polyadenylation signal**, followed by a $5'$-YA-$3'$, where Y=pyrimidine, in the next 11–20 nt (Figure 2a). Downstream, a GU-rich sequence is often present. Collectively, these sequence elements make up the **polyadenylation site**.

A number of specific protein factors recognize these sequence elements and bind to the pre-mRNA to form a cleavage and polyadenylation complex. These factors are initially carried by the transcribing RNA Pol II enzyme and transferred onto the poly(A) site once it has been transcribed. The $3'$ processing reaction they carry out is linked to transcription termination (Section H4). **Cleavage and polyadenylation specificity factor** binds to the sequence $5'$-AAUAAA-$3'$. One of its subunits is the endonuclease that cleaves the pre-mRNA. **Cleavage stimulation factor** then binds to the GU-rich region. **Poly(A) polymerase** (**PAP**) is the third factor required and is the enzyme that synthesizes the poly(A)

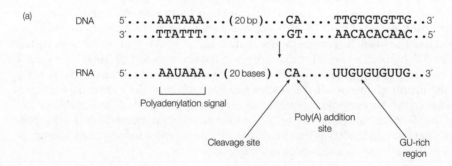

Figure 1. The 5′ cap structure of eukaryotic mRNA.

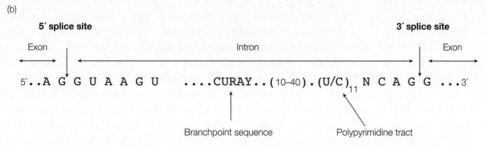

(a)

DNA 5′....AATAAA...(20 bp)...CA....TTGTGTGTTG..3′
3′....TTATTT...........GT....AACACACAAC..5′

RNA 5′....AAUAAA..(20 bases).CA....UUGUGUGUUG..3′

Polyadenylation signal

Cleavage site

Poly(A) addition site

GU-rich region

(b)

5′ splice site   **3′ splice site**

Exon   Intron   Exon

5′..A G G U A A G U   ....CURAY..(10–40).(U/C)$_{11}$ N C A G G ...3′

Branchpoint sequence   Polypyrimidine tract

Figure 2. Sequences of (a) a typical polyadenylation site and (b) the splice site consensus.

tail. When the complex has assembled, cleavage takes place first and then PAP adds up to 250 A residues to the 3′-end of the cleaved pre-mRNA. The nuclear **poly(A) binding protein** (**PABII**) binds to the growing poly(A) tail and accelerates the process. Once the mature mRNA has been exported, PABII is replaced by **PABI** in the cytoplasm. The poly(A) tail on (pre-)mRNA stabilizes these molecules since PABI and PABII binding prevents 3′-exonuclease action. In addition, the poly(A) tail helps in the translation of the mature mRNA in the cytoplasm. Histone pre-mRNAs do not normally get polyadenylated, but are cleaved between a hairpin structure and a sequence that is bound by the **U7 snRNP** called the **histone downstream element** to generate their mature 3′-ends.

Splicing

In eukaryotes, some internal sequences in the pre-mRNA transcript are also removed during creation of the mature mRNA. These intervening sequences, or **introns**, interrupt the **exons,** which are usually the protein-coding regions of the mRNA and which will then become adjacent regions in the mature mRNA. The process of cutting the pre-mRNA to remove the introns and joining together of the exons is called **splicing**. Like the polyadenylation process, it takes place in the nucleus before the mature mRNA can be exported to the cytoplasm. Splicing generally takes place cotranscriptionally and introns near the 5′-end are usually removed before introns located nearer the 3′-end. However, some introns may not be removed until after 3′-cleavage and polyadenylation. Indeed, the temporal coupling of these pre-mRNA processing events is underlined by the observations that both splicing and polyadenylation factors tend to be carried along by the RNA Pol II complex (see below and Section H4) and transferred to exon-intron junctions (or poly(A) sites) when these sequences have been transcribed and are emerging from the transcription complex.

As with 3′-processing, splicing also requires certain specific sequences to be present (Figure 2b). The 5′-end of almost all introns has the sequence 5′-GU-3′ while the 3′-end is usually 5′-AG-3′. The AG at the 3′-end is preceded by a pyrimidine-rich sequence called the **polypyrimidine tract**. About 10–40 residues upstream of the polypyrimidine tract is a sequence called the **branchpoint sequence**, which is 5′-CURAY-3′ in vertebrates (where R=purine and Y=pyrimidine) but is the more specific sequence 5′-UACUAAC-3′ in yeast. Splicing involves a **two-step reaction** (Figure 3a). First, the bond in front of the G at the 5′-end of the intron at the so-called **5′-splice site** is attacked by the 2′-hydroxyl group of the A residue of the branchpoint sequence to create a tailed circular molecule called a **lariat** and free 5′ exon (exon 1 in Figure 3a). In the second step, cleavage at the **3′-splice site** occurs after the G of the AG as the two exon sequences are joined together. The intron is released in the lariat form and is eventually degraded.

The splicing process is catalyzed by the U1, U2, U4, U5, and U6 snRNPs, as well as other splicing factors. The RNA components of these snRNPs form base pairs with the various conserved sequences at the 5′- and 3′-splice sites and the branchpoint (Figure 3b). Early in splicing, the 5′-end of the U1 snRNP binds to the intron sequence at the 5′-splice site and then U2 binds to the branchpoint. The tri-snRNP complex of U4, U5, and U6 can then bind, and in so doing the intron is looped out and the 5′- and 3′-exons are brought into close proximity. The snRNPs interact with one another forming a complex that folds the pre-mRNA into the correct conformation for splicing. This complex of snRNPs and pre-mRNA that forms to hold the upstream and downstream exons close together while looping out the intron is called a **spliceosome**. After the spliceosome forms, a rearrangement takes place before the two-step splicing reaction can occur with release of the intron as a lariat. The complete spliceosome is around 12×10^6 Da ($5 \times$ the mass of a ribosome). The

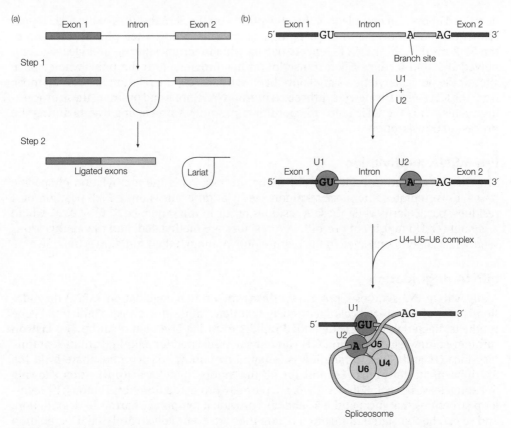

Figure 3. Splicing of eukaryotic pre-mRNA. (a) The two-step reaction; (b) involvement of snRNPs in spliceosome formation.

five snRNAs and their protein components make up nearly half the total mass, with some 70 protein splicing factors and 30 other proteins accounting for the rest. Assembly of the spliceosome proceeds through an ordered series of six steps from the early E complex (or **commitment complex**) to the catalytic C2 complex. When the intron has been removed, a protein complex called the **exon junction complex** forms where the two exons have been joined. This complex plays a role in both **mRNA export** and **mRNA surveillance** (Section L3).

Although most eukaryotic introns are removed by the standard spliceosome described above (i.e. are **U2-dependent**), one exception is a minor class of intron that usually has the sequences AU and AC at the ends, rather than the normal GU and AG, as well as a variant and less degenerate branchpoint sequence. These minor introns are spliced by a variant of the normal spliceosome in which the RNA components of U1 and U2 are replaced by U11 and U12 respectively. Despite other minor differences, the splicing mechanism for these **U12-dependent introns** is essentially identical.

Role of Pol II CTD in processing

The phosphorylated C-terminal domain (CTD) of RNA Pol II (Section H1) is involved in recruiting various factors that carry out capping, 3′-cleavage and polyadenylation, and splicing. First, the capping enzymes bind to the CTD to carry out the capping reactions described above. On completion, the phosphates on the Ser5 residues of the CTD repeat sequences are removed and the capping enzymes dissociate. Then phosphorylation of the Ser2 residues in the CTD repeats occurs which recruits splicing and cleavage and polyadenylation factors. When transcription has progressed to the point where splice sites or the polyadenylation signal have been synthesized, the relevant factors can move from the CTD on to the newly synthesized pre-mRNA regions and promote these processing events. Thus the CTD helps to coordinate pre-mRNA processing events during the process of transcription.

Pre-mRNA methylation

The final modification undergone by many pre-mRNAs is the methylation of specific bases. In vertebrates, the most common methylation event is on the N6 position of A residues, particularly when these A residues occur in the sequence 5′-RRACX-3′, where X is rarely G. Up to 0.1% of pre-mRNA A residues are methylated, and the methylations seem to be largely conserved in the mature mRNA, though their function is unknown.

mRNA degradation

Along with mRNA synthesis, protein synthesis and protein degradation, mRNA degradation is part of the overall cellular strategy that determines the levels of individual gene products. Bacterial mRNA is degraded rapidly from the 5′-end and so the first **cistron** (protein-coding region, Section G1) can only be translated for a limited amount of time (Sections G1 and G2). Degradation is initiated by removal of pyrophosphate from the 5′-triphosphate terminus (Section J4) by the 'nudix' hydrolase **RppH**, thus allowing 5′-exonucleases to act. Some mRNAs are partially protected from RppH attack by stem-loop structures that form at the 5′-end and provide a temporary barrier to degradation, and so can be translated more often before they are eventually hydrolysed. Degradation also involves endonucleases and 3′-exonucleases.

The 5′-cap and 3′-poly(A) present on a mature eukaryotic mRNA protect it from attack by 5′→3′ and 3′→5′ exonucleases. As sequence-independent endonucleases are not abundant in eukaryotes, most mRNA is degraded by one of two major pathways that both involve removal of all but about 10 A residues from the poly(A) tail by the enzyme **poly(A) nuclease**. In the first pathway, the oligo(A)-containing mRNA is decapped by a **decapping complex** that includes another 'nudix' hydrolase, **DCP2**, and then degraded by **XRN1**, a 5′→3′ exonuclease. In the other major pathway, 3′→5′ degradation is carried out by the **exosome**. There are four other degradation pathways that target specific subsets of mRNAs such as the nonpolyadenylated histone mRNAs. The most significant of these pathways involves miRNAs (Sections J2 and L4).

J4 Alternative mRNA processing

Key Notes

Alternative processing

Alternative mRNA processing is the conversion of pre-mRNA species into more than one type of mature mRNA. This can result from the use of different splice sites, different poly(A) sites and from RNA editing, and greatly extends the protein-coding potential of eukaryotic genomes.

Alternative promoters

Many eukaryotic genes make use of alternative promoters to produce alternative transcripts and protein products. The promoters may have different strengths and be used in a developmental stage- or tissue-specific manner. The alternative protein products may have different properties, such as subcellular targeting.

Alternative splicing

Different mature mRNAs can be generated from a single pre-mRNA by varying the use of 5′- and 3′-splice sites. This can be achieved by using different promoters, by using different poly(A) sites, by retaining certain introns, by retaining or removing certain exons in either a mutually exclusive or combinatorial manner, or by the use of alternative 5′- or 3′-splice sites within an exon to vary the length of the exon included in the product. Most cases involve the binding of members of the SR and/or hnRNP protein families to *cis*-acting sequences in either exons or introns to enhance or silence particular alternative splicing patterns.

Alternative poly(A) sites

Some pre-mRNAs contain more than one poly(A) site and these may be used under different circumstances (e.g. in different cell types) to generate different mature mRNAs. In some cases, factors will bind near to and activate or repress a particular site.

RNA editing

This is a form of RNA processing in which the nucleotide sequence of the primary transcript is altered either by changing, inserting, or deleting residues at specific points along the molecule. In the case of human Apo-B protein, intestinal cells make a truncated protein by creating a stop codon in the mRNA by editing a C to a U. RNA editing can involve guide RNAs.

Related topics

(H4) RNA Pol II genes: promoters and enhancers

(J3) mRNA processing, hnRNPs and snRNPs

Alternative processing

In eukaryotes a particular pre-mRNA species can give rise to more than one type of mature mRNA and it is now believed that >90% of the protein-coding gene transcripts from the human genome fall into this category. This can occur when some of the exons are also removed by splicing (**alternative** or **differential splicing**) and so are not retained in the mature mRNA product. Additionally, if alternative poly(A) sites can be used, different 3'-ends can be present in the mature mRNAs (**alternative**, or **differential poly(A) processing**). Changing the actual base sequence of the primary transcript after synthesis (**RNA editing**) also gives rise to variation. Together, alternative pre-mRNA processing, along with the use of **alternative transcriptional promoters**, and **alternative translational start codons** (Section L3), allows a mammalian genome of 20–25 000 genes to encode up to a predicted 10^6 different proteins.

Alternative promoters

Although not strictly a form of alternative processing, the use of alternative RNA Pol II promoters to yield transcripts with different 5'-exons from the same gene is widespread among mammals: at least 50% of human genes may use alternative promoters. The resulting transcripts may encode different protein isoforms with different N-termini that target them to different subcellular locations (Section L4), or they may vary only in their 5'-UTRs, which could influence their stability, processing (e.g. alternative splicing), regulation or translational efficiency. Alternative promoters may be of different strengths and may be used differentially during development or on a tissue-specific basis, depending on the presence of the required transcription factors. For example, the gene for the digestive enzyme α-amylase has a weak downstream promoter that is active in the liver, and an upstream alternative strong promoter active in salivary glands. The difference between these two promoters accounts for the >100-fold difference in expression of α-amylase in salivary glands compared to liver and also leads to alternative splicing (see below). The erroneous activation of an alternative promoter at the wrong time or in the wrong place can have a serious phenotypic effect, such as cancer.

Alternative splicing

Eight types of alternative splicing are summarized in Figure 1. In Figure 1a, it is the choice of promoter that forces the pattern of splicing, as happens in the α-amylase and myosin light chain genes. The exon transcribed from the **upstream promoter** has the stronger 5'-splice site, which out-competes the downstream one for use of the first 3'-splice site. This happens to α-amylase pre-mRNA in salivary glands where specific transcription factors cause transcription from the upstream promoter. In the liver, the **downstream promoter** is used and the weaker (second) 5'-splice site is used by default. Alternative splicing caused by differential use of poly(A) sites (see below) is shown in Figure 1b. The stronger 3'-splice site is only present if the downstream poly(A) site is used and thus the penultimate 'exon' will be removed. When the upstream poly(A) site is used (such as in a different cell or at a different stage of development), splicing occurs by default using the weaker (upstream) 3'-splice site. In the case of immunoglobulins, use of a downstream poly(A) site includes exons encoding membrane-anchoring regions, whereas when the upstream site is used these regions are not present and the secreted form of immunoglobulin is produced. In some situations, introns can be retained, as shown in Figure 1c. If the intron contains a stop codon then a truncated protein will be produced on translation. This can give rise to an inactive protein, or one with a different activity, as in the case of the P element transposase in *Drosophila* somatic cells (Section E4). In germ cells, the

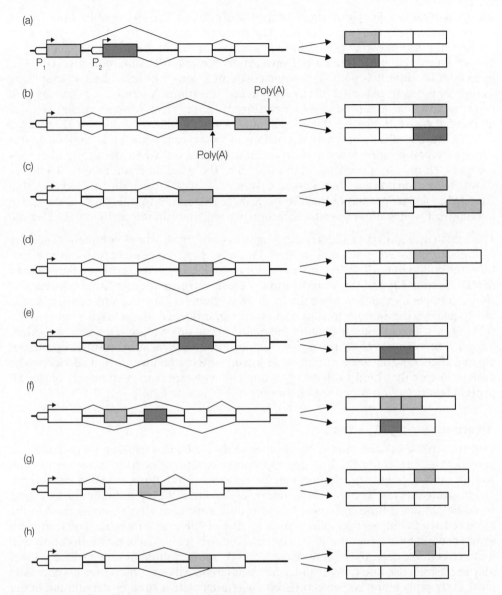

Figure 1. Modes of alternative splicing. (a) Alternative selection of promoters P₁ or P₂; (b) alternative selection of cleavage/polyadenylation sites; (c) retention of an intron; (d) exon skipping; (e) mutually exclusive exons; (f) combinatorial exons; (g) alternative 5′-splice sites; (h) alternative 3′-splice sites. Empty boxes are exons, filled boxes are alternative exons and thin lines are introns.

lack of a specific factor present in somatic cells causes the correct splicing of the intron and a longer mRNA is made which is translated into a functional transposase enzyme only in these cells.

Figure 1d illustrates that some exons can be retained or removed in different circumstances. A likely reason is the existence of a factor in one cell type that either promotes the use of a particular splice site or prevents the use of another. The rat troponin-T pre-mRNA can be differentially spliced in this way. The **mutually exclusive** use of exons (Figure 1e) is where only one of two or more alternative exons appears in the product, depending on the cell type or circumstances. The **Down syndrome cell-adhesion molecule (DSCAM)** gene from *Drosophila* is perhaps the most extreme example of such alternative splicing as only 24 exons from a total of 115 are retained in various combinations. The DSCAM gene has 95 exons organized into four clusters of 12, 48, 33, and two mutually exclusive alternative exons. Hence the theoretical number of alternative products is $12 \times 48 \times 33 \times 2$ $= 38\,016$. The complexity of synaptic junctions in the nervous system may depend on this variety. The **combinatorial use of exons** is shown in Figure 1f where the alternative exons are included only in a particular subset of all of the possible combinations. Figure 1g shows that within an exon there can be a choice of which 5′-splice site is used such that a longer or shorter section of the 5′-exon is included. Similarly, in Figure 1h, the use of alternative 3′-splice sites can cause inclusion of a longer or shorter section of the 3′-exon.

The most common types of alternative splicing are those where complete exons are included or excluded. About one-third of alternatively spliced products contain a premature stop codon that will target the mRNA for degradation by **nonsense mediated decay (NMD)** (Section L3) and this would seem to be a mechanism for regulating gene expression. No single mechanism can explain all these different forms of alternative splicing, but sequences within both introns and exons can influence the patterns of alternative splicing by using splicing regulatory factors to stimulate or repress the use of particular splice sites. These *cis*-acting sequences in exons are known as **exon splicing enhancers** and **silencers** and those in introns as **intron splicing enhancers** and **silencers**. The protein factors that bind to these *cis*-acting elements are usually members of the **SR protein family** (serine–arginine rich) or the hnRNP protein family.

Alternative poly(A) sites

Some pre-mRNAs contain more than one set of the sequences required for cleavage and polyadenylation (Section J3). If an upstream site is used then downstream sequences that control mRNA stability or location may be removed in the portion that is cleaved off. Thus mature mRNAs with the same coding region, but differing stabilities or locations, could be produced from one gene. In some situations, both sites could be used in the same cell at a frequency that reflects their relative efficiencies (strengths) and so the cell would contain both types of mRNA. The efficiency of a poly(A) site may reflect how well it matches the consensus sequences (Section J3). In other situations, one cell (e.g. liver) may exclusively use one poly(A) site, while a different cell (e.g. brain) uses another. The most likely explanation is that in one cell the stronger site is used by default, but in the other cell a factor is present that activates the weaker site so that it is used exclusively, or that prevents the stronger site from being used. In some cases, the use of alternative poly(A) sites can cause different patterns of splicing to occur (see above).

RNA editing

An unusual form of RNA processing in which the sequence of the primary transcript is altered is called **RNA editing**. Several examples exist. In man, editing in intestinal cells causes a single base change from C to U in the apolipoprotein B pre-mRNA, creating a stop codon in the mRNA at position 6666 in the 14 500-nt molecule. The unedited RNA in the liver makes apolipoprotein B100, a 512-kDa (4563 aa) protein, but in the intestine editing causes the truncated apolipoprotein B48 (241 kDa, 2153 aa) to be made. The process is carried out by the **apolipoprotein B editing complex**, which recognizes a 26-nt sequence near the editing site. Similarly, a single A to G base change in the glutamate receptor pre-mRNA gives rise to an altered form of the receptor in neuronal cells. RNA editing in the ciliated protozoan *Leishmania* is much more extensive. When the cDNA for the mitochondrial cytochrome *b* gene was cloned, it had a coding region corresponding to the protein sequence. However, although the gene was known to be located in the mitochondrial genome, no corresponding sequence could be found. Eventually, some cDNA clones were obtained that had sequences corresponding to intermediates between the mature mRNA and the genomic sequence. It seems that the primary transcript is edited successively by introducing U residues at specific points. Many cycles of editing eventually produce the mature mRNA, which is then translated. Short RNA molecules called **guide RNAs** are involved. Their sequences are complementary to regions of the genomic DNA and the edited RNA. Several other types of RNA editing are known, including those involving the deamination of adenosine to inosine. Once thought to be confined to tRNAs (Section L1) and a few protein-coding regions, this modification seems to be quite common, particularly in the noncoding regions of primary transcripts and miRNAs (Section J2).

K1 The genetic code

Key Notes

Principle	The genetic code defines the way in which the nucleotide sequence in nucleic acids specifies the amino acid sequence in proteins. It is a triplet code, where the codons (groups of three nucleotides) are adjacent (nonoverlapping) and are not separated by punctuation (comma-less).
Deciphering	The standard genetic code was deciphered by adding homopolymers, copolymers, or synthetic nucleotide triplets to cell extracts that were capable of limited translation. Sixty-one codons specify the 20 amino acids and there are three stop codons.
Degeneracy, universality, and ambiguity	The code is degenerate, as 18 of the 20 amino acids are specified by multiple (or synonymous) codons that are grouped together in the genetic code table. Usually they differ only in the third codon position. If this is a pyrimidine, then the codons always specify the same amino acid. If a purine, then this is usually also true. The standard genetic code was originally considered universal; however, deviations are now known to occur in mitochondria and in some unicellular organisms. Stop codons can sometimes be used to specify an amino acid.
Effect of mutation	The grouping of synonymous codons means that the effects of point mutations are minimized. Transitions in the third position often have no effect, as do transversions more than half the time. Mutations in the first and second position often result in a chemically similar type of amino acid being used.
ORFs	Open reading frames are suspected coding regions usually identified by computer in newly sequenced DNA. They are continuous groups of adjacent codons between a start codon and a stop codon. They show the protein-coding potential of a genome.
Overlapping genes	These occur when the coding region of one gene partially or completely overlaps that of another. Thus one reading frame encodes one protein, and one of the other possible frames encodes part or all of a second protein. Some small viral genomes use this strategy to increase the coding capacity of their genomes.
Related topics	(E2) Mutagenesis (L2) Mechanism of protein synthesis (J4) Alternative mRNA processing (L3) Initiation in eukaryotes (K2) tRNA structure and function

Principle

The **genetic code** is the correspondence between the sequence of the four bases in nucleic acids and the sequence of the 20 standard amino acids in proteins. It has been shown that the code is a triplet code, where three nucleotides encode one amino acid, and this agrees with mathematical argument as being the minimum necessary $[(4^2=16)<20<(4^3=64)]$. However, since there are only 20 amino acids to be specified and potentially 64 different triplets, most amino acids are specified by more than one triplet, and so the genetic code is said to be **degenerate**, or to have **redundancy**. From a fixed start point, each group of three bases in the coding region of the mRNA represents a **codon**, which is recognized by a complementary triplet, or **anticodon**, on a particular tRNA molecule (Section K2). The triplets are read in nonoverlapping groups and there is no punctuation between the codons to separate or delineate them. They are simply decoded as adjacent triplets once the process of decoding has begun at the correct start point. A sequence made up of base triplets could be decoded in three different ways (three **reading frames**) to yield three entirely different amino acid sequences depending at which of the three bases reading commences. Usually only one reading frame is used, so part of the initiation process involves selecting this (Section L2). As more gene and protein sequence information has been obtained, it has become clear that the genetic code is very nearly, but not quite, universal. This supports the hypothesis that all life has evolved from a single common origin.

Deciphering

In 1968, Marshall Nirenberg, Gobind Khorana, and Robert Holley received the Nobel prize for deciphering the genetic code. Previous genetic experiments had suggested it was a triplet code but the assignment of specific codons to specific amino acids began with experiments involving synthetic mRNAs containing only one or two bases, e.g. the **homopolymers** poly(U), poly(C), poly(A), and poly(G), which were synthesized by the enzyme **polynucleotide phosphorylase**. When added to an extract of *E. coli*, poly(U) directed the synthesis of polyphenylalanine, hence UUU must encode Phe. Similarly, poly(C) coded for polyproline and poly(A) for polylysine. Poly(G) did not work because it formed a complex secondary structure. Further assignments were possible by analyzing the protein products made using **random copolymers** containing a mixture of two nucleotides. If G and U are present at unequal ratios such as 1:3, then the triplet GGG will be the rarest and UUU the most common. Triplets with two Us and one G will be the next most frequent, etc.

The remaining assignments followed the discovery that **synthetic trinucleotides** could attach to the ribosome and bind their corresponding aminoacyl-tRNAs (Section K2) from a mixture. Using radioactive amino acids, it was possible to determine which triplet bound which amino acid in a large complex that could be trapped on a membrane. In this way, a total of 61 codons were shown to code for amino acids. Three codons, UAA, UAG, and UGA, specified no amino acid; these are 'stop' codons, used to terminate translation (Table 1; see Section A3, Figure 2 for the one-letter and three-letter amino acid codes). The codon AUG has a dual function. It specifies the amino acid methionine but is also used as a 'start' codon to initiate protein synthesis (Section L1). Note that Table 1 shows the codons in mRNA. The corresponding codons in DNA have T in place of U, e.g. methionine/start is ATG.

Degeneracy, universality, and ambiguity

The genetic code is degenerate (redundant). This is because 18 out of 20 amino acids can be specified by more than one codon, called **synonymous codons**. Only methionine and

Table 1. The universal genetic code

First position (5′-end)	Second position								Third position (3′-end)
	U		C		A		G		
U	Phe	UUU	Ser	UCU	Tyr	UAU	Cys	UGU	U
	Phe	UUC	Ser	UCC	Tyr	UAC	Cys	UGC	C
	Leu	UUA	Ser	UCA	**Stop**	UAA	**Stop**	UGA	A
	Leu	UUG	Ser	UCG	**Stop**	UAG	Trp	UGG	G
C	Leu	CUU	Pro	CCU	His	CAU	Arg	CGU	U
	Leu	CUC	Pro	CCC	His	CAC	Arg	CGC	C
	Leu	CUA	Pro	CCA	Gln	CAA	Arg	CGA	A
	Leu	CUG	Pro	CCG	Gln	CAG	Arg	CGG	G
A	Ile	AUU	Thr	ACU	Asn	AAU	Ser	AGU	U
	Ile	AUC	Thr	ACC	Asn	AAC	Ser	AGC	C
	Ile	AUA	Thr	ACA	Lys	AAA	Arg	AGA	A
	Met	AUG	Thr	ACG	Lys	AAG	Arg	AGG	G
G	Val	GUU	Ala	GCU	Asp	GAU	Gly	GGU	U
	Val	GUC	Ala	GCC	Asp	GAC	Gly	GGC	C
	Val	GUA	Ala	GCA	Glu	GAA	Gly	GGA	A
	Val	GUG	Ala	GCG	Glu	GAG	Gly	GGG	G

tryptophan have single codons. The synonymous codons are not positioned randomly, but are grouped in the table. Generally, they differ only in their third position (the **wobble** position, Section L1). In all cases, if the third position is a pyrimidine, then the codons specify the same amino acid (are synonymous). In most cases, if the third position is a purine the codons are also synonymous. If the second position is a pyrimidine then generally the amino acid specified is hydrophilic. If the second position is a purine then generally the amino acid specified is polar.

For a long time after the genetic code was deciphered, it was thought to be universal, i.e. the same in all organisms. However, mitochondria, which have their own small genomes, utilize a genetic code that differs slightly from the standard code. Indeed, it is now known that some unicellular eukaryotes also have a variant genetic code. Table 2 lists some of these variations. In addition, the amino acid **selenocysteine** (Section A3) is specified by certain UGA stop codons when followed by a specific sequence element called **SECIS** (selenocysteine insertion sequence), while in some archaea the stop codon UAG can be translated as **pyrrolysine**, a modified form of lysine. This context-dependent dual meaning of stop codons is an example of the **ambiguity** of the code. Another example is the occasional use by some organisms of GUG (Val) and UUG (Leu) as alternative start codons, in which case they specify Met instead (Section L1).

Even though the code is 'nearly' universal, synonomous codons do not occur with the same frequency in all organisms. Thus, of the six arginine codons, AGG and AGA are preferred in the human genome but are rarely used by E. coli, which favors CGC and CGU.

Table 2. Modifications of the genetic code

Codon	Usual meaning	Alternative	Organelle or organism
AGA	Arg	**Stop**, Ser	Some animal mitochondria
AGG			
AUA	Ile	Met	Mitochondria
CGG	Arg	Trp	Plant mitochondria
CUN	Leu	Thr	Yeast mitochondria
AUU	Ile	Start	Some prokaryotes
GUG	Val		
UUG	Leu		
UAA	**Stop**	Glu	Some protozoans
UAG			
UGA	**Stop**	Trp	Mitochondria, mycoplasma

This is known as **codon usage bias** and can cause practical difficulties when expressing genes from one organism in another (Sections A5 and S5). Even within one organism, the genes for highly expressed proteins tend to use those codons from within a synonomous group for which the corresponding tRNAs (Section K2) are more abundant. Codon bias also extends to stop codons; UAG is used in only about 10% of *E. coli* mRNAs.

Effect of mutation

It is generally considered that the genetic code evolved in such a way as to minimize the effect of mutations (Section E2). The most common type of mutation is a **transition**, where either a purine is mutated to the other purine or a pyrimidine is changed to the other pyrimidine. **Transversions** are where a pyrimidine changes to a purine or vice versa. In the third position, transitions usually have no effect, but can cause changes between Met and Ile, or Trp and stop. Just over half of transversions in the third position have no effect and the remainder usually result in a similar type of amino acid being specified, for example Asp or Glu. In the second position, transitions will usually result in a similar chemical type of amino acid being used, but transversions will change the type of amino acid. In the first position, mutations (both transition and transversions) usually specify a similar type of amino acid, and in a few cases it is the same amino acid.

ORFs

Computer scanning of DNA sequences, such as those obtained by genome sequencing projects, will identify long, continuous groups of adjacent codons that start with ATG and end with TGA, TAA, or TAG and which therefore have the potential to encode a protein. These are referred to as **open reading frames**, or **ORFs**, when there is no known protein product. When a particular ORF is known to encode a certain protein, the ORF is usually referred to as a coding region. Hence, an ORF is a **suspected coding region**. Scanning for ORFs allows the minimum number of proteins encoded by a newly sequenced genome to be predicted. In the case of eukaryotic genomes, ORFs can only be identified after prediction of splice sites and removal of introns *in silico* (i.e. by computer).

Overlapping genes

Although it is generally true that one gene encodes one polypeptide, and the evolutionary constraints on having more than one protein encoded in a given region of sequence are great, many examples of overlapping genes are known in plasmids, viruses, mitochondria, and bacteria. One reason is the need for greater information storage density in small genomes. For example, the phage ΦX174 (Section M2) makes 11 proteins of combined molecular mass 262 kDa from a 5386-nt genome. Without overlapping genes, this genome could encode at most 200 kDa of protein. Three proteins are encoded within the coding regions for longer proteins; the ribosomes have to find the second start codon to be able to translate the overlapping gene and they may achieve this without detaching from the RNA. Examples of overlapping eukaryotic nuclear genes are also known, most involving overlapping transcripts arising from the two DNA strands of the gene (sense and antisense), although there are examples of same-strand overlap as well.

K2 tRNA structure and function

Key Notes

tRNA primary structure

The linear sequence (primary structure) of tRNAs is 60–95 nt long, most commonly 76. There are many modified nucleosides present, notably, thymidine, pseudouridine, dihydrouridine and inosine. There are 15 invariant and eight semi-variant residues in tRNA molecules.

tRNA secondary structure

The cloverleaf structure is a common secondary structural representation of tRNA molecules that shows the base pairing of various regions to form four stems (arms) and three loops. The 5'- and 3'-ends are largely base-paired to form the amino acid acceptor stem, which has no loop. Working in the 5'→3' direction, there is the D-arm, the anticodon arm and the T-arm. Most of the invariant and semi-variant residues occur in the loops not the stems.

tRNA tertiary structure

Nine hydrogen bonds form between the bases (mainly the invariant ones) in the single-stranded loops and fold the secondary structure into an L-shaped tertiary structure, with the anticodon and amino acid acceptor stems at opposite ends of the molecule.

tRNA function

When charged by attachment of a specific amino acid to their 3'-end to become aminoacyl-tRNAs, tRNA molecules act as adaptor molecules in protein synthesis.

Aminoacylation of tRNAs

tRNA molecules become charged (aminoacylated) in a two-step reaction. First, the aminoacyl-tRNA synthetase attaches AMP to the carboxyl group of the amino acid to create an aminoacyl adenylate intermediate. Then the appropriate tRNA displaces the AMP.

Aminoacyl-tRNA synthetases

The synthetase enzymes can be monomers, dimers or one of two types of tetramer. They use certain regions of the tRNAs, called identity elements, to distinguish one tRNA from another.

Proofreading

Proofreading occurs when a synthetase carries out step 1, or even step 2, of the aminoacylation reaction with the wrong, but chemically similar, amino acid. It either then does not carry out step 2 but hydrolyzes the aminoacyl adenylate instead, or it hydrolyzes the misacylated tRNA.

Related topics

(A2) Nucleic acid structure and function

(J2) tRNA and other small RNA processing

tRNA primary structure

tRNAs are the **adaptor** molecules that deliver amino acids to the ribosome and decode the information in mRNA. Their **primary structure** (i.e. the linear sequence of nucleotides) is 60–95-nt long, but most commonly 76. They have many **modified bases**, accounting for up to 20% of the total bases in any one tRNA molecule. Over 50 different types of modified base have been found in the tRNA molecules characterized to date, all of which are created post-transcriptionally by specific tRNA modifying enzymes. Seven of the most common are shown as their **nucleosides** in Figure 1. Four of these, **ribothymidine (T)**, which unusually contains the DNA base thymine, **pseudouridine (Ψ)**, **dihydrouridine (D)**, and **inosine (I)**, which contains the base hypoxanthine, are very common in tRNA, all but the last being present in similar positions in the sequences of nearly all tRNA molecules. In the tRNA primary structure, there are 15 invariant nucleotides and eight semivariant that are either purines (R) or pyrimidines (Y). The positions of these nucleotides are important in maintaining either the secondary or tertiary structure (see below and Figure 2).

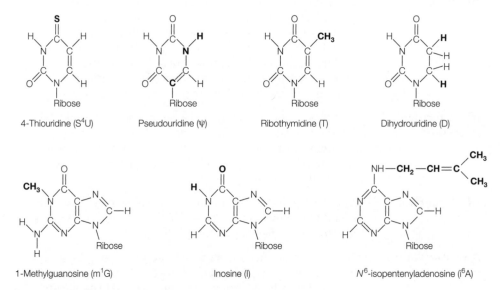

Figure 1. Modified nucleosides found in tRNA.

tRNA secondary structure

All tRNAs have a common **secondary structure** (i.e. base pairing of different regions to form stems and loops), most easily visualized as a **cloverleaf** (Figure 2a). This structure has a 5′-phosphate terminus formed by RNase P cleavage (Section J2), rather than the 5′-triphosphate or cap found in mRNA (Section J3). It has a 7-bp stem formed by base-pairing between the 5′- and 3′-ends; however, the invariant residues 74–76 (i.e. the terminal 5′-CCA-3′), which are added during processing in eukaryotes (Section J2), are not included in this base-paired region. This stem is called the **amino acid acceptor stem**. Working 5′→3′ (anticlockwise), the next secondary structural feature is called the D-arm, which is composed of a 3- or 4-bp stem and a loop called the **D-loop** (DHU-loop) usually containing the modified base dihydrouracil. The next structural feature consists of a 5-bp stem and a seven-residue loop in which there are three adjacent nucleotides called the **anticodon** that are complementary to the codon sequence (a triplet in the mRNA) that the tRNA recognizes. The presence of inosine in the anticodon gives a tRNA the ability to base-pair to more than one codon sequence (Section L1). Next there is a **variable arm** that can have between three and 21 residues and may form a stem of up to 7 bp. The other positions of length variation in tRNAs are in the D-loop, shown as dashed lines in Figure 2a. The final major feature of secondary structure is the **T-arm** or TΨC-arm which is composed of a 5-bp stem ending in a loop, the TΨC-loop (or T-loop), containing the invariant residues GTΨC. Note that the majority of the invariant residues in tRNA molecules are in the loops and do not play a major role in forming the secondary structure. However, several of them help to form the tertiary structure.

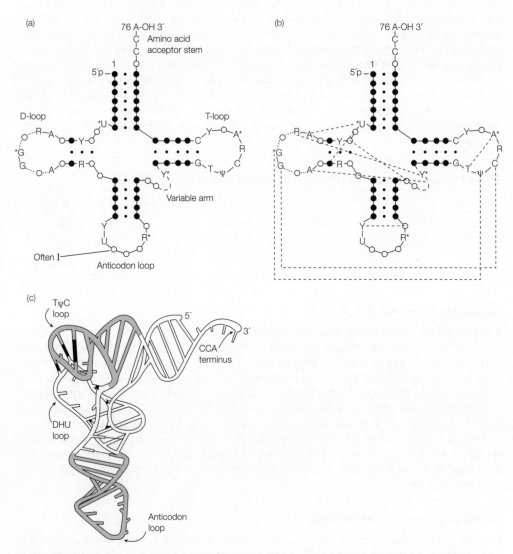

Figure 2. tRNA structure. (a) Cloverleaf structure showing the invariant and semi-variant nucleotides, where I=inosine, Ψ=pseudouridine, R=purine, Y=pyrimidine and * indicates a modification. (b) Tertiary hydrogen bonds between the nucleotides in tRNA are shown as dashed lines. (c) The L-shaped tertiary structure of yeast tRNA^Tyr. Part (c) adapted from Brown (2011) *Introduction to Genetics: A Molecular Approach*, Garland Science, Oxford.

tRNA tertiary structure

Nine hydrogen bonds (**tertiary hydrogen bonds**) help to form the three-dimensional structure of tRNA molecules. They mainly involve base pairing between invariant bases and are shown in Figure 2b. Base-pairing between residues in the D- and T-arms fold the tRNA molecule into an **L-shape**, with the anticodon at one end and the amino acid acceptor site at the other. This tertiary structure is strengthened by base stacking interactions (Figure 2c and Section B1).

tRNA function

tRNAs are joined to amino acids to become **aminoacyl-tRNAs** (charged tRNAs) in a reaction called aminoacylation (see below). It is these charged tRNAs that are the adaptor molecules in protein synthesis. Special enzymes called aminoacyl-tRNA synthetases carry out the joining reaction, which is extremely specific (i.e. a specific amino acid is joined to a specific tRNA). These pairs of specific amino acids and tRNAs, or tRNAs and aminoacyl-tRNA synthetases, are called **cognate pairs**, and the nomenclature used is shown in Table 1. The levels of the different tRNAs for synonomous codons reflects the codon usage bias of the organism (Section K1). Curiously, tRNAs have a number of other functions outside protein synthesis, including cell wall remodeling and retroviral replication (Section M4).

Table 1. Nomenclature of tRNA-synthetases and charged tRNAs

Amino acid	Cognate tRNA	Cognate aminoacyl-tRNA synthetase	Aminoacyl-tRNA
Serine	$tRNA^{Ser}$	Seryl-tRNA synthetase	Seryl-tRNA^{Ser}
Leucine	$tRNA^{Leu}$	Leucyl-tRNA synthetase	Leucyl-tRNA^{Leu}
	$tRNA^{Leu}_{UUA}$		Leucyl-tRNA$^{Leu}_{UUA}$

Aminoacylation of tRNAs

The general **aminoacylation reaction** is shown in Figure 3. In the first step, ATP is hydrolysed to AMP and pyrophosphate (PPi) and the AMP becomes linked to the carboxyl group of the amino acid giving a high-energy intermediate called an **aminoacyl-adenylate**. The hydrolysis of the released PPi to two molecules of inorganic phosphate drives this reaction forward. In the second step, the aminoacyl-adenylate reacts with the **appropriate** uncharged tRNA to give the aminoacyl-tRNA and AMP. Some synthetases join the amino acid to the 2′-hydroxyl of the ribose and some to the 3′-hydroxyl, but once joined the two species can interconvert. The formation of an aminoacyl-tRNA helps to drive protein synthesis, as the aminoacyl-tRNA bond is of a higher energy than a peptide bond and thus peptide bond formation is a favorable reaction once this energy-consuming step has been performed.

Aminoacyl-tRNA synthetases

Despite the fact that they all carry out the same reaction of joining an amino acid to a tRNA, the different aminoacyl-tRNA synthetases vary in structure. They are classified into two groups: **class I synthetases** are usually monomeric and have structurally similar N-terminal active sites, whereas **class II synthetases** are mostly dimeric, or tetrameric, and their active sites are nearer their C-terminal ends. The subunit composition can be α, α_2, α_4, or $\alpha_2\beta_2$.

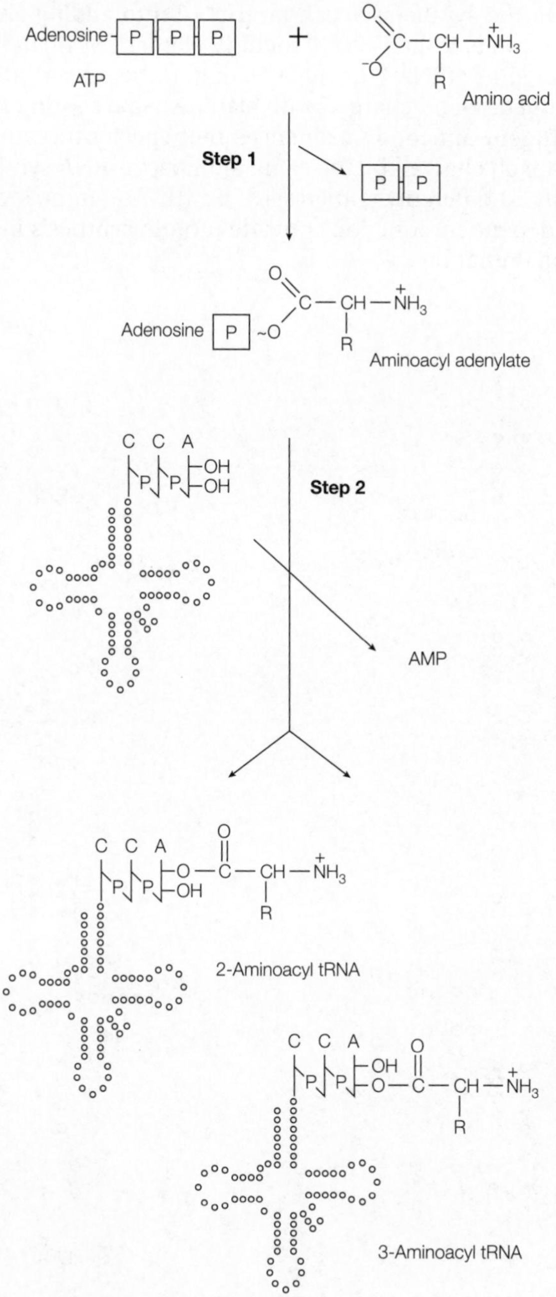

Figure 3. Formation of aminoacyl-tRNA.

These synthetases have to be able to distinguish between about 40 similarly shaped but different tRNA molecules in the cell, and they do so by recognizing specific regions of the tRNAs called **identity elements** (Figure 4). These do not always include the anticodon sequence, which does differ between tRNAs, and often involves base-pairs in the acceptor stem, especially base 73 (the **discriminator base**). If these elements are swapped between tRNAs then the synthetases can be tricked into adding an amino acid to the wrong tRNA. For example, if the G3:U70 identity element of tRNA^{Ala} is used to replace the 3:70 base pair of either tRNA^{Cys} or tRNA^{Phe}, then these altered tRNAs are recognized by alanyl-tRNA synthetase and charged with alanine. **Isoaccepting tRNAs** (those different tRNAs with different anticodon sequences that specify the same amino acid, e.g. the leucyl-tRNAs) are all charged by the same aminoacyl-tRNA synthetase and so most organisms have only 18 different synthetases. Like tRNAs, aminoacyl-tRNA synthetases have numerous, independent functions outside protein synthesis including the control of apoptosis and inflammation.

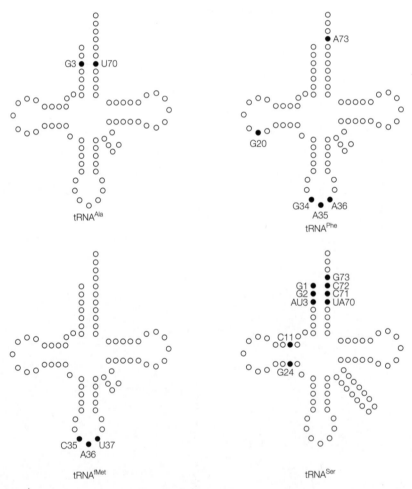

Figure 4. Identity elements in various tRNA molecules.

Proofreading

Synthetases that have to distinguish between two chemically similar amino acids need to carry out a proofreading step. Usually, if they accidentally carry out step 1 of the amino-acylation reaction with the wrong amino acid, then they will not carry out step 2. Instead they will hydrolyze the aminoacyl-adenylate. This is known as **pre-transfer editing**.

Some synthetases carry out **post-transfer editing** by hydrolyzing the misacylated tRNA after step 2. This proofreading ability is only necessary when a single recognition step is not sufficiently discriminating. However, discrimination between the amino acids Phe and Tyr can be achieved in one step because of the -OH group difference on the benzene ring, so in this case there is no need for proofreading.

L1 Aspects of protein synthesis

Key Notes	
Codon–anticodon interaction	In the cleft of the ribosome, the three bases of the codon on the mRNA and the anticodon on the tRNA pair in an antiparallel fashion. If the 5′-anticodon base is modified, the tRNA can usually interact with more than one codon.
Wobble	The wobble hypothesis explains how nonstandard base pairs can form between modified 5′-anticodon bases and 3′-codon bases. When the wobble nucleoside is inosine, the tRNA can base pair with three codons – those ending in A, C or U.
Ribosome binding site	The ribosome-binding site (RBS) is a sequence just upstream of the initiation codon in bacterial mRNA that base-pairs with a complementary sequence near the 3′-end of the 16S rRNA to position the ribosome for initiation of protein synthesis. It is also known as the Shine–Dalgarno sequence after its discoverers.
Initiator tRNA	A special tRNA (initiator tRNA) recognizes the first AUG codon and is used to initiate protein synthesis in both prokaryotes and eukaryotes. In bacteria, the initiator tRNA is first charged with methionine by methionyl-tRNA synthetase. The methionine residue is then converted to *N*-formylmethionine by transformylase. In eukaryotes, the methionine on the initiator tRNA is not modified. There are structural differences between the *E. coli* initiator tRNA and the tRNA that inserts internal Met residues.
Polysomes	Polyribosomes (polysomes) form on an mRNA when successive ribosomes attach, begin translating, and then move 5′→3′ along the mRNA. A polysome is a complex of multiple ribosomes at various stages of translation on one mRNA molecule.
Related topics	(J1) rRNA processing and ribosomes (J2) tRNA and other small RNA processing	(K2) tRNA structure and function (L2) Mechanism of protein synthesis

Codon–anticodon interaction

The **anticodon** in one of the stem-loops of the tRNA interacts with a complementary triplet of bases on the mRNA, the **codon**, when both are brought together in the cleft of the ribosome (Section J1). The interaction is **antiparallel** in nature (Figure 1). When some highly purified tRNA molecules were found to interact with more than one codon,

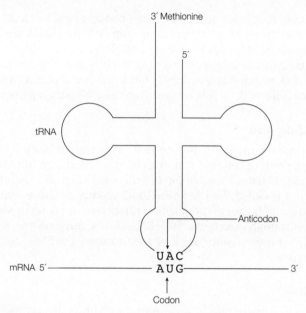

Figure 1. Codon–anticodon interaction.

this ability was correlated with the presence of modified nucleosides in the 5′-anticodon position, particularly inosine (Section K2). Inosine (I) is formed by post-transcriptional processing (Section J2) of adenosine when it occurs at this position. This is carried out by **tRNA-specific adenosine deaminase**, which converts the 6-amino group of the base adenine to a keto group (hypoxanthine).

Wobble

The wobble hypothesis was suggested by Crick to explain the redundancy of the genetic code. He realized by model building that the 5′-anticodon base was able to undergo more movement than the other two bases and could thus form **nonstandard base-pairs** as long as the distances between the ribose units were close to normal. His specific predictions are shown in Table 1 along with actual observations. No purine–purine or pyrimidine–pyrimidine base pairs are allowed as the ribose distances would be incorrect.

Table 1. Original wobble predictions

5′ Anticodon base	Predicted 3′ codon base read	Observations
A	U	A converted to I by anticodon deaminase
C	G	No wobble, normal base-pairing
G	C and U	G, and modified G, can pair with C and U
U	A and G	U not found as 5′-anticodon base
I	A and	Wobble as predicted. Inosine (I) can
	C and	recognize 3′ -A, -C, or -U
	U	

No single tRNA could recognize more than three codons, hence at least 32 tRNAs would be needed to decode the 61 codons, excluding stop codons. tRNAs can recognize either one, two, or three codons, depending on their wobble base (the 5′-anticodon base). If this is C it will recognize only the codon ending in G. If it is G, it will recognize the two codons ending in U or C. If U, which is subsequently modified, it will pair with either A or G. The wobble nucleoside is never A, as this is converted to I which then pairs with A, C, or U.

Ribosome binding site

In bacterial mRNAs a conserved, purine-rich sequence, usually 5′-AGGAGGU-3′, is found 8–13 nt upstream of the first codon to be translated (the initiation codon). This sequence base-pairs with the 3′-end of the 16S rRNA in the small subunit of the ribosome (5′-ACCUCCU-3′). It is called the **ribosome binding site**, or **Shine–Dalgarno sequence** after its discoverers, and is thought to position the ribosome correctly with respect to the initiation codon. Although eukaryotic 18S rRNA shows conservation with bacterial 16S rRNA, this ribosome binding sequence is absent. Instead, the 5′ cap may serve a similar function in eukaryotes.

Initiator tRNA

The first amino acid incorporated into a protein chain is always methionine in both prokaryotes and eukaryotes, though in bacteria and in mitochondria and chloroplasts this Met is modified to **N-formylmethionine**. The AUG initiation codon is recognized by a special **initiator tRNA** that differs from the one that pairs with AUG codons in the rest of the coding region. In *E. coli*, there are subtle differences between these two tRNAs (Figure 2). There are unpaired bases at the end of the amino acid acceptor stem that are required for formylation. There are also three G-C base-pairs in the anticodon arm that are required for entry into the ribosomal P-site (Section L2). Finally, the initiator tRNA

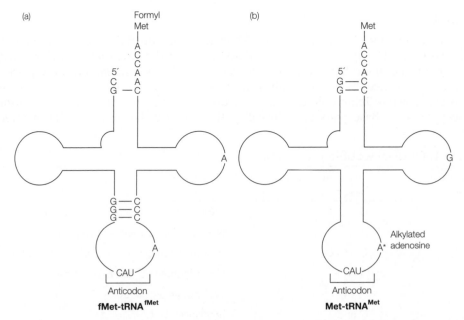

Figure 2. The *E. coli* methionine-tRNAs. (a) The initiator tRNA fMet-tRNA[fMet]; (b) the methionyl-tRNA Met-tRNA[Met].

allows more flexibility in base-pairing (wobble) because it lacks the alkylated A in the anticodon loop and hence it can recognize not only AUG as an initiation codon, but also GUG and very rarely UUG, which occur at much lower frequencies in bacterial mRNAs. The noninitiator tRNA is less flexible and can only pair with AUG codons. Both tRNAs are charged with Met by the same **methionyl-tRNA synthetase** (Section K2) to give the methionyl-tRNA (Met-tRNA), but only the initiator Met-tRNA (Met-tRNA^{fMet}) is modified by the enzyme **transformylase** to give N-formylmethionyl-tRNA^{fMet} (fMet-tRNA^{fMet}). The link to the N-formyl group resembles a peptide bond and may help this initiator tRNA to enter the P-site of the ribosome, an essential requirement for initiation, whereas all other tRNAs enter the A-site (Section L2).

Polysomes

When a ribosome has begun translating an mRNA molecule (Section L2), and has moved about 70–80 nt downstream from the initiation codon, a second ribosome can assemble at the initiation site and start to translate the mRNA. When this second ribosome has moved away, a third can begin, and so on. Multiple ribosomes on a single mRNA are called **polysomes** (short for polyribosomes) and there can be as many as 50 on some mRNAs, although they cannot be positioned closer than about 80 nt apart.

L2 Mechanism of protein synthesis

Key Notes	
Overview	There are three stages in protein synthesis:
	• Initiation – the assembly of a ribosome on an mRNA
	• Elongation – repeated cycles of amino acid delivery, peptide bond formation and movement along the mRNA (translocation)
	• Termination – the release of the polypeptide chain
Initiation	In bacteria, initiation requires the large and small ribosomal subunits, the mRNA, the initiator tRNA, three initiation factors (IFs), and GTP. IF1 and IF3 bind to the 30S subunit and prevent the large subunit binding. IF_2+GTP can then bind and will help the initiator tRNA to bind later. This small subunit complex can now attach to an mRNA via its ribosome-binding site. The initiator tRNA can then base-pair with the AUG initiation codon, which releases IF_3, thus creating the 30S initiation complex. The large subunit then binds, displacing IF_1 and IF_2+GDP, giving the 70S initiation complex, which is the fully assembled ribosome at the correct position on the mRNA.
Elongation	Elongation involves the three elongation factors (EFs), EF-Tu, EF-Ts and EF-G as well as GTP, charged tRNAs, and the 70S initiation complex (or its equivalent). It consists of three steps:
	• A charged tRNA is delivered as a complex with EF-Tu and GTP. The GTP is hydrolyzed and EF-Tu–GDP is released, which can be re-used with the help of EF-Ts and GTP (via the EF-Tu–EF-Ts exchange cycle).
	• Peptidyl transferase, an activity of the large subunit rRNA, makes a peptide bond by transferring the polypeptide in the P-site to the aminoacyl-tRNA in the A-site without the need for further energy input.
	• Using the energy of GTP hydrolysis, translocase (EF-G) moves the ribosome one codon along the mRNA, ejecting the uncharged tRNA and transferring the growing polypeptide back to the P-site.
Termination	Release factors (RF1 or RF2) recognize the stop codons and, helped by RF3, cause peptidyl transferase to transfer the finished polypeptide to a water molecule, thus releasing it.

	Ribosome recycling factor helps to dissociate the ribosome subunits from the mRNA.	
Related topics	(J2) tRNA and other small RNA processing (K2) tRNA structure and function	(L1) Aspects of protein synthesis (L3) Initiation in eukaryotes

Overview

Not all of the mRNA sequence is translated into protein, only the region from (and including) the initiation codon up to the termination codon. The regions upstream and downstream of these codons are called the **5′-untranslated (5′-UTR)** and **3′-untranslated (3′-UTR) regions** respectively and are important for the regulation of translation.

The actual mechanism of protein synthesis can be divided into three stages:

- **Initiation** – the assembly of a ribosome on an mRNA molecule

- **Elongation** – repeated cycles of amino acid addition

- **Termination** – the release of the new protein chain

These are illustrated in Figures 1–3 and involve the activities of a number of factors. In bacteria, the factors are abbreviated as IF or EF for **initiation** and **elongation factors** respectively, whereas in eukaryotes they are called eIF and eEF. There are distinct differences of detail between the mechanism in bacteria and eukaryotes, most of which occur at the initiation stage. For this reason, this section will describe the mechanism in bacteria and the following section (L3) will detail the differences that occur in eukaryotes. Archaeal protein synthesis is a mosaic, showing features of both.

Initiation

The purpose of the initiation step is to assemble a complete ribosome onto an mRNA molecule at the correct start point, the **initiation codon**. The components involved are the large and small **ribosomal subunits**, the **mRNA**, the **initiator tRNA** in its charged form (Section L1), **three initiation factors**, and **GTP**. The initiation factors IF_1, IF_2, and IF_3 are all just over one-tenth as abundant as ribosomes. The overall sequence of events (Figure 1) is as follows:

- IF_1 and IF_3 bind to a free 30S subunit. IF_3 helps to prevent the formation of an inactive ribosome by stopping a large subunit binding in the absence of mRNA. Later it assists the correct charged initiator tRNA to enter the appropriate binding site (the P-site). IF_1 blocks the site at which noninitiator tRNAs would bind.

- IF_2 complexed with GTP then binds to the small subunit to assist the charged initiator tRNA to bind.

- The 30S subunit and associated IFs attach to an mRNA molecule by recognizing the ribosome-binding site (RBS) on the mRNA (Section L1).

- The initiator tRNA then binds to the complex in the P-site (see below) by base-pairing of its anticodon with the first AUG codon on the mRNA. At this point, IF_3 can be released as its roles in keeping the subunits apart, helping the mRNA to bind, and directing the initiator tRNA are fulfilled. This complex is called the **30S initiation complex**.

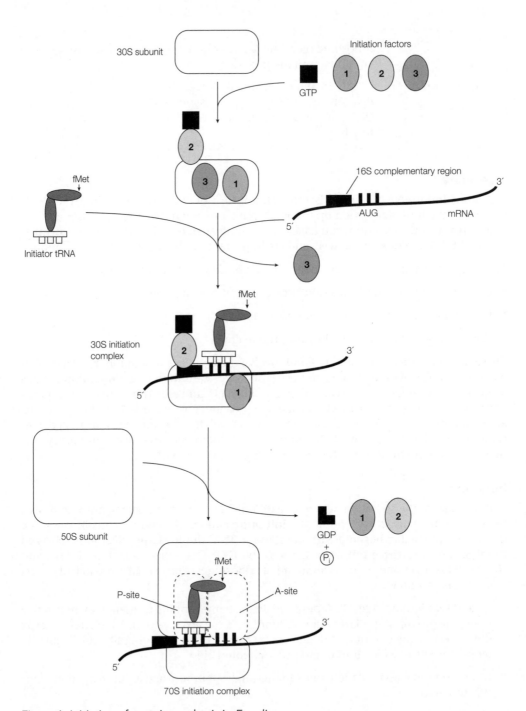

Figure 1. Initiation of protein synthesis in *E. coli*.

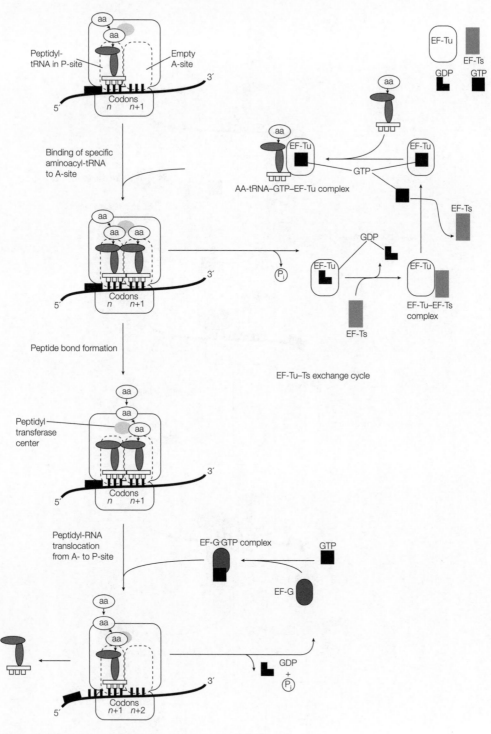

Figure 2. Elongation stage of protein synthesis in *E. coli*.

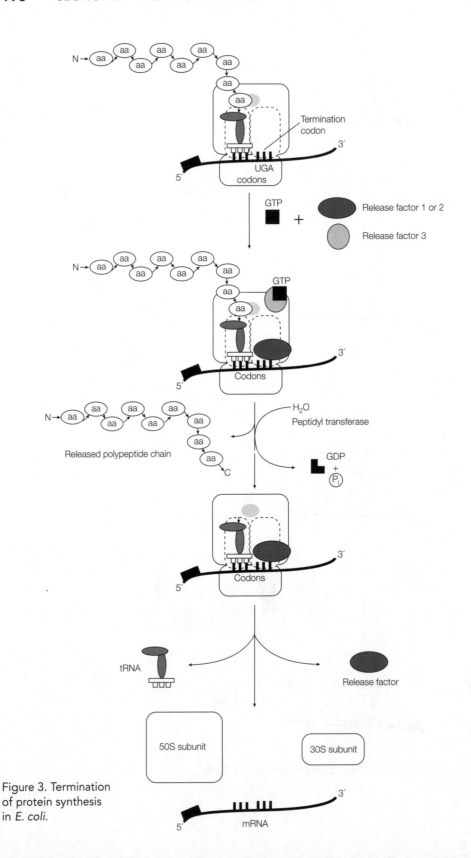

Figure 3. Termination of protein synthesis in *E. coli*.

- The 50S subunit can now bind, displacing IF_1 and IF_2. The GTP is hydrolyzed in this step. The complex formed at the end of the initiation phase is called the **70S initiation complex**.

As shown for simplicity in Figures 1–3, the assembled ribosome has two tRNA-binding sites. These are called the **A- and P-sites**, for aminoacyl and peptidyl sites. The A-site is where incoming aminoacyl-tRNA molecules bind, and the P-site is where the growing polypeptide chain is usually found. These sites are in the cleft of the small subunit (Section J1) and contain adjacent codons that are being translated. One major outcome of initiation is the placement of the initiator tRNA in the P-site. It is the only tRNA that does this, as all others must enter the A-site (Section L1). Correct positioning of the initiating AUG codon in this way fixes the **reading frame** for the rest of the translation process (Section K1).

Elongation

With the formation of the 70S initiation complex, the **elongation cycle** can begin. It can be subdivided into three steps as follows: (i) **aminoacyl-tRNA delivery**, (ii) **peptide bond formation**, and (iii) **translocation**. These are shown in Figure 2, beginning with an occupied P-site and an empty A-site. It involves **three elongation factors**, **EF-Tu**, **EF-Ts**, and **EF-G**, all of which bind GTP or GDP. EF-Ts and EF-G are about as abundant as ribosomes, but EF-Tu is nearly 10 times more abundant.

(i) Aminoacyl-tRNA delivery. EF-Tu is required to deliver the aminoacyl-tRNA to the A-site and energy is consumed in this step in the form of GTP hydrolysis to GDP+Pi. The released **EF-Tu-GDP** complex is regenerated with the help of EF-Ts. In the **EF-Tu–EF-Ts exchange cycle**, EF-Ts displaces the GDP and subsequently is displaced itself by GTP. The resulting EF-Tu-GTP complex is now able to bind another aminoacyl-tRNA and deliver it to the ribosome. All aminoacyl-tRNAs can form this complex with EF-Tu, except the initiator tRNA.

(ii) Peptide bond formation. After aminoacyl-tRNA delivery, the A- and P-sites are both occupied and the two amino acids that are to be joined are in close proximity. The **peptidyl transferase** activity of the 50S subunit can now form a peptide bond between these two amino acids without the input of any more energy, since energy in the form of ATP was used to charge the tRNA (Section K2). Interestingly, the catalytic site of peptidyl transferase is part of the 23S (or 28S) rRNA in the large subunit and does not involve a protein. This rRNA is therefore a **ribozyme** (Section J2).

(iii) Translocation. A complex of EF-G (**translocase**) and GTP binds to the ribosome and, in an energy-consuming step, the discharged tRNA is ejected from the P-site, the peptidyl-tRNA is moved from the A-site to the P-site and the mRNA moves by one codon relative to the ribosome. GDP and EF-G are released, the latter being re-used. A new codon is now present in the vacant A-site. In bacteria the discharged tRNA is first moved to an **E-site** (**exit site**), located mainly on the 50S subunit, and is ejected when the next aminoacyl-tRNA binds. In this way, the ribosome maintains contact with the mRNA via 6 bp, which may well reduce the chances of **frameshifting**, i.e. slipping by one or two bases, which would change the reading frame and the resulting protein sequence (Section E2). There is no evidence for an E-site in eukaryotes.

Once one cycle of the three-step elongation process has been completed, the cycle is repeated until one of the three termination codons (stop codons) appears in the A-site. In this way, the mRNA is translated 5'→3' and the protein is synthesized from its N-terminus to its C-terminus.

Termination

Apart from the few specialized cases where they can be interpreted as an amino acid (Section K1), stop codons are not normally recognized by tRNA species. Instead, proteins called **release factors** interact with these codons and cause release of the completed polypeptide chain (Figure 3). The class 1 release factors RF1 and RF2 recognize the codons UAA and UAG, and UAA and UGA, respectively. They cause peptidyl transferase to transfer the polypeptide to water rather than to an aminoacyl-tRNA, and so the new protein is released. The class 2 release factor RF3 promotes the dissociation of RF1 or RF2 from the ribosome. To remove the uncharged tRNA from the P-site and release the mRNA, EF-G together with **ribosome recycling factor** (**RRF**) are needed for the complete dissociation of the subunits. IF_3 can now bind the small subunit to prevent inactive 70S ribosomes reforming.

Ribosomes can stall on an mRNA if a base is damaged and unrecognizable (Section E1), if the message has no stop codon, or if there is insufficient of the correct tRNA for a particular codon. Mechanisms exist to deal with the problems of damaged or truncated (shortened) mRNAs being translated into defective proteins. In bacteria, a small RNA called **tmRNA** (transfer-messenger RNA) that has properties of both tRNA and mRNA is used to free the stalled ribosome and ensure degradation of the defective protein. First, the tmRNA behaves like a tRNA by delivering alanine to the A-site and allowing peptide bond formation to take place. Then translocation occurs, placing another part of the tmRNA in the A-site where it behaves like an mRNA directing the translation of 10 codons and a stop codon contained within its own sequence. The released protein, therefore, has a tag of 10 amino acids encoded by the tmRNA at its C-terminus that target it for rapid degradation. Different mechanisms exist in eukaryotes (Section L3).

L3 Initiation in eukaryotes

Key Notes

Overview
Most of the differences in the mechanism of protein synthesis between bacteria and eukaryotes occur at the initiation stage. In addition, eukaryotes have just one release factor (eRF) and the eukaryotic initiator tRNA does not become N-formylated as in bacteria.

Scanning
The eukaryotic 40S ribosome subunit complex binds to the 5′-cap of the mRNA and then moves 5′→3′ scanning the mRNA for the AUG start codon. This is not necessarily the first AUG, as it must be in the appropriate sequence context. More than one AUG can be used to initiate, leading to different proteins.

Initiation
Initiation comprises four major steps involving at least 12 eIFs that fall into different functional groups. These assemble the 43S pre-initiation complex, bind to the mRNA, or recruit the 60S subunit by displacing other factors. In contrast to bacteria, initiation involves binding of the initiator tRNA to the 40S subunit before it binds to the mRNA. Important controls include phosphorylation of eIF2, which delivers the initiator tRNA, and eIF4E binding proteins, which inhibit eIF4G binding thereby blocking recruitment of the 43S complex.

Elongation
This stage of protein synthesis is essentially identical to that described for bacteria (Section L2). The factors EF-Tu, EF-Ts, and EF-G have direct eukaryotic equivalents called eEF1α, eEF1βγ, and eEF2 respectively, which carry out the same roles.

Termination
Eukaryotes have only one class 1 release factor (eRF1) for termination of protein synthesis. It can recognize all three stop codons. The class 2 release factor, eRF3, causes eRF1 to leave the ribosome. Surveillance mechanisms operate to release stalled ribosomes or degrade defective mRNAs.

Related topics
(J2) tRNA and other small RNA processing
(K2) tRNA structure and function

(L2) Mechanism of protein synthesis

Overview

Apart from protein synthesis in the mitochondria and chloroplasts (which are thought to originate from symbiotic bacteria), the details of eukaryotic translation differ somewhat from that described in Section L2. Most of these differences are in the initiation phase where a greater number of **eukaryotic initiation factors** (**eIFs**) are involved. Finding the correct start codon involves a **scanning process** as there is no ribosome binding sequence. Although there are two different tRNA species for methionine, one of which is the initiator tRNA, the attached methionine does not become converted to *N*-**formyl-methionine**. A comparison of the factors involved in bacteria and eukaryotes is given in Table 1.

Table 1. Comparison of protein synthesis factors in bacteria and eukaryotes

Bacterial	Eukaryotic	Function
Initiation factors		
IF1, IF3	eIF1, eIF1A, eIF3, eIF5/eIF2, eIF2B	Binding to small subunit/initiator tRNA delivery
IF2	eIF4B, eIF4F, eIF4H	Binding to mRNA
	eIF5B	Displacement of other factors and large subunit recruitment
Elongation factors		
EF-Tu	eEF1α	Aminoacyl tRNA delivery to ribosome
EF-Ts	eEF1βγ	Recycling of EF-Tu or eEF1α
EF-G	eEF2	Translocation
Termination factors		
RF1	eRF1	Stop codon recognition and polypeptide chain release (class 1)
RF2		
RF3	eRF3	Dissociation of class 1 factors (class 2)

Scanning

The lack of a recognizable RBS in eukaryotic mRNA means that the mechanism of selecting the start codon must be different. Marilyn Kozak proposed a **scanning model** in which the 40S subunit, already carrying the initiator tRNA, attaches to the capped 5′-end of the mRNA and scans in a 5′→3′ direction until it finds an appropriate AUG. This is not always the first one as it must be in the correct **sequence context** (5′-gccRcc**AUG**G-3′), where R=purine and lower case indicates slight preference. The closeness of the actual sequence surrounding the AUG to this **Kozak consensus sequence** determines the likelihood of initation taking place. Thus it is possible to initiate at more than one AUG with different probabilities leading to the synthesis of related proteins with different N-terminal sequences.

Initiation

Figure 1 shows the steps and factors involved in the initiation of protein synthesis in eukaryotes. There are four major steps: (i) the assembly of the **43S pre-initiation complex** via a **multifactor complex** (**MFC**); (ii) recruitment of the 43S pre-initiation complex by the mRNA through interactions at its 5′-end; (iii) scanning for the initiation codon; and (iv) recruitment of the 60S subunit to form the 80S initiation complex. There are at least 12 reasonably well-defined initiation factors involved in eukaryotic protein synthesis, and

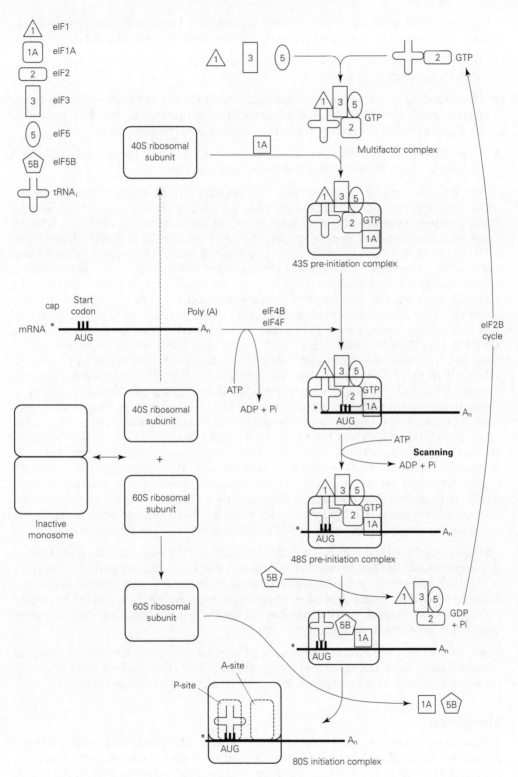

Figure 1. Initiation of protein synthesis in eukaryotes.

some have analogous functions to the three bacterial IFs. They can be grouped in various ways but it is logical to group them according to the steps at which they act as follows:

- Those involved in assembly of the 43S pre-initiation complex, such as eIF1, eIF1A, eIF2, eIF3, and eIF5.

- Those binding to the mRNA to recognize the 5'-cap and to melt secondary structure, such as eIF4B and eIF4F. eIF4F is a heterotrimer complex of an RNA helicase called eIF4A, a cap binding protein called eIF4E, and a scaffold protein, eIF4G.

- Those that recruit the 60S subunit by displacing other factors such as eIF5B, which releases five other factors so that the 60S subunit can bind.

The following events take place, starting with the **eIF2-GTP binary complex**, formed by eIF2B recycling the eIF2-GDP released late during initiation. The initiator $tRNA_i$ joins to make a **ternary complex** of three components, the **initiator $tRNA_i$**, **eIF2**, and **GTP**. In yeast the ternary complex then forms part of an MFC containing **eIF1**, **eIF2-GTP-tRNA$_i$**, **eIF3**, and **eIF5**. The binding of the MFC to a free **40S** subunit is assisted by **eIF1A** and the resulting complex is called the **43S pre-initiation complex**. In some eukaryotes the ternary complex may bind later, i.e. to the 40S subunit containing eIF1, eIF1A, eIF3, and eIF5. Note this different order of assembly in eukaryotes where the initiator $tRNA_i$ is bound to the small subunit before the mRNA binds (compare Section L2, Figure 1). Before this large complex can bind to the mRNA, the latter must have interacted with **eIF4B** and **eIF4F** (which recognizes the 5'-cap via its subunit **eIF4E**), and using energy from ATP, have been unwound and have had secondary structure removed by subunit **eIF4A**. **eIF4H** may help in this. The **eIF4G** scaffold subunit of eIF4F can bind to **PABI**, the cytoplasmic poly(A) binding protein (Section J3), and cause stimulation of translation by circularizing the mRNA (not covalently, but spatially through protein–protein contacts). This PABI stimulation of translation also provides a check that the ribosome is assembling on an intact mRNA that has both a 5'-cap and a 3'-poly(A) tail. The second major step occurs when the 43S pre-initiation complex has bound to the mRNA complex via interactions between **eIF4G** and **eIF3**. In the third step, ATP is hydrolyzed as the mRNA is scanned to find the AUG start codon. This is usually, but not always, the first one (see above). In the fourth step, **eIF5B** must displace eIF1, eIF2, eIF3, and eIF5 to allow the **60S** subunit to bind, and GTP is hydrolyzed. eIF1A and eIF5B are released when the latter has promoted binding of the 60S subunit to form the complete **80S initiation complex**.

The released eIF2-GDP complex is recycled by **eIF2B** and the rate of recycling (and hence the rate of initiation of protein synthesis) is regulated by phosphorylation of the α-subunit of eIF2. Certain events, such as viral infection and the resultant production of interferon, cause an inhibition of protein synthesis by promoting **phosphorylation of eIF2**. Another point of regulation involves **eIF4E binding/inhibitory proteins** that can block complete assembly of eIF4F on some mRNAs (Section L4).

Some RNA viruses use a cap-independent process to assemble ribosomes at **internal ribosome entry sites** (**IRESs**) on their polycistronic mRNAs (Sections M1 and M4). IRESs are also found in some cellular genes, especially those involved in stress survival.

Elongation

The protein synthesis elongation cycle in bacteria and eukaryotes is quite similar. Three factors are required with properties similar to their bacterial counterparts (Table 1). eEF1α, eEF1βγ, and eEF2 have the equivalent roles described in Section L2 for EF-Tu, EF-Ts, and EF-G.

Termination

In eukaryotes, a **single class 1 release factor**, **eRF1**, recognizes all three stop codons and performs the roles carried out by RF1 (or RF2) in bacteria. The **class 2 release factor eRF3** causes release of eRF1 and requires GTP for activity.

mRNA surveillance mechanisms operate in eukaryotes to release stalled ribosomes or degrade defective mRNAs that may have arisen due to mutation (Section E2). **Nonstop decay** releases stalled ribosomes that have translated an mRNA lacking a stop codon and become stuck at the 3′-end. The translation product is defective in that it will have polylysine at its C-terminus due to translation of the poly(A) tail. A protein factor (Ski7) helps to dissociate the ribosome and recruit a 3′→5′ exonuclease to degrade the defective mRNA. The defective polylysine-tagged protein is also rapidly degraded.

In NMD, mRNAs containing premature stop codons are recognized due to the presence of protein complexes deposited at exon–exon junctions following splicing in the nucleus (Section J3). In normal mRNAs, these complexes are displaced by the first ribosome to translate the mRNA and the stop codon is reached later. In defective mRNAs, the premature stop codon is reached before all the complexes are displaced and this causes decapping of the mRNA and consequent 5′→3′ degradation. NMD is also used for regulating the expression of normal transcripts.

L4 Translational control and post-translational events

Key Notes

Translational control in bacteria	In bacteria, the level of translation of different cistrons can be affected by: (i) the binding of proteins that prevent ribosome access, (ii) the relative stability to nucleases of regions of the polycistronic mRNA, and (iii) the binding of short antisense RNA molecules.
Translational control in eukaryotes	In eukaryotes, protein binding can also mask the mRNA and prevent translation, and repeats of sequences like 5'-AUUUA-3' can make the mRNA unstable and less frequently translated. *Trans*-acting microRNAs in the RNA-induced silencing complex can target specific mRNAs for degradation by binding to complementary sequences in the 3'-UTR. *Cis*-acting sequences such as the iron response element in transferrin and ferritin mRNAs are important for the control of iron transport and storage.
Polyproteins	A single translation product that is cleaved to generate two or more separate proteins is called a polyprotein. Many viruses produce polyproteins.
Protein targeting	Short terminal or internal peptide sequences in proteins determine the subcellular location of the protein, e.g. nucleus, mitochondrion, etc. The signal sequence of secreted proteins causes the translating ribosome to bind factors that make the ribosome dock with a membrane and transfer the protein through the membrane as it is synthesized. The signal sequence is usually then removed by a signal peptidase.
Protein folding and modification	Chaperones assist newly synthesized proteins to fold by preventing misfolding and aggregation. The most common alterations to nascent polypeptides are those of cleavage and chemical modification. Cleavage occurs to remove signal peptides, to release mature fragments from polyproteins, to remove internal peptides, and to trim both N- and C-termini. All but six of the amino acid side chains can undergo post-translational chemical modifications, of which there are many.
Protein degradation	Damaged, modified or inherently unstable proteins are marked for degradation by having multiple molecules of ubiquitin covalently attached. The ubiquitinylated protein is then degraded by a 26S protease complex, the proteasome. Defects in protein degradation are responsible for several human diseases.

Related topics	(H4) RNA Pol II genes: promoters and enhancers (J3) mRNA processing, hnRNPs, and snRNPs	(J4) Alternative mRNA processing (L2) Mechanism of protein synthesis (L3) Initiation in eukaryotes

Translational control in bacteria

Because of differences between bacterial and eukaryotic mRNAs (e.g. polycistronic vs. monocistronic, Section G1) and the absence of a nuclear membrane in the former, translation is controlled in different ways. At its simplest, translation of a polycistronic bacterial mRNA involves movement of the ribosome from the 5′- to the 3′- end. The 30S ribosomal subunit remains attached to the mRNA throughout, with full ribosome assembly, initiation, and termination of each individual encoded protein occurring sequentially as each RBS, start and stop codon is encountered. However, RBSs can be obstructed by regulatory RNA-binding proteins that bind nearby, while secondary structures such as stem-loops can temporarily obscure the RBS of a downstream cistron; in the latter case, this inhibitory structure is disrupted when a ribosome translates the upstream cistron, allowing a 30S subunit to bind internally. The formation of stem-loop structure can inhibit the action of enzymes whose function is to degrade the mRNA, especially at the 5′-end, where they inhibit the action of RppH, the enzyme that initiates degradation (Section J3). Bacteria also employ small regulatory RNA molecules with partial sequence complementarity to regions of specific mRNAs, e.g. the RBS. The resulting duplex structure inhibits translation. Mechanisms such as these allow different amounts of translation products to be made from the cistrons of a polycistronic mRNA. Several operons encoding ribosomal proteins in *E. coli* show an interesting form of translational control where a region of the mRNA has a tertiary structure that resembles the binding site for a ribosomal protein encoded by the mRNA. If there is insufficient rRNA available for the translation product to bind to, it will bind to its own mRNA and prevent further translation.

Translational control in eukaryotes

Global control of translation in eukaryotes can be achieved by phosphorylation of key factors involved in the initiation of translation. For example, global repression of eukaryotic translation is caused when the α-subunit of eIF2 is phosphorylated (Section L3). Repression is also caused by the binding of a protein **eIF4EBP1** (a member of the **4E-BP** group of translational repressors) to eIF4E, thus preventing active ribosome assembly (Section L3). Growth factors, hormones, and cell division stimulators can activate a **kinase** (**mTor**) that phosphorylates eIF4EBP1, causing it to dissociate from eIF4E, thus activating translation. Although eukaryotes generally control the amount of specific proteins by varying the level of transcription of the gene (Section H4) and/or by RNA processing (Sections J3 and J4), specific translational control also occurs in the cytoplasm. The presence of multiple copies of **AU-rich elements** such as **5′-AUUUA-3′**, usually in the 3′-noncoding region, marks the mRNA for rapid degradation, often by the **exosome** (Section J3), and so limits translation. Another form of translational control involves proteins of the 4E-BP group binding indirectly to a specific mRNA and preventing translation. This RNA is called '**masked mRNA**'. In appropriate circumstances, the mRNA can be translated when the protein dissociates.

Translational regulation of individual or groups of mRNAs involves *trans*-acting small ncRNAs such as **microRNAs (miRNAs)**, **short interfering RNAs (siRNAs)**, and **piwi-associated RNAs (piRNAs)**. These RNAs are responsible at least in part for the phenomenon of **RNA silencing**, the down-regulation of gene expression by exogenous double-stranded RNAs (dsRNAs) first observed in plants. miRNAs and siRNAs share many properties and functions and the distinction between them is not completely clear. **Piwis** are regulatory, nucleic acid-binding proteins and piRNAs are a special class of miRNA found in germ cells. There may be at least 1000 miRNA genes in the human genome, some of which are located in the introns of the protein-coding genes they regulate, while others are found between structural genes. Some are evolutionarily conserved and developmentally regulated. Most miRNAs are synthesized by RNA Pol II as larger precursors, which undergo processing such as capping, polyadenylation, and cleavage (Section J2). These **primary miRNAs (pri-miRNAs)** have extensive secondary structure and some can yield several mature miRNAs, which are short (21–23 nt) and double-stranded. These RNAs inhibit translation by binding to a complementary region in the target mRNA, usually in the 3′-UTR, a process known as **RNA interference** (**RNAi**). In plants, the match is near perfect, whereas in animals, it is only partial, meaning that one animal miRNA can often target multiple mRNAs. The miRNA is part of an RNP complex called the **RISC** that includes the processing protein Dicer (Section J2) and **argonaute** proteins that promote repression of translation either by inhibition, or by cleaving the mRNA directly or targeting it for enhanced degradation (Section J3). It has been estimated that miRNAs regulate the expression of more than half the genes in the human genome, and overexpression of some miRNAs has been correlated with certain types of cancer. RNAi may originally have evolved to protect eukaryotic cells from RNA viruses (Section M4) but has now been adapted for a much wider role in controlling gene expression. It is also now being exploited experimentally to discover the functions of specific genes by synthetic **RNAi knockdown** (Section S2).

An interesting example of translational control in mammals involving *cis*-acting noncoding sequences within mRNA is the regulation of cellular iron content. Iron is essential for the activity of some proteins, but is harmful in excess. Iron is transported into cells by the **transferrin receptor** and is stored bound to **ferritin**. The mRNAs for each of these proteins contain a noncoding sequence called the **iron response element** (**IRE**) that can form stem-loop structures to which an iron sensing protein (ISP) can bind. However, the position of the IRE and the action of the bound ISP is very different for each mRNA. The IRE in the transferrin receptor mRNA is in the **3′-UTR** and when the ISP binds, which it does when iron is scarce, it stabilizes the mRNA, increasing translation of the receptor in order to improve iron uptake. However, when iron is abundant the ISP dissociates from the 3′-IRE and unmasks destabilizing sequences that are then attacked by nucleases, leading to mRNA degradation, reduced translation, and reduced iron uptake. Conversely, under these same conditions of high iron, the translation of the ferritin mRNA is increased. This occurs because at low iron levels the IRE located in the **5′-UTR** of the ferritin mRNA binds the ISP, reducing the ribosome's ability to translate the ferritin mRNA. When iron levels rise, the ISP dissociates from the 5′-IRE and so translation proceeds, increasing the level of ferritin needed for storage. This translational control system rapidly and responsively regulates intracellular iron levels. There are many other examples of translational control that depend on the binding of proteins in the vicinity of destabilizing sequences, leading to suppression of their effect.

Polyproteins

Bacteriophage and viral transcripts (Section M2) and many mRNAs for hormones in eukaryotes (e.g. pro-opiomelanocortin) are translated to give a single polypeptide chain that is cleaved subsequently by specific proteases to produce multiple mature proteins from one translation product. The parent polypeptide is called a **polyprotein**.

Protein targeting

The ultimate cellular location of proteins is usually determined by relatively short, specific amino acid sequences within the proteins themselves. These sequences determine whether proteins are secreted, imported into the nucleus or targeted to other organelles. **Protein secretion** by both prokaryotes and eukaryotes involves a **signal sequence** in the nascent protein, and specific exporting proteins. In bacteria, several secretion systems are involved, some of which are used by pathogens to secrete toxins within the infected organism. For eukaryotic secreted proteins, the signal sequence is recognized by an RNP particle, the **signal recognition particle** (**SRP**), which contains **7SL RNA** (Figure 1). When a cytosolic ribosome starts to translate an mRNA encoding a protein that is to be secreted, SRP binds to the ribosome and to the emerging polypeptide and arrests translation by preventing binding of elongation factors. SRP recognizes ribosomes with a nascent protein chain containing a signal sequence (**signal peptide**) composed of about 13–36 amino acids with at least one positively charged residue followed by a hydrophobic core of 7–13 residues followed by a small, neutral residue, often Ala. SRP (attached to the arrested ribosome) then binds to the **SRP receptor** (**docking protein**) on the cytosolic side of the **endoplasmic reticulum** (**ER**) and is then released for re-use when the ribosome becomes attached to **ribosome receptor proteins** of the translocation complex (**translocon**) on the ER. The ribosome is able to continue translation, and the nascent polypeptide chain is pushed through into the **lumen** (inside) of the ER. As it passes

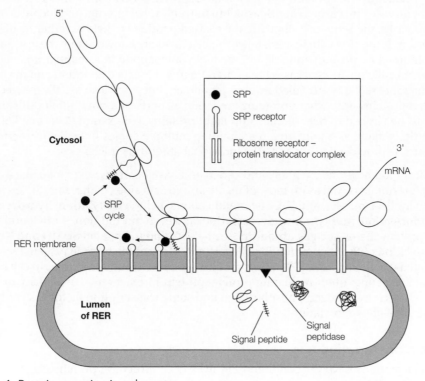

Figure 1. Protein secretion in eukaryotes.

through, **signal peptidase** removes the signal peptide. When the protein is released into the ER it is usually modified, often by **glycosylation**; different patterns of glycosylation seem to control the final destination of the protein. The formation of the correct disulfide bonds (Section A3), which are most commonly found in secreted proteins, is assisted by **protein disulfide isomerase** in the ER.

Other peptide sequences are responsible for the intracellular location of proteins. Different N- or C-terminal sequences can cause proteins to be imported into mitochondria, chloroplasts, and peroxisomes while the internal sequence -Lys-Lys-Lys-Arg-Lys (or any five consecutive positive amino acids) can act as a **nuclear localization signal** causing the protein containing it (e.g. histone) to be imported into the nucleus.

Protein folding and modification

A newly translated polypeptide does not always immediately generate a functional protein (Section A3). Folding into the correct tertiary structure is required and this begins before synthesis is complete. Although the tertiary structure is inherent in the primary sequence and many polypeptides will correctly fold spontaneously *in vitro*, the cotranslational nature of folding and the high protein concentration in the cell requires that **chaperone** proteins assist the folding process, primarily to prevent misfolding due to aggregation of newly exposed hydrophobic sequences during translation. Many chaperones are also **HSPs** (e.g. **Hsp70, GroEL/GroES**) as they both prevent and reverse aggregation caused by protein denaturation at elevated temperatures.

Apart from correct folding and the possible formation of disulfide bonds, a number of other alterations may be required for activity. These **post-translational modifications** include **cleavage** and covalent modifications. Cleavage is very common, especially trimming by **amino-** and **carboxypeptidases**, but the removal of internal peptides also occurs. In most cases, the sequences flanking the excised peptide are held together by disulfide linkages, as is the case with insulin; however, instances are known where these sequences are combined by **protein splicing** into a single polypeptide in a manner analogous to pre-mRNA splicing. In such cases the excised fragment is called an **intein** and the joined, informational sections are called **exteins**. Signal sequences are also usually cleaved from secreted proteins and, where proteins are made as parts of a polyprotein, this precursor must be cleaved to release the component proteins. For example, the small 8.5-kDa **ubiquitin** is made as a polyprotein containing multiple copies linked end-to-end, and this must be processed to generate the individual ubiquitin molecules.

Chemical modifications are many and varied and have been shown to take place on the N and C termini, as well as on most of the 20 amino acid side chains, with the exception of Ala, Gly, Ile, Leu, Met, and Val. The modifications include **acetylation, hydroxylation, phosphorylation, methylation, glycosylation**, and even the addition of **nucleotides** and small protein chains such as **ubiquitin** (see below), and the related **SUMO and NEDD8**. Hydroxylation of Pro is common in collagen, and some of the histone proteins are often acetylated (Sections C2 and C3). These modifications have numerous roles in cell regulation and signaling, often mediating protein–protein interactions. The activity of many enzymes, such as glycogen phosphorylase and some transcription factors, is controlled by reversible phosphorylation.

Protein degradation

Different proteins have very different half-lives ($t_{1/2}$). Structural proteins such as collagen have $t_{1/2}$ measured in weeks or months while regulatory proteins tend to turn over rapidly ($t_{1/2}$=minutes or hours). Cells must also be able to degrade and dispose of faulty and damaged proteins. In eukaryotes, the **N-terminal residue** has been proposed to play a critical role in inherent stability. Eight N-terminal amino acids (Ala, Cys, Gly, Met, Pro, Ser, Thr, Val) correlate with stability ($t_{1/2}$>20 h), eight (Arg, His, Ile, Leu, Lys, Phe, Trp, Tyr) with short $t_{1/2}$ (2–30 min) and four (Asn, Asp, Gln, Glu) are destabilizing following chemical modification. A protein that is damaged, modified, or has an inherently destabilizing N-terminal residue becomes **ubiquitinylated** by the covalent linkage of Lys residues in the protein to the C-terminal glycines of ubiquitin molecules. The ubiquitinylated protein is recognized and digested by a **26S protease complex** (**proteasome**) in a reaction that requires ATP, releasing intact ubiquitin for re-use. The amino acids released can be reused to make new proteins. Defects in protein degradation can cause insoluble protein aggregates to accumulate in cells and have been associated with the pathogenesis of several cancers and neurodegenerative disorders such as Alzheimer's, Parkinson's and Huntington's disease (Section A3).

M1 Introduction to viruses

Key Notes

Viruses	Viruses are extremely small (20–300 nm) parasites, incapable of replication, transcription, or translation outside of a host cell. Viruses of bacteria are called bacteriophages. Virus particles (virions) essentially comprise a nucleic acid genome and protein coat or capsid. Some viruses have a lipoprotein outer envelope, and some also contain nonstructural proteins essential for transcription or replication soon after infection.
Virus genomes	Viruses can have genomes consisting of either RNA or DNA, which may be double-stranded or single-stranded, and, for single-stranded genomes, positive, negative, or ambi-sense (defined relative to the mRNA sequence). The genomes vary in size from around 1 kb to nearly 300 kb, and replicate using combinations of viral and cellular enzymes.
Replication strategies	Viral replication strategies depend largely on the type and size of genome. Small DNA viruses may make more use of cellular replication machinery than large DNA viruses, which often encode their own polymerases. RNA viruses, however, require virus-encoded RNA-dependent polymerases for their replication. Some RNA viruses use an RNA-dependent DNA polymerase (reverse transcriptase) to replicate via a DNA intermediate.
Virus virulence	Many viruses do not cause any disease, and often the mechanisms of viral virulence are accidental to the viral life cycle, although some may enhance transmission. Some viruses are associated with tumor formation.
Related topics	(M2) Bacteriophages (M4) RNA viruses (M3) DNA viruses

Viruses

It is difficult to give a precise definition of a virus. The word originally simply meant a toxin, and was used by Jenner in the 1790s when describing the agents of cowpox and smallpox. Virus particles (**virions**) are sub-microscopic and can replicate only inside a host cell. As they are not free-living and lack the property of **homeostasis** – the ability to regulate their internal environment to maintain stability and constancy – they are regarded as nonliving entities. Virtually all viruses rely entirely on the host cell for translation, while many also rely on the infected host for various transcription and replication factors. As well as being useful tools in practical molecular biology, e.g. as cloning vectors, they provide superb examples of how the basic processes of information flow and processing can be modified and manipulated in different ways for different purposes and so are discussed here in some detail. Viruses of bacteria are called **bacteriophages** or **phages**.

Viruses essentially consist of a nucleic acid genome of single- or double-stranded DNA, or RNA surrounded by a virus-encoded protein coat, the **capsid**. It is the capsid and its interaction with the genome that largely determine the structure of the virus (see Figure 1 for examples of the structure of different viruses). Viral capsids tend to be composed of protein subunits assembled into larger structures during the formation of mature particles, a process that may require interaction with the genome (the complex of genome and capsid is known as the **nucleocapsid**). Capsid proteins may also have nonstructural roles in virus transcription and replication proteins and are often important antigens of great interest to those designing vaccines.

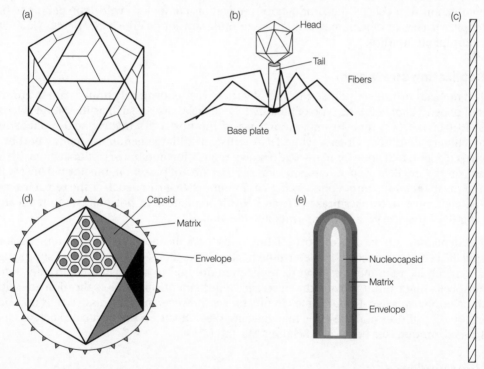

Figure 1. Some examples of virus morphology (not drawn to scale). (a) Icosahedral virion (e.g. poliovirus); (b) complex bacteriophage with icosahedral head, and tail (e.g. bacteriophage T4); (c) helical virion (e.g. bacteriophage M13); (d) enveloped icosahedral virion (e.g. herpesviruses); (e) a rhabdovirus's typical bullet-shaped, helical, enveloped virion.

Many types of virus also have an outer bilayer lipoprotein **envelope**. The envelope is derived from host cell membranes by **budding** and sometimes contains host cell proteins. Virus-encoded **envelope glycoproteins** are important for the assembly and structure of the virion and often bind to specific **receptors** on the appropriate host cell. Many RNA viruses and some DNA viruses also contain nonstructural proteins within the virion, necessary for immediate transcription or genome replication after infection of the cell.

Virus genomes

The genome of a virus is defined by its state in the mature virion and varies between virus families. Unlike the genomes of true organisms, the virus genome can consist of DNA or RNA, which may be double- or single-stranded. In some viruses, the genome consists of a single, linear, or circular molecule of nucleic acid but in others it is **segmented** or diploid. Single-stranded viral genomes are described as **positive sense** (i.e. the same nucleotide sequence as the mRNA), **negative sense**, or **ambi-sense** (in which genes are encoded in both senses, often overlapping; Section K1). These genomic mRNAs are therefore **poly-cistronic** as they encode several viral proteins in a single molecule (Section G1).

Not all virions have a complete or functional genome; indeed the ratio of the number of virus particles in a virus preparation (as counted by electron microscopy) to the number of infectious particles (determined in cell culture) is usually greater than 100 and often many thousands. Genomes unable to replicate by themselves may be **rescued** during coinfection of a cell by the products of replication-competent **wild-type** genomes of **helper viruses**, or of genomes with different mutations or deletions in a process known as **complementation**.

Replication strategies

The replication/transcription strategies of viruses vary enormously from group to group and depend largely on the type of genome. Small DNA viruses such as the **papovavirus** SV40 (Section M3) tend to make more use of the host cell's nucleic acid replication machinery than RNA viruses, while DNA viruses with large genomes such as **herpesvirus** (Section M3) code for many of their own replication and transcription factors. RNA viruses require RNA-dependent polymerases that are not present in the normal host cell and must, therefore, be encoded by the virus. Some RNA viruses such as the **retroviruses** encode a **reverse transcriptase** (**RT**, an RNA-dependent DNA polymerase) to replicate their RNA genome via a DNA intermediate (Section M4).

The dependence of viruses on host cell functions for replication and the requirements for specific cell-surface receptors determine host cell specificity. Cells capable of supplying the metabolic requirements of virus replication are said to be **permissive** to infection. Host cells that cannot provide the necessary requirements for virus replication are said to be **nonpermissive**. Under some circumstances, however, nonpermissive cells may be infected by viruses and the virus may have marked effects on the host cell, such as cell **transformation** (see below and Sections M3 and N2).

Virus virulence

Some viruses damage the cells in which they replicate, and if enough cells are damaged then the consequence is disease. It is important to realize that viruses do not exist in order to cause disease; this is simply a consequence of their replication. In many circumstances **virulence** (the capacity to cause disease) may be selectively disadvantageous (i.e. it may decrease the capacity for viral replication) but in others it may aid transmission. The evolution of virulence often results from a trade-off between damaging the host and maximizing transmission. Virulence can include cellular transformation and tumor formation (Section N2). For example, papillomaviruses can cause benign skin warts but also cervical cancer, some papovaviruses such as SV40 and polyoma can cause tumors in nonpermissive hosts (Section M3), Epstein–Barr virus can cause nasopharyngeal carcinoma, while many animal retroviruses are associated with sarcomas, leukemias, and other tumors (Section M4).

M2 Bacteriophages

Key Notes	
General properties	Bacteriophages infect bacteria. Although some phages have small genomes and a simple life cycle, others have large genomes and complex life cycles involving regulation of both viral and host cell metabolism.
Lytic and lysogenic infections	In lytic infection, virions are released from the cell by lysis. However in lysogenic infection, viruses integrate their genomes into that of the host cell, and may be stably inherited through several generations before returning to lytic infection.
Bacteriophage M13	Bacteriophage M13 has a small single-stranded DNA genome, replicates via a double-stranded DNA replicative form, and can infect cells without causing lysis. Its simplicity led to its use as a model system for studying DNA replication.
Bacteriophage λ (lambda)	Probably the best-studied lysogenic phage is bacteriophage λ. Temporally regulated expression of various groups of genes enables the virus to either undergo rapid lytic infections, or, if environmental conditions are adverse, undergo lysogeny as a prophage integrated into the host cell's genome. Expression of the lambda repressor, the product of the *cI* gene, is an important step in the establishment of a lysogenic infection.
Transposable phages	Some phages, for example bacteriophage Mu, routinely integrate into the host cell and replicate by replicative transposition.
Related topics	(D1) DNA replication: an overview (F4) Transcription initiation, elongation, and termination (M1) Introduction to viruses	(M3) DNA viruses (M4) RNA viruses (P2) Bacteriophages, cosmids, YACs, and BACs

General properties

Bacteriophages, or **phages**, are viruses that infect bacteria. Their genomes can be of RNA or DNA and range in size from around 2.5 to 150 kb. They can have simple lytic life cycles or more complex, tightly regulated life cycles involving **integration** in the host genome or even **transposition** (Section E4).

Bacteriophages have played an important role in the history of both virology and molecular biology. They have been studied intensively as model viruses and were important tools in the original identification of DNA as the genetic material, the determination of the genetic code, the existence of mRNA, and many more fundamental concepts of molecular biology. Since phages parasitize bacteria, they often have significant sequence similarity to their hosts, and have, therefore, also been used extensively as simple models for various aspects of bacterial molecular biology. Some phages are also used as cloning vectors (Section P2).

Lytic and lysogenic infections

Some phages replicate extremely quickly: infection, replication, assembly, and release by lysis of the host cell may all occur within 20 minutes. In such cases, replication of the phage genome occurs independently of the bacterial genome. Sometimes, however, replication and release of new virus can occur without lysis of the host cell (e.g. in bacteriophage M13 infection). Other phages alternate between a **lytic** phase of infection, with DNA replication in the cytosol, and a **lysogenic** phase in which the viral genome is integrated into that of its host (e.g. bacteriophage λ). Yet another group of phages replicate while integrated into the host cell genome via a combination of replication and **transposition** (e.g. bacteriophage μ, Mu).

Bacteriophage M13

Bacteriophage M13 has a small (6.4 kb) single-stranded, positive-sense, circular DNA genome (Figure 1). M13 particles attach specifically to *E. coli* through a minor coat protein (g3p) located at one end of the particle. Binding of the minor coat protein induces a structural change in the major capsid protein. This causes the whole particle to shorten, injecting the viral DNA into the host cell. Host enzymes then convert the viral single-stranded genome into a dsDNA **replicative form**. The genome has ten tightly packed genes and a small intergenic region which contains the origin of replication. Transcription occurs, again using host cell enzymes, from any of several promoters, and continues until it reaches one of two terminators (Figure 1). Multiple copies of the replicative form are produced by normal, double-stranded DNA replication, except that initiation of replicative form replication involves elongation of the 3′-OH group of a nick made in the (+) strand by a viral endonuclease (the product of gene 2), rather than RNA priming (Section D1). Finally, multiple single-stranded (+) strands for packaging into new phage particles are made by continuous replication of each replicative form, with the synthesis of the complementary (−) strand being blocked by coating the new (+) strands with the phage gene 5 protein. These packaging precursors are transported to the cell membrane

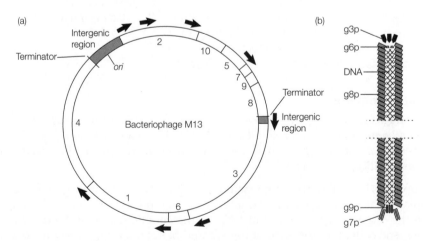

Figure 1. Overview of (a) the genome and (b) virion structure of M13. Arrows in (a) are promoters. In (b), gene 3 protein=g3p, etc.

and, there, the DNA binds to the major capsid protein. At the same time, new virions are extruded from the cell's surface without lysis. M13-infected cells continue to grow and divide (albeit at a reduced rate), giving rise to generations of cells each of which is also infected and continually releasing M13 phage. What is more, the amount of DNA found in any particle is highly variable (giving rise to the variable length of the particles): virions containing multiple genomes and virions containing only partial genomes are found in any population.

Because they rely almost entirely on host cell factors for their replication, M13 and related phages such as ΦX174 were used as early **model systems** for the study of bacterial DNA replication because they are much easier to handle *in vitro* than the large, fragile *E. coli* chormosome. M13 was also employed extensively for cloning small ssDNA fragments for sequencing, but this use has now been largely supplanted by newer technologies (Section R2).

Bacteriophage λ (lambda)

One of the best studied bacteriophages is bacteriophage λ, which has been much studied as a model for regulation of gene expression. Derivatives have been used as cloning vectors (Section P2). The λ phage virion consists of an icosahedral head containing the 48.5 kb linear dsDNA genome, and a long flexible tail. The phage binds to specific receptors on the outer membrane of *E. coli*, and the viral genome is injected through the phage tail into the cell. Although the viral genome is linear within the virion, its termini are single-stranded and complementary. These **cos sites** (Section P2) are commonly known as cohesive or 'sticky' ends and rapidly bind to each other once in the cell, producing a circular genome with staggered single-strand breaks, which are sealed by cellular DNA ligase to give a covalently closed circle. Within the infected cell, the λ phage may either undergo **lytic** or **lysogenic** life cycles. In the lysogenic life cycle, the bacteriophage genome becomes integrated as a linear copy, or **prophage**, in the host cell's genome.

There are three classes of λ genes, which are expressed at different times after infection. First, **immediate-early** and then **delayed-early** gene expression result in genome replication. Subsequently, **late** expression produces the structural proteins necessary for the assembly of new virus particles and lysis of the cell. Circularization of the genome is followed rapidly by the onset of immediate-early transcription. This is initiated at two promoters, *pL* and *pR*, and leads to the transcription of the immediate-early *N* and *Cro* genes. The two promoters are transcribed to the left (*pL*) and to the right (*pR*), using different strands of the DNA as templates for RNA synthesis (Figure 2). The terminators of both the *N* and *Cro* genes depend on transcription termination by **rho** (Section F4). The N protein acts as a transcription antiterminator, which enables the RNA polymerase to read through transcription termination signals of the *N* and *Cro* genes. As a result, mRNA transcripts are made from both *pL* and pR, which continue transcription, to the right through the replication genes *O* and *P* and into the *Q* **gene**, and to the left, through genes involved in recombination and enhancement of replication. This leads to replication of the genome, which at this stage involves bidirectional DNA replication by host enzymes (Section D1), but initiated at the *ori* (origin of replication) site by a complex of the proteins pO and pP, and host cell helicase (DnaB). Later, build up of the *gam* gene product leads to conversion to **rolling circle replication** and the production of concatamers (mutiple length copies) of linear genomes (Figure 3).

Transcription from *pR* results in the build-up of the **Cro protein**. Cro protein binds to sites overlapping *pL* and *pR*, and inhibits transcription from these promoters. As a result, early transcription is shut down. **Q protein**, like **N protein**, has an antitermination function

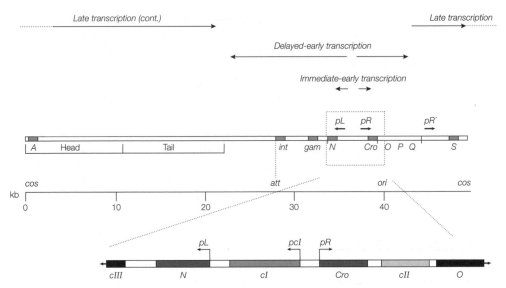

Figure 2. Simplified map of the bacteriophage λ genome (linearized).

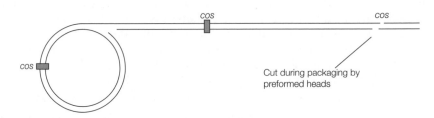

Figure 3. Rolling circle replication of bacteriophage λ.

(they can be compared with the HIV Tat protein, Section I2). The build-up of Q protein allows late transcription from *pR'* to occur (Figure 2). The late genes encode the structural proteins of the virion head and tail, a protein to cleave the *cos* ends to produce a linear genome, and a protein that allows host cell lysis and viral release.

The above lytic cycle can be completed in around 35 minutes with the release of about 100 particles. For the lytic cycle to proceed, it requires the expression of the late genes. **Lysogeny** depends on the synthesis of a protein called the **lambda repressor**, which is the product of the *cI* gene. The delayed-early genes include replication and recombination genes, but also encode three regulators. One of these regulators, Q protein (described above), is responsible for the expression of the late genes. Two further regulator genes, ***cII*** and ***cIII***, are expressed from *pR* and *pL* respectively. The *cII* gene product, which can be stabilized by the *cIII* gene product, binds to and activates the promoters of the ***int* gene**, which is responsible for lysogenic λ integration into the host cell genome by site-specific recombination at the ***att*** (*attP*) site (Section E4), and the ***cI* gene**, which encodes the λ repressor. The λ repressor represses both *pR* and *pL*, and thereby all early expression (including *Cro* and Q expression). Consequently, it represses both late gene expression and the lytic cycle, and this leads to lysogeny. The balance between the lytic and lysogenic pathways is determined by the concentration of the Cro protein (which inhibits early expression and *cI* expression) and Q protein (which activates the late genes) which favor lysis and, on the other hand, by the **cII** and **cIII proteins** and λ repressor protein, which establish the lysogenic pathway.

Lysogenic infection can be maintained for many generations, during which the prophage is replicated like any other part of the bacterial genome. Transcription during lysogeny is largely limited to the *cI* gene: transcription of *cI* is from its own promoter which is enhanced by, but does not require, the cII protein, since the *cI* product can also regulate its own transcription. Escape from lysogeny occurs particularly in situations when the infected cell is itself under threat or if damage to DNA occurs (e.g. through ionizing radiation). Such situations induce the host cell to express RecA (Section E4), whose associated protease activity cleaves the λ repressor, enabling progression into the lytic cycle.

Transposable phages

Transposable phages are found mainly in Gram-negative bacteria, particularly *Pseudomonas* species. One of the best-studied examples is bacteriophage Mu (μ). Transposable phages have lytic and lysogenic phases of infection similar to those of λ phage, except that the method of genome replication is different. In the 'early' lytic phase, a complex process involving viral transposase, bacterial DNA polymerases, and other viral and bacterial enzymes mediates both replication and transposition of the copy genome to elsewhere in the host cell's genome, without the original viral genome having to leave the host cell's genome (Section E4). Only after several rounds of **replicative transposition** are viral genomes, along with regions of adjacent cell DNA, excised from the cell's genome and encapsidated, causing degradation of the host cell's genome and lytic release of new phage particles.

M3 DNA viruses

Key Notes

DNA genomes: replication and transcription

DNA virus genomes can be double-stranded or single-stranded. Almost all eukaryotic DNA viruses replicate in the host cell's nucleus and make use of host cellular replication and transcription as well as translation. Large dsDNA viruses often have more complex life cycles, including temporal control of transcription, translation and replication of both the virus and the cell. Viruses with small DNA genomes may be much more dependent on the host cell for replication.

Small DNA viruses

One example of a small DNA virus family is the Papovaviridae. Papovaviruses, such as SV40 and polyoma, rely on overlapping genes and splicing to encode six genes in a small, 5-kb double-stranded genome. These viruses can transactivate cellular replicative processes that mediate not only viral but cellular replication; hence they can cause tumors in their hosts.

Large DNA viruses

Examples of large DNA viruses include the family Herpesviridae. Herpesviruses infect a range of vertebrates, causing a variety of important diseases.

Herpes simplex virus-1

Herpes simplex virus-1 (HSV-1) has over 70 ORFs and a genome of around 150 kb. After infection of a permissive cell, three classes of genes, the immediate-early (α), early (β) and late (λ) genes are expressed in a defined temporal sequence. These genes express a cascade of *trans*-activating factors that regulate viral transcription and activation. This virus has the ability to undergo latent infection.

Related topics

(Section H) Transcription in eukaryotes	(M4) RNA viruses
(M1) Introduction to viruses	(N2) Oncogenes
(M2) Bacteriophages	(P3) Eukaryotic vectors

DNA genomes: replication and transcription

The 'large' DNA viruses (e.g. herpesviruses or adenoviruses) have double-stranded genomes encoding up to 200 genes, and complex life cyles that can involve not only regulation of their own replication but sometimes that of the life cycle and functions of their host cells. However, the enormous mimivirus, which infects *Acanthamoeba*, has a 1.2-Mb genome and nearly 1000 genes – more than some prokaryotes – and begins to blur the distinction between nonliving viruses and living organisms. At the other extreme, papovaviruses have an extremely small, double-stranded circular genome encoding only a few genes, and rely on their host for most replication functions.

Most eukaryotic DNA viruses replicate in the host cell's nucleus, where even the largest and most complex viruses can make use of cellular DNA metabolic pathways. This means that there is often considerable similarity between the sequences of viral promoters and those of their host cell. For this reason, the strong promoters of DNA viruses are often used in mammalian expression vectors (Section P3).

Small DNA viruses

The papovaviruses include **simian virus 40 (SV40)**, a monkey virus. This virus is well studied because it is tumorigenic (Sections M1 and N2). SV40 has a 5-kb, double-stranded circular genome, which is supercoiled and packaged with cell-derived histones within a 45-nm, icosahedral virus particle. In order to pack five genes into so small a genome, the genes are found on both strands and overlap each other. The genes are separated into two overlapping transcription units, the early genes and the late genes (Figure 1). The different proteins are produced by a combination of overlapping reading frames and differential splicing. SV40 depends on host cell enzymes for transcription and replication, but the early genes produce transcription activators (known as **large T-antigen** and **small t-antigen**) that stimulate both viral and host cell transcription and replication. These are responsible for the tumorigenic properties of this virus. The late genes produce three proteins, VP1, VP2, and VP3, which are required for virion production. **Adenoviruses** have linear dsDNA genomes of about 25–45 kb.

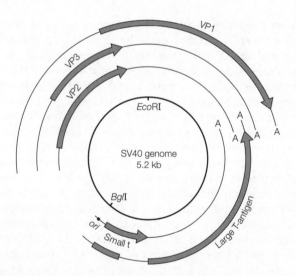

Figure 1. Structure of the SV40 virus genome. Outer lines indicate transcripts and coding regions; A=poly(A) tail.

Large DNA viruses

Herpesviruses provide a good example of the complex ways in which a 'large' DNA virus and its host cell can interact. Herpesviruses infect vertebrates. The virions are large, icosahedral, and enveloped (Section M1), and contain a double-stranded, linear DNA genome of up to 270 kb encoding around 100 ORFs. They are divided into three subfamilies, based on biological characteristics and genomic organization. There are well over 100 species, most of which are fairly host specific. Human examples include: herpes simplex virus-1 (HSV-1), the cause of cold sores; varicella zoster virus, the cause of chickenpox and shingles; and Epstein–Barr virus, a cause of infectious mononucleosis (glandular fever) and certain tumors. Other examples of large DNA viruses that have also found a use as cloning vectors are the insect **baculoviruses** (80–180 kb) and **poxviruses** such as vaccinia (190 kb), which is related to the cowpox and smallpox viruses.

Herpes simplex virus-1

HSV-1 is particularly well studied. Its genome is around 150 kb and contains over 70 ORFs. Genes can be found on both strands of DNA, sometimes overlapping each other. The genome (Figure 2) can be divided into two parts ('short' and 'long'), each consisting of a unique section (U_L and U_S) with inverted repeats at the internal ends of the regions (IR_L and IR_S) and at the termini (TR_L and TR_S). These inverted repeats consist of sequences b and b' and c and c'. In addition, a short sequence (a) is repeated a variable number of times (a_n and a_m).

Figure 2. Genome structure of herpes simplex virus-1

The transcription and replication of the herpesvirus genome are tightly controlled temporally. After infection of a permissive cell (Section M1), the genome circularizes, and a group of genes located largely within the terminal repeat regions, the **immediate-early** or **α genes**, are transcribed by cellular RNA polymerase II (Section H4). Transcription of α genes is, however, *trans*-activated by a virus-encoded protein (**α-*trans*-inducing factor**, or **α-TIF**). In common with around one-third of the gene products of HSV-1, α-TIF is not essential for replication. Part of the mature virion's matrix, it interacts with cellular transcription factors after entry into the cell, binds to specific sequences upstream of α promoters and enhances their expression. The α mRNAs, some of which are spliced, encode *trans*-activators of the **early** or **β genes**. The β genes encode most of the nonstructural proteins used for further transcription and genome replication, and a few st-ructural proteins. Early gene products include enzymes involved in nucleotide synthesis (e.g. thymidine kinase, ribonucleotide reductase), DNA polymerase, inhibitors of immediate-early gene expression and other products that can down-regulate various aspects of host cell metabolism. The promoters of β genes are similar to those of their hosts. They have an obvious TATA box 20–25 bases upstream of the mRNA transcription start site, and CCAAT box and transcription factor-binding sites (Section H4) – indeed herpesvirus β gene promoters function extremely well if incorporated into the host cell's genome, and have long been studied as model RNA polymerase II promoters.

Replication appears to be initiated at one of several possible origin (ORI) sites within the circular genome, and involves an ORI-binding protein, helicase–primase complexes, and a polymerase–DNA-binding protein complex, which are all virus encoded. DNA synthesis is semi-discontinuous (Section D1) and results in concatamers (mutiple length copies) with multiple replication complexes and forks.

Late or **γ genes** (some of which can only be expressed after DNA replication) largely encode structural proteins, or factors that are included in the virion for use immediately after infection (such as α-TIF). Virus assembly takes place in the nucleus: empty capsids apparently associate with a free genomic terminal repeat 'a' sequence and one genome equivalent is packaged and cleaved, again at an 'a' sequence. The envelope is derived from modified nuclear membrane, and contains several viral glycoproteins important for attachment and entry.

In addition to this tightly regulated replication cycle, herpesviruses can undergo **latent infection**, i.e. they can down-regulate their own transcription to such an extent that the circularized genome can persist extra-chromosomally in an infected cell's nucleus without replication. Latent infection can last the life of the cell, with only periodic reactivation (virus replication). HSV-1 undergoes latency mainly in neurons, but other herpesviruses can latently infect other tissues, including lymphoid cells. The precise mechanisms of herpesvirus latency and reactivation are still not fully understood, but may involve the transcription of specific RNAs called **latency-associated transcripts** encoded within the terminal repeat regions. One of these appears to exert an anti-apoptotic effect (Section N4) and prevents the immune system from destroying the infected cells.

Recently, the HSV-1 genome has been found to encode at least 16 miRNAs (Sections A2 and L4), some of which are antisense transcripts of viral genes and which may, therefore, regulate their expression. Whether others can target host cell mRNAs or regulatory ncRNAs to control productive or latent infection and whether host miRNAs can mediate an antiviral defense against HSV infection remain to be established.

M4 RNA viruses

Key Notes

RNA genomes: general features

Viral RNA genomes may be single- or double-stranded, positive or negative sense, and have a wide variety of mechanisms of replication. All, however, rely on virus-encoded RNA-dependent polymerases, the inaccuracy of which in terms of making complementary RNA is much higher than that of DNA-dependent polymerases. This significantly affects the evolution of RNA viruses by increasing their ability to adapt, but limits their size.

Retroviruses

Retroviruses have diploid, positive sense RNA genomes, and replicate via a dsDNA intermediate. This intermediate, called the provirus, is inserted into the host cell's genome. Retroviruses share many properties with eukaryotic retrotransposons such as the yeast Ty elements and their RTs are similar enough to suggest that they have evolved from a common ancestor

Oncogenic retroviruses

Insertion of the retrovirus into the host genome may cause either deregulation of host cell genes or, occasionally, may cause recombination with host cell genes (and the acquisition of those genes into the viral genome). This may give rise to cancer if the retrovirus alters the expression or activity of a critical cellular regulatory gene called an oncogene.

Retroviral genome structure and expression

Retroviruses have a basic structure of *gag*, *pol* and *env* genes flanked by 5'- and 3'-LTRs. The retroviral promoter is found in the U3 region of the 5' LTR and this promoter is responsible for all retroviral transcription. The viral transcripts are polyadenylated and may be spliced. In HIV, Tat regulates transcriptional elongation from the viral promoter and Rev regulates the transport of unspliced RNAs to the cell cytoplasm.

Retroviral mutation rates

The RTs of some retroviruses can have a high error rate of up to one mutation per 10 000 nt. Defective genomes may be rescued by complementation and recombination. This, combined with the rapid turnover of virus (10^9–10^{10} new virions per day in the case of HIV), enables it to adapt rapidly to selective pressure.

Related topics

(H4) RNA Pol II genes: promoters and enhancers
(I2) Examples of transcriptional regulation

(M1) Introduction to viruses
(M3) DNA viruses
(N2) Oncogenes
(Q2) cDNA libraries

RNA genomes: general features

Depending on the family, the genome of an RNA virus may be single- or double-stranded; if single-stranded it may be positive or negative sense (Section M1). As host cells do not contain RNA-dependent RNA polymerases, these must be encoded by the virus genome. The polymerase gene (*pol*) may encode other nonstructural functions and so is usually the largest gene found in the genome of an RNA virus. This makes RNA viruses true parasites of translation, and often totally independent of the host cell nucleus for replication and transcription. Thus, unlike eukaryotic DNA viruses, many RNA viruses replicate in the cytoplasm.

RNA-dependent polymerases are not as accurate as DNA-dependent polymerases and, as a rule, are not capable of proofreading (Section E2). Therefore, RNA viruses have mutation rates of around one per 10^3–10^4 bases per replication cycle, compared with rates as low as 10^{-8}–10^{-11} in large DNA viruses. This has three main consequences:

(i)   If the virus has a rapid replication cycle, significant changes in antigenicity and virulence can develop very rapidly through mutation, i.e. RNA viruses can evolve rapidly and quickly adapt to a changing environment, new hosts, etc..

(ii) Some RNA viruses mutate so rapidly that they exist as **quasispecies**, i.e. as populations of different genomes within any individual host; these often replicate through complementation and can only be molecularly defined in terms of a majority or average sequence.

(iii) As many of the mutations are deleterious to viral replication, the mutation rate puts an upper limit on the size of an RNA genome at around 10^4 nt, the inverse of the mutation rate (i.e. the size that, on average, would give one mutation per genome).

Some positive sense RNA viruses such as poliovirus have IRESs in their genomes that allow cap-independent translation of their genes, a probable advantage during infection (Section L3). These sites form recognizable secondary and tertiary structures.

Retroviruses

Retroviruses have a single-stranded RNA (ssRNA) genome. Two copies of the sense strand of the genome are present within the viral particle. When they infect a cell, the ssRNA is converted into a dsDNA copy (the **provirus**) by the viral-encoded enzyme **RT**. Curiously, the RT uses a host-encoded tRNA as a primer for DNA synthesis. The concept of **reverse transcription** and the use of RTs in the laboratory are covered in Sections A1 and Q2. Although the different viruses that make use of reverse transcription can have very different morphologies, genome structures, and life cycles, the amino acid sequences of their RTs and core proteins are sufficiently similar that they probably evolved from a common ancestor. Their genomes also have obvious structural similarities (and some sequence homology) to the retrotransposons found in many eukaryotes, such as the yeast Ty retrotransposons (Section E4).

Viral replication and transcription occur from the dsDNA provirus, which is integrated into the host cell genome by a viral **integrase** enzyme. Retroviruses vary in complexity (Figure 1). At one extreme there are those with relatively simple genomes that differ from retrotransposons essentially only by having an *env* gene, which encodes the envelope glycoproteins essential for infectivity. At the other extreme, lentiviruses, such as the **HIVs**, have larger genomes that also encode various *trans*-acting factors and *cis*-acting sequences, which are active at different stages of the replication cycle and are involved in regulation of both viral and cellular functions.

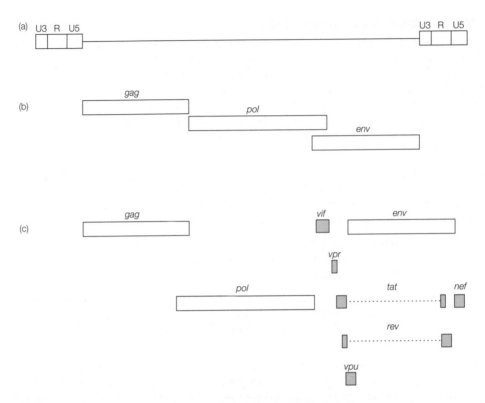

Figure 1. The structure of retrovirus proviral genomes. (a) The basic structure of the retroviral provirus, with LTRs consisting of U3, R, and U5 elements; (b) avian leukosis virus (ALV), a simple retrovirus; (c) HIV-1, a complex retrovirus, with several extra regulatory factors.

Oncogenic retroviruses

Retroviruses sometimes recombine with host genomic material and such viruses are usually replication deficient. The retrovirus may either disrupt the regulated expression of a host gene, or it may recombine with the host DNA to insert the gene into its own genome. If the cellular gene encodes a protein involved in regulation of cell division, it may act as an **oncogene** (Section N2) and cause cancer. Many oncogenic retroviruses expressing human oncogenes are known.

Retroviral genome structure and expression

Examples that represent the extreme ranges of retroviral genomes are shown in Figure 1. All retroviral genomes have a similar basic structure. The *gag* gene encodes the core proteins of the icosahedral capsid, the *pol* gene encodes the enzymatic functions involved in viral replication (i.e. RT, RNase H, integrase, and protease), and the *env* gene encodes the envelope proteins. At either end of the viral genome are unique elements (U5 and U3) and repeat elements that are involved in replication, host cell integration and viral gene expression.

Transcription of integrated provirus depends on the host's RNA polymerase II, directed by promoter sequences in the U3 region of the 5′ LTR. RNA transcripts are polyadenylated and may be spliced. Unspliced RNA is translated on cytoplasmic ribosomes to produce both a gag **polyprotein** and a gag-pol polyprotein (processed to pol proteins) through, for example, **translational frameshifting** (slippage of the ribosome on its template RNA by one reading frame to avoid a stop codon). The spliced mRNA is translated on membrane-bound ribosomes to produce the envelope glycoproteins. Some retroviruses, e.g. the lentiviruses, also produce other multiply or differently spliced RNAs that are translated to produce various factors such as Tat and Rev that regulate transcription and mRNA processing respectively. The Tat protein regulates elongation of transcripts originating from the 5′ LTR of the HIV genome (Section I2). The Rev protein enhances nuclear to cytoplasmic export of unspliced viral mRNAs that encode a full range of structural viral proteins.

Retroviral mutation rates

The RTs of some retroviruses (e.g. HIV-1) have a high error rate of around 10^{-4} (see above). Thus many genomes are replication defective, although this may be overcome in any virion through complementation by the second genome (two RNA genomes are packaged into each retroviral particle). Furthermore, recombination can occur between the two different genomes within any virion during reverse transcription. These two features, combined with the rapid turnover (10^9–10^{10} new virions per day) of HIV-1 enable it to adapt rapidly to new environments (e.g. under selective pressure from antibodies or drug treatment), a property that is increasingly recognized as central to our understanding of the pathogenesis of infections such as AIDS. The ability of retroviruses to integrate into the host cell's genome, and the relative ease with which their ability to replicate can be genetically modified in the laboratory, has made them prime candidates as vectors both for the creation of novel cell lines and in whole organisms for gene therapy (Sections P3 and S5).

N1 The cell cycle

Key Notes

The cell cycle

The cell cycle involves DNA replication followed by cell division to produce two daughter cells from one parent. It is an ordered process and is controlled by the cell cycle machinery.

Cell cycle phases

There are four main phases: G1, S, G2, and M phase. S phase is the DNA synthesis phase, whereas M phase is the cell division phase. These two phases are separated by two gap phases, G1 and G2. M phase can be divided into further phases: prophase, metaphase, anaphase, telophase, and cytokinesis. G1, S, and G2 comprise the interphase. Cells can enter a nonproliferative phase from G1, which is called G0 phase or quiescence.

Checkpoints and their regulation

The cell cycle is regulated in response to the cell's environment and to avoid the proliferation of damaged cells. Checkpoints are stages at which the cell cycle may be halted if the circumstances are not right for cell division. Principal checkpoints occur at the end of the G1 and G2 phases, but also during S and M phases. At the R point in G1 phase, cells starved of mitogens withdraw from the cell cycle into the G0 phase.

Cyclins and cyclin-dependent kinases

The cell cycle is controlled through protein phosphorylation catalyzed by multiple protein kinase complexes. These complexes consist of cyclins, the regulatory subunits, and CDKs, the catalytic subunits. Different cyclin-CDK complexes control different phases of the cell cycle. In turn, their activity is regulated through transcriptional control of their synthesis, alteration of their enzyme activity by inhibitor proteins, and by regulation of their proteolytic destruction.

Regulation by E2F and Rb

G1 progression is controlled by activation of a family of transcription factors called E2Fs, which regulate expression of genes required for later phases of the cell cycle. E2F-1 is inhibited in early G1 by the binding of hypophosphorylated Rb. Phosphorylation of Rb by G1 cyclin-CDK complexes frees E2F-1, which can then activate transcription.

Cell cycle activation, inhibition, and cancer

The CIP and INK4 classes of proteins halt cell cycle progression by inhibition of the activity of cyclin-CDK complexes. The G1 to S phase transition is regulated by proto-oncogenes and by tumor suppressor proteins. B cell tumors are associated with over-expression of the cyclin D1 gene. Rb, a critical regulator of G1 to S progression, and the

INK4 p16 protein are tumor suppressor proteins. The CIP protein p21/Waf1/Cip1 is activated by the tumor suppressor p53 in response to DNA damage.

Related topics	(C3) Eukaryotic chromosome structure	(I2) Examples of transcriptional regulation
	(D3) Eukaryotic DNA replication	(N2) Oncogenes
		(N3) Tumor suppressor genes
	(H4) RNA Pol II genes: promoters and enhancers	(N4) Apoptosis

The cell cycle

The two main stages of cell division are replication of cellular DNA and the division of the cell into two daughter cells. The repeated process of cell division, where daughter cells continue to divide to form further generations of cells, can be thought of as a repeated cycle of DNA replication followed by cell division. This is known as the **cell cycle**.

The cell cycle is an ordered and regulated process. It could be catastrophic if a cell attempted to divide without first replicating its DNA, or alternatively if it attempted to re-replicate all or part of its DNA before cell division had occurred. Thus, we can consider cell proliferation as a cyclic process, coordinated by the cell cycle machinery to ensure the correct ordering of the cellular processes.

Cell cycle phases

Since the processes of DNA replication and cell division occur at distinct and regulated time intervals, the cell cycle can be conceptually divided into four phases (Figure 1).

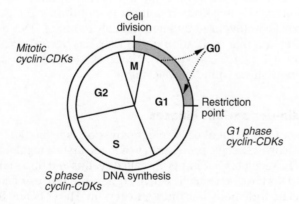

Figure 1. Schematic diagram of the phases of the cell cycle. The regulatory complexes which are important in each phase are shown in italics.

G1: the longest phase (called a **gap phase**) during which the cells are preparing for DNA replication;

S: or **DNA synthesis phase** during which the DNA is replicated and a complete copy of each of the chromosomes is made (Section D3);

G2: a short **gap phase** occuring after S phase and before mitosis;

M: the **mitotic** phase during which the new chromosomes are segregated equally between the two daughter cells (**mitosis**) prior to cell division (**cytokinesis**).

Mitosis itself can be subdivided into **prophase**, during which the chromosomes condense; **metaphase**, during which the sister chromatids attached at the centromere (Section C3) and become aligned in the centre of the cell; **anaphase**, where the sister chromatids separate and segregate towards opposite ends of the cell; and finally **telophase**, when the separated chromosomes decondense back to chromatin.

Together, G1, S and G2 comprise the **interphase**. After mitosis, proliferating cells will enter the G1 phase of the next cell cycle. Cells can also exit the cell cycle after mitosis and enter a nonproliferative resting state, **G0**, which is also called **quiescence**.

Checkpoints and their regulation

The initiation of a cell division cycle requires the presence of extracellular growth factors, or **mitogens**. In the absence of mitogens, cells withdraw from the cell cycle in G1 and enter the G0 resting phase (Figure 1). The point in G1 at which information regarding the environment of the cell is assessed, and the cell decides whether to enter another division cycle, is called the **restriction point** (or **R point**). Cells starved of mitogens before reaching the R point enter G0 and fail to undergo cell division. Cells that are starved of mitogens after passing through the R point continue through the cell cycle to complete cell division before entering G0. In most cell types, the R point occurs a few hours after mitosis. The R point is clearly of crucial importance in understanding the commitment of cells to undergoing a cell division cycle. The interval in G1 between mitosis and the R point is the period in which multiple signals coincide and interact to determine the fate of the cell.

Those parts of the cell cycle, such as the R point, where the process may be stopped are known as **checkpoints**. The major checkpoints operate during the gap phases. These ensure that the cell is competent to undergo another round of DNA replication (at the R point in G1 phase) and that replication of the DNA has been successfully completed before cell division (G2 phase checkpoint). However, there is also an **intra-S-phase** checkpoint and a **metaphase** (**spindle**) checkpoint.

Cyclins and cyclin-dependent kinases

A major mechanism for control of cell cycle progression is regulation by protein phosphorylation. This involves specific protein kinases made up of a regulatory subunit and a catalytic subunit. The regulatory subunits are called **cyclins** and the catalytic subunits are called **CDKs**. The CDKs have no catalytic activity unless they are associated with a cyclin, and each can associate with more than one type of cyclin. The CDK and the cyclin present in a specific CDK-cyclin complex jointly determine which target proteins are phosphorylated by the protein kinase.

There are three different classes of cyclin-CDK complexes, which are associated with the G1, S or M phases of the cell cycle respectively. The **G1 CDK complexes** prepare the cell for S phase by activating proteins including transcription factors that cause expression of enzymes required for DNA synthesis and the genes encoding S phase CDK complexes.

The **S phase CDK complexes** stimulate the onset of organized DNA synthesis (Section D3). The machinery ensures that each chromosome is replicated only once. The **mitotic CDK complexes** induce chromosome condensation and ordered chromosome separation into the two daughter cells.

The activity of the CDK complexes is regulated in three ways:

(i)   By control of the transcription of the CDK complex subunits.

(ii) By inhibitors that reduce the activity of the CDK complexes. For example, the mitotic CDK complexes are synthesized in S and G2 phase, but their activity is repressed until DNA synthesis is complete.

(iii) By organized proteolysis of the CDK complexes at a defined stage in the cell cycle where they are no longer required.

Regulation by E2F and Rb

Cell cycle progression through G1 and into S phase is in part regulated by activation (and in some cases inhibition) of gene transcription, whereas progression through the later cell cycle phases appears to be regulated primarily by post-transcriptional mechanisms. Passage through the key G1 R point critically depends on the activation of a transcription factor, **E2F**. E2F stimulates the transcription and expression of genes encoding proteins required for DNA replication and deoxyribonucleotide synthesis as well as for cyclins and CDKs required in later cell cycle phases. The activity of E2F is inhibited by binding of the protein **Rb** (the **retinoblastoma tumor suppressor protein**, Section N3) and related proteins. When Rb is hypophosphorylated (under-phosphorylated), E2F activity is inhibited. The phosphorylation of Rb by cyclin-CDK complexes during middle and late G1 phase frees E2F so that it can activate transcription (Figure 2).

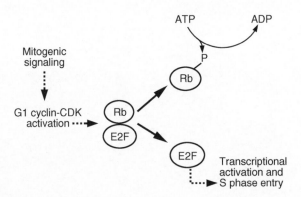

Figure 2. Schematic diagram showing the regulation of E2F by Rb and G1 cyclin-CDK complexes.

Cell cycle activation, inhibition, and cancer

Small inhibitor proteins can delay cell cycle progression by repression of the activity of cyclin-CDK complexes. There are two classes of these inhibitors, CIP proteins and INK4 proteins. For example, during skeletal muscle cell differentiation, the differentiating cells are induced to withdraw from the cell cycle via activation of one of the CIP protein genes, encoding the p21/Waf1/Cip1 protein. This is controlled by the master regulator transcription factor MyoD (Section I2).

There is a fundamental link between regulation of the cell cycle and cancer. This is supported by the observation that proto-oncogenes (Section N2) and tumor suppressor genes (Section N3) regulate the passage of cells from G1 to S phase. One of the G1 cyclins, cyclin D1, is often over-expressed in tumors of immunoglobulin-producing B cells. This is caused by a chromosomal translocation that brings the immunoglobulin gene enhancer (Section H4) to a chromosomal position adjacent to the cyclin D1 gene. The INK4 p16 protein has been shown to have all the characteristics of a tumor suppressor. Importantly, the products of the two most significant tumor suppressor genes in human cancer, Rb and p53 (Section N3), are both critically involved in cell cycle regulation. Rb is involved in direct regulation of E2F activity (see above), and p53 is involved in the inhibition of the cell cycle in response to DNA damage. This occurs through p53-induced transcriptional activation of the *p21/Waf1/Cip1* gene in response to DNA damage.

N2 Oncogenes

Key Notes

Oncogenes

Oncogenes are genes whose activity causes cells to become cancerous. They are derived from normal cellular proto-oncogenes but their products behave in an uncontrolled fashion compared to those of the proto-oncogene, over which they are genetically dominant.

Oncogenic retroviruses

Oncogenic retroviruses were the source of the first oncogenes to be isolated. Retroviruses become oncogenic either by expressing mutated versions of cellular growth-regulatory genes or by stimulating the overexpression of normal cellular genes.

Discovery of cellular oncogenes

The isolation of oncogenes was aided by the development of an assay that tests the ability of DNA to transform the growth pattern of NIH-3T3 mouse fibroblasts. This assay has many practical advantages and allowed the isolation of many oncogenes, but also has limitations to its use.

Relationship between oncogenes and proto-oncogenes

An oncogene may differ from the proto-oncogene either by possessing a mutation that alters the activity of its gene product or by being expressed at an abnormally high level, often as a result of a change in genomic location.

Oncogenes and growth factors

Many oncogenes code for proteins involved in the pathways by which cells respond to growth factors. Such oncogenes include *c-ErbB*, *int-2*, *c-Fms*, and *c-Ras* and cause the cancer cell to behave as though it were being continuously stimulated to divide by a growth factor.

Nuclear oncogenes

Other oncogenes code for nuclear transcription factors that regulate the expression of genes controlling cell division. These include *c-Myc*, *c-Fos*, *c-Jun*, and *c-ErbA*.

Cooperation between oncogenes

Usually, both a growth factor-related and a nuclear oncogene are required in order to convert normal cells to full malignancy. This reflects the fact that carcinogenesis *in vivo* is a multistep process requiring several changes to occur in a cell.

Related topics

(I1) Eukaryotic transcription factors
(M4) RNA viruses

(N1) The cell cycle
(N3) Tumor suppressor genes

Oncogenes

Oncogenes are genes whose abnormal activity causes cells to become cancerous. They are derived from normal cellular genes called **proto-oncogenes** either by **mutation** or **dysregulation** but their gene products behave in an uncontrolled and usually overactive

fashion compared to those of the proto-oncogene, over which they are genetically dominant, i.e. only one copy of an oncogene is sufficient to cause a change in the cell's behavior.

Oncogenic retroviruses

The basic concepts of oncogenes were discovered from studies on oncogenic viruses,
particularly **oncogenic retroviruses.** Retroviruses are RNA viruses that replicate via a
DNA intermediate (the **provirus**) that inserts itself into the cellular DNA and is transcribed into new viral RNA by cellular RNA polymerases (Section M4). Oncogenic viruses
are an important cause of cancer in animals, although only a few rare forms of human
cancer have been linked to them. Many oncogenic retroviruses were found to contain an
extra gene that was not present in closely related but nononcogenic viruses. This extra
gene was shown to be an oncogene by transfecting it into noncancerous cells, which then
became **tumorigenic** (able to form tumors when injected into animals). Different oncogenic viruses contain different oncogenes.

Discovery of cellular oncogenes

The isolation of oncogenes depends upon the use of a specific cellular assay(s). One such
assay is **DNA transfection** (Section S5) of **NIH-3T3 cells**. These mouse fibroblast cells
(connective tissue cells that grow particularly well *in vitro*) do not normally give rise to
tumors when injected into immune-deficient mice. The growth pattern of NIH-3T3 cells
in culture is that of normal, noncancerous cells. If NIH-3T3 cells are transfected with DNA
from a cancer cell that contains a suspected oncogene, a few cells will take up and express
the foreign DNA. If this DNA contains an oncogene, the growth pattern of the cells in
culture may be changed to have the generally disorganized characteristics of cancerous
cells (Figure 1). This straightforward assay has been extremely successful in discovering
oncogenes, although not all can be identified in this way.

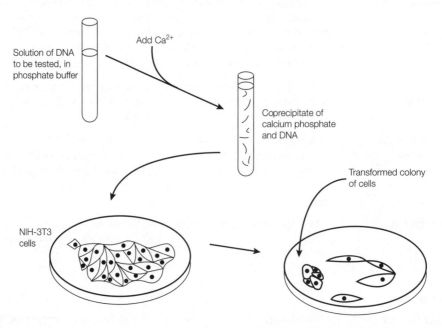

Figure 1. Testing for the presence of an oncogene in DNA by revealing its ability to cause a
change in the growth pattern of NIH-3T3 cells.

Relationship between oncogenes and proto-oncogenes

The first oncogenes to be isolated were those present in oncogenic retroviruses. When these were cloned, an important discovery was made – genes with DNA sequences homologous to retroviral oncogenes were present in the DNA of normal cells. It was then realized that retroviral oncogenes must have originated from normal cellular proto-oncogenes that became incorporated into the viral genome when the provirus integrated itself close to the proto-oncogene in the cellular genome. Subsequently, similar (sometimes the same) oncogenes were isolated from nonvirally caused cancers. In oncogene terminology, the name of a cellular oncogene is prefixed with '*c-*' whereas the related viral oncogene is prefixed by '*v-*'. In all cases, the cancer-causing oncogene differs from the normal proto-oncogene in one of two main ways.

- A change in oncogene sequence: the coding sequence is altered, for example by deletion or point mutation, so that the protein product is functionally different (Section E2). Normally the oncogene protein is hyperactive.

- A change in oncogene expression: the coding sequence of the gene may be unaltered; however, the gene may be under the control of a viral promoter/enhancer (Section H4) or may have become translocated to a new site in the genome, where it is transcribed at a higher rate than normal.

These processes are best exemplified by oncogenes that encode proteins involved in the cellular response to **growth factors**.

Oncogenes and growth factors

Growth factors are proteins secreted by many cell types that bind to specific **cell-surface receptors** on nearby tissue cells (of the same or a different type from the secretory cell). They stimulate a response that frequently includes an increase in cell proliferation. Oncogenes whose actions have been shown to depend upon their growth factor-related activity include:

- The ***c-ErbB*** oncogene, which codes for a **truncated** version of the receptor protein that binds **epidermal growth factor**. Because the missing region is responsible for binding the growth factor, the oncogene version is always active, permanently sending signals to the nucleus (Figure 2). This inappropriate signal causes the cell to proliferate in an uncontrolled manner.

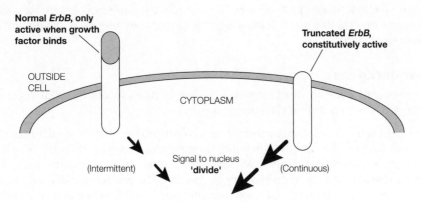

Figure 2. The oncogene version of c-*ErbB* codes for a constitutively active growth factor receptor.

- The mouse **int-2** (*fgf-3*) gene, which codes for a **fibroblast growth factor**. Breast cancer in mice can be caused by **mouse mammary tumor** (**retro**) **virus** (**MMTV**). If the MMTV provirus inserts itself close to the mouse *int-2* gene, the viral enhancer sequences (Section H4) overstimulate the transcription of *int-2* and the excess of growth factor causes the cells to divide continuously (Figure 3).

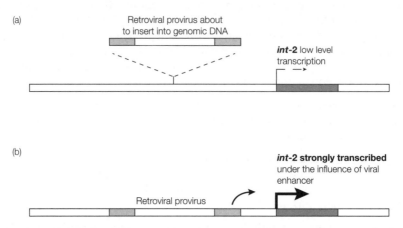

Figure 3. The mechanism by which MMTV causes cancer in mouse mammary cells. (a) The mouse *int-2* gene before integration of the MMTV provirus; (b) integration of the provirus results in overexpression of *int-2*.

- The **c-Fms** oncogene, which codes for a mutated version of the receptor for **colony-stimulating factor-1** (**CSF-1**), a growth factor that stimulates bone marrow cells during blood cell formation. The 40 amino acids at the C-terminus of the normal CSF-1 receptor are replaced by 11 unrelated amino acids in the Fms protein due to a frameshift mutation. As a result, the Fms protein is constitutively active no matter whether CSF-1 is present or absent.

- The various **c-Ras** oncogenes, which encode members of the **G-protein** family of plasma membrane signaling proteins. These transmit signals from many cell surface receptors to enzymes that synthesize small molecules called **second messengers** that initiate further signaling pathways. When activated, normal G-proteins bind GTP but are then inactivated by their own GTPase activity. *c-Ras* oncogenes possess point mutations that inhibit their GTPase activity so that they remain activated for longer than normal (Figure 4).

Nuclear oncogenes

Other oncogenes code for nuclear transcription factors (Section I1) that regulate the expression of genes controlling cell division.

- Expression of the **c-Myc** gene in normal cells is induced by a variety of **mitogens** (agents that stimulate cells to divide), including platelet-derived growth factor (PDGF). The c-Myc protein binds to specific DNA sequences and controls the expression of many genes, some of which are required for cell division. Overexpression of *c-Myc* in cancer cells can occur by: (i) increased transcription under the influence of a viral enhancer; (ii) translocation of the coding sequence from its normal site on chromosome 8 to a site on chromosome 14, which places it under the control of the active promoter

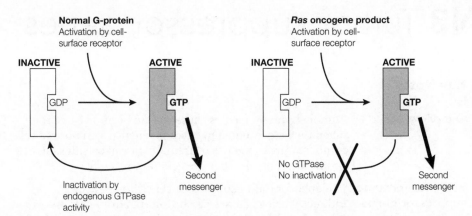

Figure 4. The *c-Ras* oncogene codes for a signal transmission protein that is mutated so that it has lost the ability to inactivate itself.

for the immunoglobulin heavy chain (in **Burkitt's lymphoma**); or (iii) deletion of the 5'-noncoding sequence of the mRNA, which increases the mRNA stability.

- The *c-Fos* and *c-Jun* oncogenes code for subunits of a normal transcription factor, **AP-1**. In normal cells, expression of *c-Fos* and *c-Jun* occurs only transiently, immediately after mitogenic stimulation. The normal cellular concentrations of the *c-Fos* and *c-Jun* gene products are regulated not only by the rate of gene transcription, but also by the stability of their mRNA. In cancer cells, both processes may be increased. The *c-ErbA* oncogene codes for a truncated version of the **nuclear receptor** for **thyroid hormone**. Thyroid hormone receptors act as transcription factors regulating the expression of specific genes when they are activated by binding the hormone (Section I2). The c-ErbA protein lacks the C-terminal region of the normal receptor so that it cannot bind the hormone and cannot stimulate gene transcription. However, it can still bind to the same sites on the DNA and appears to act as an antagonist of the normal thyroid hormone receptor.

Cooperation between oncogenes

Transformation of a normal cell into a fully malignant cancer cell is a multi-step process involving alterations in the expression of several genes. Transfection of any one of the oncogenes described above will, in most cases, cause oncogenic transformation of the NIH-3T3 cell-line. However, simultaneous introduction of both the *c-Ras* and the *c-Myc* oncogenes into cells is required for full transformation of normal fibroblasts. A variety of pairs of oncogenes are able to achieve this effect. Interestingly, the pair must usually include one growth factor-related oncogene and one nuclear oncogene in order to fully convert a completely normal cell into a malignant cancer cell. Alternatively, a third type of oncogene – one whose normal function is to suppress **apoptosis** (Section N4) – may be required in addition to one other.

N3 Tumor suppressor genes

Key Notes

Overview	Tumor suppressor genes cause cells to become cancerous when they are mutated to become inactive. In a normal cell, tumor suppressor genes restrain the rate of cell division or promote apoptosis. They are generally recessive.
Evidence for tumor suppressor genes	Evidence for tumor suppressor genes is varied and indirect. Their existence can be inferred from patterns of inheritance of certain cancers and loss of heterozygosity for chromosomal markers in tumor cells.
RB1 gene	The *RB1* gene was the first tumor suppressor gene to be isolated. Loss or mutation of the *RB1* gene causes retinoblastoma, a childhood tumor of the eye, but has also been detected in breast, colon, and lung cancers.
p53 gene	*p53* is the tumor suppressor gene that is mutated in the largest number of different types of tumor. When first identified, it appeared to have characteristics of both an oncogene and a tumor suppressor gene. It is now known to be a tumor suppressor gene that can act in a dominant-negative manner to interfere with the function of the remaining, normal allele.
Related topics	(N1) The cell cycle (N4) Apoptosis
	(N2) Oncogenes

Overview

Tumor suppressor genes act in a fundamentally different way from oncogenes. Whereas proto-oncogenes are converted to oncogenes by processes that increase or alter their normal activity, tumor suppressor genes become oncogenic as the result of mutations that eliminate their activity. For example, the normal version of a tumor suppressor gene may act to prevent a cell from entering mitosis and cell division; removal of this negative control allows a cell to divide. Other tumor suppressor genes may promote the apoptosis of damaged cells (Section N4); their loss allows damaged cells to survive, accumulate more mutations, and become tumorigenic. An important consequence of this mechanism of action is that both copies (alleles) of a tumor suppressor gene have to be inactivated to remove all restraint, i.e. tumor suppressor genes act in a genetically recessive fashion.

Evidence for tumor suppressor genes

Because there are no assays for tumor suppressor genes equivalent to the NIH-3T3 assay for oncogenes, their basic importance for the development of cancer was not appreciated for a long time. However, examination of the inheritance of certain **familial cancers** suggested that they result from recessive mutations, while in many cancer cells there is a consistent loss of characteristic regions of certain chromosomes. This '**loss of heterozygosity**' is believed to indicate the loss of the remaining active allele of a tumor suppressor gene in a cell that has already lost the other allele, either by previous mutation or by inheritance (Figure 1). Thus the cell is already heterozygous for the active tumor suppressor gene.

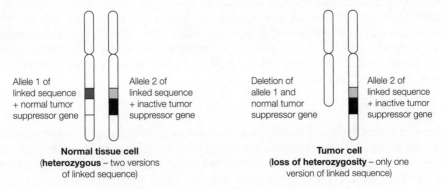

Figure 1. Loss of heterozygosity is the process whereby a cell loses a portion of a chromosome that contains the only active allele of a tumor suppressor gene.

Retinoblastoma is a childhood tumor of the eye, and is the classic example of a cancer caused by loss of a tumor suppressor gene. Retinoblastoma takes two forms: familial (40% of cases), which exhibits the inheritance pattern for a recessive gene and which frequently involves both eyes, and sporadic, which does not run in families and usually only occurs in one eye. Retinoblastoma results from two mutations that inactivate both alleles of the gene. In the familial form of the disease, one mutated allele is inherited in the germ line. On its own this is harmless, but the occurrence of a mutation in the remaining normal allele in a retinoblast cell causes a tumor. Since there are 10^7 retinoblasts per eye, all at risk, the chances of a tumor are relatively high. In the sporadic, noninherited form of the disease, both inactivating mutations have to occur in the same cell during the lifetime of the individual, so the likelihood is very much less and only one eye is usually affected (Figure 2). It should be noted that, whilst familial retinoblastoma constitutes the minority of cases, it is responsible for the majority of tumors. The 'two-hit' hypothesis for retinoblastoma is also supported by evidence for loss of heterozygosity.

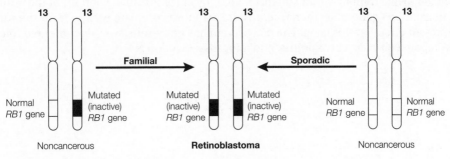

Figure 2. Retinoblastoma results from the inactivation of both copies of the *RB1* gene on chromosome 13. This can occur by mutation of both normal copies of the gene (sporadic retinoblastoma) or by inheritance of one inactive copy followed by an acquired mutation in the remaining functional copy (familial retinoblastoma).

RB1 gene

The **retinoblastoma gene** (*RB1*) is located on human chromosome 13. DNA sequencing in patients with the familial disease (i.e. who are heterozygous for *RB1*) has shown a deletion in the region of the remaining *RB1* gene in retinoblastoma tumor cells, but not in nontumor cells. *RB1* codes for the 110-kDa cell-cycle regulator **Rb** (Section N1), which binds to DNA and inhibits the transcription of proto-oncogenes such as *Myc* and *Fos*. *RB1* mRNA was found to be absent or abnormal in retinoblastoma cells. The role of *RB1* in retinoblastoma was established definitively when it was shown that retinoblastoma cells growing in culture reverted to a nontumorigenic state when they were transfected with a cloned, normal *RB1* gene. Unexpectedly, *RB1* mutations have also been detected in breast, colon, and lung tumors.

p53 gene

Similar techniques have subsequently been used to identify/isolate tumor suppressor genes associated with other cancers, but the gene that really put tumor suppressors on the map is one called *p53*. The gene for p53 is located on the short arm of chromosome 17, and deletions of this region have been associated with nearly 50% of human cancers. The mRNA for p53 is 2.2–2.5 kb and codes for a 52-kDa nuclear protein. The protein is found at a low level in most cell types and has a very short half-life (6–20 min). Confusingly, although usually classified as a tumor suppressor, *p53* has some of the properties of both an oncogene and a tumor suppressor gene.

- Many mutations (point mutations, deletions, insertions) have been shown to occur in the *p53* gene, and all cause it to become oncogenic. Mutant forms of *p53*, when cotransfected with the *c-Ras* oncogene, will transform normal rat fibroblasts. In cancer cells, the p53 protein has an extended half-life (4–8 h), resulting in elevated levels. All this seems to suggest that *p53* is an oncogene.

- A consistent deletion of the short arm of chromosome 17 has been seen in many tumors. In brain, breast, lung and colon tumors, where a *p53* gene was deleted, the remaining allele was found to be mutated. This suggests that *p53* is a tumor suppressor gene!

The explanation seems to be that p53 acts as a dimer. When a mutant (inactive) p53 protein is present it dimerizes with the wild-type protein to create an inactive complex (Figure 3). This is known as a **dominant-negative** effect. However, inactivation of the normal *p53* gene by the mutant gene would not be expected to be 100%, since some normal–normal dimers would still form. Loss (by chromosomal deletion) of the remaining normal *p53* gene may, therefore, result in a more complete escape from the tumor suppressor effects of this gene. The p53 protein has numerous, well-documented roles in DNA repair, cell cycle checkpoints, and apoptosis (Section N4).

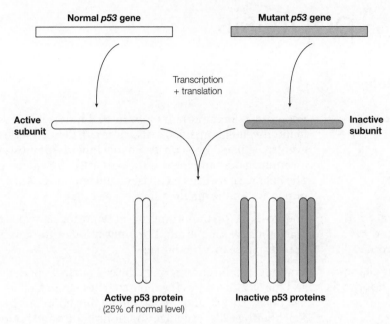

Figure 3. The dominant-negative effect of a mutated *p53* gene results from the ability of the protein to dimerize with and inactivate the normal protein.

N4 Apoptosis

Key Notes

Apoptosis	Apoptosis is an important pathway that results in cell death in multi-cellular organisms. It occurs as a defined series of events, regulated by a conserved machinery. Apoptosis is essential in development for the removal of unwanted cells. The balance between cell division and apoptosis is very important for the maintenance of cell number.
Removal of damaged or dangerous cells	Apoptosis has an important role in removing damaged or dangerous cells, for example in prevention of autoimmunity or in response to DNA damage.
Cellular changes during apoptosis	In apoptosis, the chromatin in the cell nucleus condenses and the DNA becomes fragmented. The cells detach from neighbors, shrink, and then fragment into apoptotic bodies. Neighboring cells recognize apoptotic bodies and remove them by phagocytosis.
Apoptosis in *Caenorhabditis elegans*	The nematode worm *C. elegans* has a fixed number of 959 cells in the adult. These result from 1090 cells being formed, of which precisely 131 die through apoptosis during development. The *ced-3* and *ced-4* genes are required for cell death, which can be suppressed by the product of the *ced-9* gene, the absence of which results in excessive apoptosis.
Apoptosis in mammals	The mammalian homolog of *C. elegans ced-9* is *Bcl-2*, which acts to suppress apoptosis. Other homologs of *Bcl-2* may either suppress or enhance apoptosis to achieve a balance between cell survival and death. The mammalian homologs of the *ced-3* gene encode proteases called caspases, which are important for the execution of apoptosis.
Apoptosis in disease and cancer	Defects in apoptosis are important in disease and cancer. Some proto-oncogenes such as *Bcl-2* prevent apoptosis, reflecting the role of apoptosis-suppression in tumor formation. The *c-Myc* proto-oncogene has a dual role in promoting cell proliferation, as well as triggering apoptosis when appropriate growth signals are not present. Cancer chemotherapy treatment uses DNA damaging drugs that act by triggering apoptosis.
Related topics	(E1) DNA damage (N1) The cell cycle (E2) Mutagenesis (N2) Oncogenes (E3) DNA repair (N3) Tumor suppressor genes

Apoptosis

Apoptosis is the mechanism by which cells normally die. It involves a defined set of programmed biochemical and morphological changes. It is a frequent and widespread process in multi-cellular organisms with an important role in the formation, maintenance and molding of normal tissues. It occurs in almost all tissues during development and also in many adult tissues. For example, apoptosis is responsible for the gaps between the human fingers, which would otherwise be webbed. It is also responsible for the loss of the tadpole's tail during amphibian metamorphosis. It seems that in many cell types, apoptosis is a built-in self-destruct pathway, often called **programmed cell death**, which is automatically triggered when growth signals that normally give the signal for the cell to survive are absent. Thus, homeostasis of tissue mass in an adult organism is maintained not only by the proliferation, differentiation, and migration of cells, but also, in large part, by the controlled loss of cells through apoptosis.

Although apoptosis is the major pathway of physiological cell death it is not the only way in which cells can die. In **necrosis**, the cell membrane loses its integrity and the cell lyses releasing its contents. The release of cell contents in an intact organism is generally undesirable as they may be toxic to extracellular structures and cause inflammation and tissue damage.

Removal of damaged or dangerous cells

In the thymus, over 90% of cells of the immune system undergo apoptosis. This process is very important for the removal of self-reactive T lymphocytes, which would otherwise cause auto-immunity by turning the immune system against the organism's own cells. When T cells kill other cells, they do so by activating the apoptotic pathway and so induce the cells to die. Many virus-infected cells undergo apoptosis as a mechanism for limiting the spread of the virus within the organism. The tumor suppressor protein p53 (Section N3) can induce apoptosis in response to excessive DNA damage, which the cell cannot repair (Section E3). The p53 protein therefore has a dual role in response to DNA damage, first in inhibiting the cell cycle (Section N1), and second in induction of apoptosis.

Cellular changes during apoptosis

During apoptosis, the nucleus shrinks and the chromatin condenses. When this happens, the DNA is often fragmented by nuclease-catalyzed cleavage between nucleosomes (Section C2). This can be demonstrated by gel electrophoresis of the DNA (Figure 1a; Section O3). The cell detaches from neighbors, rounds up, shrinks, and fragments into **apoptotic bodies** (Figure 1b), which often contain intact organelles and an intact plasma membrane. Neighboring cells rapidly engulf and destroy the apoptotic bodies, and it is thought that changes on the cell surface of the apoptotic bodies may have an important role in directing this phagocytosis. The morphological changes characteristic of apoptosis (Figure 1c) may occur within half an hour of initiation of the process.

Apoptosis in *Caenorhabditis elegans*

Every adult hermaphrodite of the nematode *C. elegans* is identical and comprises exactly 959 cells. The cell lineage of *C. elegans* is closely regulated. During the worm's development, 1090 cells are formed, of which 131 are removed by apoptosis. The products of two genes, *ced-3* and *ced-4*, are required for the death of these cells during development. If either gene is inactivated by mutation, none of the 131 cells die. The protein encoded by a further gene, *ced-9*, acts in an opposite way to suppress apoptosis. Inactivation of

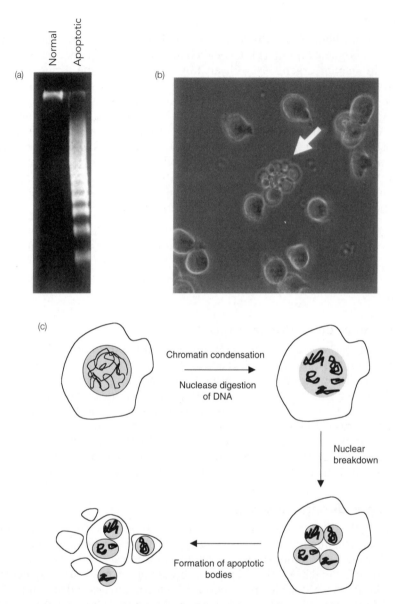

Figure 1. Apoptosis. (a) A characteristic DNA ladder from cells undergoing apoptosis; (b) a microscopic image of T-cells showing apoptotic bodies (apoptotic cell marked with an arrow, original picture from Dr D. Spiller); (c) a schematic diagram of the stages of apoptosis.

ced-9 by mutation results in excessive cell death, even in cells that do not normally die during development. Hence, *ced-9* is also required for the survival of cells that would not normally die and thus suppresses a general cellular program of cell death.

Apoptosis in mammals

The mammalian proto-oncogene **Bcl-2** is homologous to the apoptosis-suppressing nematode *ced-9* gene. *Bcl-2* was the first of a novel functional class of proto-oncogene

(Section N2) to be discovered whose members act to suppress apoptosis, rather than to promote cell proliferation. It is now known that a set of cell death-suppressing genes exists in mammals, several of which are homologous to *Bcl-2*. Another group of genes that include *Bcl-2* homologs such as *Bax* (a tumor suppressor, Section N3) promote cell death. The Bcl-2 and Bax proteins bind to each other in the cell. It therefore seems that the regulation of apoptosis by cellular signaling pathways may occur in part through a change in the relative levels of cell death-suppressor and cell death-promoter proteins in the cell. Mammalian homologs of the nematode *ced-3* killer gene have also been identified. *Ced-3* encodes a polypeptide homologous to a family of cysteine proteases of which the interleukin-1β converting enzyme (ICE) is the archetype. These **ICE proteases** are also called **caspases** and are responsible for the execution of apoptosis. Thus, it appears that apoptosis is a fundamentally important and evolutionarily conserved process.

Apoptosis in disease and cancer

Defects in the control of apoptosis appear to be involved in a wide range of diseases including neurodegeneration, immunodeficiency, cell death following a heart attack or stroke, and viral or bacterial infection. Most importantly, loss of apoptosis has a very important role in cancer. Cancer is a disease of multicellular organisms that is due to a loss of control of the balance between cell proliferation and cell death. Many proto-oncogenes (Section N2) regulate cell division, but others are known to regulate apoptosis (e.g. *Bcl-2*), reflecting the importance of the balance between these processes. The proto-oncogene *c-Myc* has a dual role since it stimulates cell division but can also act as a trigger of apoptosis. *c-Myc* triggers apoptosis when growth factors are absent, or the cell has been subjected to DNA damage. Mutations in genes that result in the absence or relative down-regulation of the apoptosis pathway may therefore result in cancer. On the other hand, over-expression of genes such as *Bcl-2*, which normally inhibits the apoptotic pathway, may also result in cancer.

Many cancer treatments involve the use of DNA-damaging drugs that kill dividing cells (for principles see Section E1). Rather than acting nonspecifically, most of these drugs act by triggering apoptosis in the cancer cells. One of the main ways by which cancer cells develop resistance to these drugs is suppression of the apoptotic pathway. This occurs in 50% of human cancers by mutation of the tumor suppressor p53 (Section N3). As a result, p53 has been called '**the guardian of the genome**'. In response to DNA damage, p53 triggers apoptosis as well as activating DNA repair pathways (Section E3) and inhibiting progression through the cell cycle (Section N1). When the damage is too great to be repaired, induction of apoptosis is crucial, not only for removing potential cancer cells, but also for inhibiting proliferation of mutated cells in the cell population that might otherwise accumulate further mutations that could give rise to a more aggressive tumor.

O1 DNA cloning: an overview

Key Notes

DNA cloning

DNA cloning facilitates the isolation and manipulation of fragments of an organism's genome by replicating them independently as part of an autonomous vector.

Hosts and vectors

Most of the routine manipulations involved in gene cloning use *Escherichia coli* as the host organism. Plasmids and bacteriophages may be used as cloning vectors in *E. coli*. Vectors based on plasmids, viruses, and whole chromosomes have been used to carry foreign genes into other prokaryotic and eukaryotic organisms.

Subcloning

Subcloning is the simple transfer of a cloned fragment of DNA from one vector to another; it serves to illustrate many of the routine techniques involved in gene cloning.

DNA libraries

DNA libraries, consisting of sets of random cloned fragments of either genomic or cDNA, each in a separate vector molecule, can be used in the isolation of unknown genes.

Screening libraries

Libraries may be screened for the presence of a gene sequence by hybridization with a sequence derived from its protein product or a related gene, or through the screening of the protein products of the cloned fragments. Increasingly, high throughput sequencing will allow the sequencing of an entire library.

Analysis of a clone

Once identified, a cloned gene may be analyzed by restriction mapping, or more straightforwardly by DNA sequencing, before being used in any of the diverse applications of DNA cloning.

Related topics

(Section P) Cloning vectors
(Section Q) Gene libraries and screening

(Section R) Analysis and uses of cloned DNA

DNA cloning

Classically, detailed molecular analysis of proteins or other constituents of most organisms was rendered difficult or impossible by their scarcity and the consequent difficulty of their purification in large quantities. The solution, developed around 40 years ago, was to isolate the gene(s) responsible for the expression of a particular RNA and/or protein. However, every organism's genome is large and complex (Section C), and any sequence of interest usually occurs only once or twice per cell. Hence, standard chemical or biochemical methods cannot be used to isolate a specific region of the genome for study, particularly as the required sequence of DNA is chemically identical to all the others. The solution to this dilemma was to place a relatively short fragment of a genome containing the gene or other sequence of interest in an autonomously replicating piece of DNA known as a **vector**, forming **recombinant DNA**, which could be replicated independently of the original genome, and normally in another host species altogether. Propagation of the host organism containing the recombinant DNA forms a set of genetically identical organisms, or a **clone**. Hence, this process is known as **DNA cloning**.

The technologies developed from this original idea have influenced almost every aspect of biology and beyond, from drug development to ecology. Among the exploding numbers of applications derived from DNA cloning, often collected together under the terms **genetic engineering** and **genomics**, are the following:

- DNA sequencing of partial or whole genomes, and hence the identification of genes, and the derivation of protein sequences (Section R2)

- Isolation and analysis of gene promoters and other control sequences (Section R4)

- Investigation of protein/enzyme/RNA function by large-scale production of normal and altered forms (Section A5 and R5)

- Identification of polymorphisms and mutations in humans and other organisms, for example gene defects leading to disease (Section C4)

- Biotechnology – the large-scale commercial production of proteins and other molecules of biological and medical importance, for example human insulin and growth hormone (Section A5)

- Engineering animals and plants, and gene therapy (Section S5)

- Engineering proteins to alter their properties (Section R5)

What follows in Sections O–Q is a discussion of the classical methods of gene cloning. In some cases, as indicated, these have to some extent been superceded by newer methods based on the **polymerase chain reaction** (Section R3) or **high throughput sequencing** (Section R2). Nevertheless, an appreciation of the basic methods is important to understanding the later developments.

Hosts and vectors

The initial isolation and analysis of DNA fragments is almost always carried out using the bacterium *E. coli* as the **host organism**, although the yeast *Saccharomyces cerevisiae* can be used to manipulate very large fragments of the human or other large genomes (Section P3). A wide variety of natural replicons have the properties required to allow them to act as **cloning vectors**. Vectors must normally be capable of being replicated and isolated independently of the host's genome, although some are designed to incorporate DNA into the host genome for longer-term expression of cloned genes. Vectors also incorporate a **selectable marker**, a gene that allows host cells containing the vector to be **selected** from amongst those that do not, usually by conferring resistance to a toxin (Section O2), or enabling their survival under certain growth conditions (Section P2).

The first *E. coli* vectors were extrachromosomal circular **plasmids** (Sections C1 and O2), and a number of **bacteriophages** (Section M2) have also been used in *E. coli*. Phage λ can be used to clone larger fragments than plasmid vectors, while **phage M13** allows cloned DNA to be isolated in single-stranded form if necessary (Section P2). More specialist vectors have been engineered to use aspects of plasmids and bacteriophages, such as the plasmid–bacteriophage λ hybrids known as **cosmids** (Section P2). Very large genomic fragments from humans and other species have been cloned in *E. coli* as **bacterial**

artificial chromosomes (**BACs**) and in *S. cerevisiae* as **yeast artificial chromosomes** (**YACs**; Section P2).

Plasmid and phage vectors have been used to express genes in a range of bacteria other than *E. coli*, and some phages may be used to incorporate DNA into the host genome, for example phage λ (Section P2). Plasmid vectors have been developed for use in yeast (**yeast episomal plasmids, YEps**), while in some plants, a bacterial plasmid (*Agrobacterium tumefaciens* **Ti plasmid**) can be used to integrate DNA into the genome (Section P3). In other eukaryotic cells in culture, vectors have often been based on viruses that naturally infect the required species, either by maintaining their DNA extrachromosomally or by integration into the host genome (examples include **SV40, baculovirus, retroviruses**; Section P3).

Subcloning

The simplest kind of cloning experiment, which exemplifies many of the basic techniques of classical DNA cloning, is the transfer of a fragment of cloned DNA from one vector to another, a process known as **subcloning**. This might be used to investigate a short region of a large cloned fragment in more detail, or to transfer a gene to a vector designed to express it in a particular species. In the case of plasmid vectors in *E. coli*, the most common situation, the process may be divided into the following steps, which are considered in greater detail in Sections O2–O4:

- **Isolation** of plasmid DNA containing the cloned sequence of interest (Section O2)

- **Digestion** (cutting) of the plasmid into discrete fragments with **restriction endonucleases** (Section O3)

- **Separation** of the fragments by **agarose gel electrophoresis** (Section O3)

- **Purification** of the desired **target** fragment (Section O3)

- **Ligation** (joining) of the fragment into a new plasmid vector, to form a new recombinant molecule (Section O4)

- Transfer of the ligated plasmid into an *E. coli* strain (**transformation**) (Section O4)

- **Selection** of transformed bacteria (Section O4)

- **Analysis** of recombinant plasmids (Section O4)

DNA libraries

There are two main sources from which DNA may be derived for cloning experiments designed to identify an unknown gene – bulk genomic DNA from the species of interest and, in the case of eukaryotes, bulk mRNA from a cell or tissue where the gene is known to be expressed. They are used in the formation of **genomic libraries** and **cDNA libraries** respectively (Sections Q1 and Q2).

DNA libraries are sets of DNA clones, each of which has been derived from the insertion of a different fragment into a vector followed by propagation in the host. Genomic libraries are prepared from random fragments of genomic DNA. However, genomic libraries can be an inefficient method of isolating a gene, particularly from large eukaryotic genomes, where much of the DNA is noncoding (Section C4). The alternative is to use as the source for the library the mRNA from a cell or tissue that is known to express the gene. Complementary DNA copies (cDNA) are synthesized from the mRNA by **reverse transcription** and are then inserted into a vector to form a cDNA library. cDNA libraries are efficient for cloning or identifying a gene sequence, but yield only the transcribed region and not the surrounding genomic sequences, which may be important for regulation.

Screening libraries

As it is not apparent which clone in a library contains the gene of interest, a method for screening for its presence is required. This is often based on the use of a radioactively or fluorescently labeled DNA **probe** (Section Q3) complementary or partially complementary to a region of the gene sequence, and which can be used to detect it by **hybridization** (Section B2). The probe sequence might be an oligonucleotide derived from the sequence of the protein product of the gene, if it is available, or from a related (**homologous**) gene from another species. An increasingly important method for the generation of probes is the **PCR** (Section R3). Other screening methods rely on the expression of the coding sequences of the clones in the library and identification of the protein product from its activity, or with a specific antibody (Section Q3). Increasingly, however, the most effective method might be to **sequence** the entire cDNA library and identify genes by comparison with homologous genes from other organisms (Section R2).

Analysis of a clone

Once a clone containing a target gene is identified, the structure of the cloned fragment may be investigated further using **restriction mapping**, the analysis of the fragmentation of the DNA with restriction enzymes (Section R1), or ultimately by sequencing the entire fragment (Section R2). The sequence can then be analyzed by comparison with other known sequences from databases, and the complete sequence of the protein product determined (Sections R2 and S6). The sequence is then available for manipulation in any of the applications of cloning described above.

Many enzymes are used *in vitro* in DNA cloning and analysis. The properties of the common enzymes are given in Table 1, along with the Sections where their use is discussed in more detail.

Table 1. Enzymes used in DNA cloning

Enzyme	Use
Alkaline phosphatase	Removes phosphate from 5'-ends of double- or single-stranded DNA or RNA (Sections O4, Q2, and R1)
DNA ligase (from phage T4)	Joins sugar-phosphate backbones of dsDNA with a 5'-phosphate and a 3'-OH in an ATP-dependent reaction. Requires that the ends of the DNA be compatible, i.e. blunt with blunt, or complementary cohesive ends (Sections O4, Q1, and Q2)
DNA polymerase I	Synthesizes DNA complementary to a DNA template in a 5' to 3' direction, beginning with a primer with a free 3'-OH. The Klenow fragment is a truncated version of DNA polymerase I which lacks the 5' to 3' exonuclease activity (Section R1)
S1 nuclease	Digests single-stranded nucleic acids, but will leave intact any region that is double helical. Will also cleave a strand opposite a nick on the complementary strand (Section R4)
Polynucleotide kinase	Adds phosphate to 5'-OH end of double- or single-stranded DNA or RNA in an ATP-dependent reaction. If [γ-^{32}P]ATP is used, then the DNA will become radioactively labeled (Section R1)
Restriction enzymes	Cut both strands of dsDNA within a (normally symmetrical) recognition sequence. Hydrolyze sugar-phosphate backbone to give a 5'-phosphate on one side and a 3'-OH on the other. Yield blunt or 'sticky' ends (5'- or 3'-overhang) (Sections O3 and R1)
Reverse transcriptase	RNA-dependent DNA polymerase. Synthesizes DNA complementary to an RNA template in a 5' to 3' direction, beginning with a primer with a free 3'-OH. Requires dNTPs (Section Q2, R4, and S2)
RNase A	Nuclease that digests RNA, but not DNA (Section O2 and R2)
T7, T3, and SP6 RNA polymerases	Specific RNA polymerases encoded by the respective bacteriophages. Each enzyme recognizes only the promoters from its own phage DNA, and can be used specifically to transcribe DNA downstream of such a promoter (Section R1)
Taq DNA polymerase	DNA polymerase derived from a thermostable bacterium (*Thermus aquaticus*). Operates at 72°C and is reasonably stable above 90°C. Used in PCR (Section R3)
Terminal transferase	Adds a number of nucleotides to the 3'-end of linear single- or double-stranded DNA or RNA. If only GTP is used, for example, then only Gs will be added (Sections Q2, R1, and R3)

O2 Preparation of plasmid DNA

Key Notes

Plasmids as vectors	Bacterial plasmids, small circular DNA molecules that replicate independently of the host genome and encode antibiotic resistance, are the commonest vectors for carrying cloned DNA.
Plasmid minipreparation	A plasmid may be obtained on a small scale for analysis by isolation from a few milliliters of culture, a process known as a minipreparation or miniprep.
Alkaline lysis	An alkaline solution of SDS lyses *E. coli* cells and denatures protein and DNA. Neutralization precipitates the chromosomal DNA and most of the protein, leaving plasmid DNA and RNA in solution.
Plasmid purification	Plasmid DNA is purified after alkaline lysis by binding it to a resin and washing away impurities in a process that can be automated.
Ethanol precipitation	Nucleic acid may be precipitated from solution by the addition of sodium acetate and ethanol followed by centrifugation. The method is used to concentrate the sample.
Cesium chloride gradient	A CsCl gradient can be used as part of a large-scale plasmid preparation to purify supercoiled plasmid DNA away from protein, RNA and linear or nicked DNA.
Related topics	(O1) DNA cloning: an overview (Section Q) Gene libraries and screening (O4) Ligation, transfomation, and analysis of recombinants

Plasmids as vectors

The first cloning vectors to be used in the mid 1970s were naturally occurring bacterial **plasmids**, originally from *E. coli* (Section C1). Plasmids contain an **origin of replication** (Section D1) that enables them to replicate independently of the major chromosome, although they normally rely on the polymerases and other components of the host cell's machinery. They usually carry a few genes, one of which may confer **resistance** to anti-bacterial substances. The most widely used resistance gene in plasmid vectors is the *bla*, or *amp*^r gene encoding the enzyme β-lactamase, which degrades penicillin antibiotics such as **ampicillin**. Another is the *tetA* gene, which encodes a transmembrane pump able to remove the antibiotic tetracycline from the cell.

Plasmid minipreparation

The first step in a subcloning procedure, for example the transfer of a gene from one plasmid to another (Section O1), is the isolation of plasmid DNA. Since plasmids are so much smaller than *E. coli* chromosomal DNA (Section C1), they can be separated from the latter by physico-chemical methods, such as alkaline lysis (see below). It is normally possible to isolate sufficient plasmid DNA for initial manipulation from a few milliliters of bacterial culture. Such an isolation is normally known as a **minipreparation** or **miniprep**. A sample of an *E. coli* strain harboring the required plasmid is inoculated into a few milliliters of culture broth. After growth to stationary phase (overnight), the suspension is centrifuged to yield a cell pellet.

Alkaline lysis

The most commonly used method for the purification of plasmid DNA away from chromosomal DNA and most of the other cell constituents is **alkaline lysis** (Figure 1). The

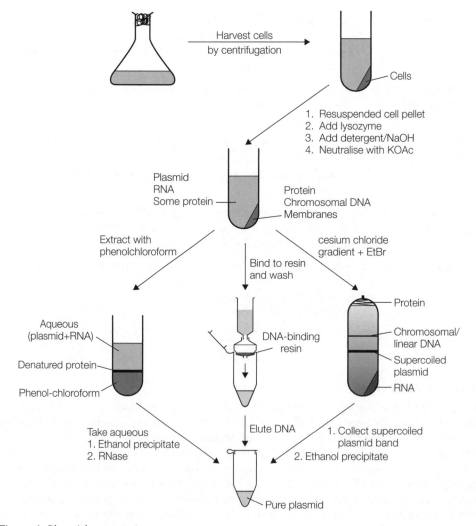

Figure 1. Plasmid preparation

cell pellet is resuspended in a buffer solution that may optionally contain lysozyme to digest the cell wall of the bacteria. The cell **lysis** solution, which contains the detergent **SDS** in an alkaline sodium hydroxide solution, is then added. The SDS disrupts the cell membrane and denatures the proteins; the alkaline conditions denature the DNA and begin the hydrolysis of RNA (Section B1). The preparation is then **neutralized** with a concentrated solution of potassium acetate (KOAc) at pH 5. This has the effect of precipitating the denatured proteins along with the chromosomal DNA and most of the detergent (potassium dodecyl sulfate is insoluble in water). The sample is centrifuged again, and the resulting supernatant (the **lysate**) now contains plasmid DNA, which, being small and closed-circular, is easily renatured after the alkali treatment along with a lot of small RNA molecules and some protein.

Plasmid purification

There are many proprietary methods for the isolation of pure plasmid DNA from the lysate above, most of which involve the selective binding of DNA to a resin or membrane and the washing away of protein and RNA (Figure 1). These steps can be carried out in a small column containing the resin, and the simplicity of this process lends itself to automation. In fact, robotic machines for the preparation of plasmid DNA, including the growth and alkaline lysis steps, are a common feature in molecular biology labs. The classical method, which is slower and involves somewhat hazardous chemicals, albeit in small amounts, involves the extraction of the lysate with **phenol** or a **phenol–chloroform** mixture. The phenol is immiscible with the aqueous layer but, when mixed vigorously with it and then allowed to separate, denatures the remaining proteins, which form a precipitate at the interface between the layers (Figure 1). **Ribonuclease A (RNase A)** may also be added to the solution to digest any remaining RNA contamination. This enzyme digests RNA but leaves DNA untouched and does not require Mg^{2+}.

Ethanol precipitation

The DNA from a miniprep may be concentrated by **ethanol precipitation**. This is a general procedure and can be used with any nucleic acid solution. If sodium acetate is added to the solution until the Na^+ concentration is more than 0.3 M, the DNA and/or RNA may be precipitated by the addition of 2–3 volumes of ethanol. Centrifugation will pellet the nucleic acid, which may then be resuspended in a smaller volume, or in a buffer with new constituents, etc. In the case of purified DNA, the pellet is taken up in tris(hydroxymethyl)aminomethane (Tris)–ethylenediamine tetra-acetic acid (EDTA) solution, the normal solution for the storage of DNA. This solution contains Tris–hydrochloride to buffer the solution (usually pH 8) and a low concentration of EDTA, which chelates any Mg^{2+} ions in the solution, protecting the DNA against degradation by nucleases, most of which require magnesium.

Cesium chloride gradient

The alkaline lysis method may also be used on a larger scale to prepare up to milligram quantities of plasmid DNA. The production of plasmids on a large scale may be required to make stocks of common cloning vectors, or if they are to be used on a large scale as substrates for enzymic reactions. Although larger-scale versions of the DNA-binding resin methods are available, **CsCl density gradient centrifugation** (Section B1) may be used as a final purification step. This is somewhat laborious, but is still the best method for the production of very pure supercoiled plasmid DNA. If a crude lysate is fractionated on a CsCl gradient (Section B1) in the presence of ethidium bromide (EtBr) (Section B3), the various constituents can all be separated (Figure 1). Supercoiled DNA binds less ethidium bromide than linear or nicked DNA (Section B3) and so has a higher density (it is less unwound). Hence, supercoiled plasmid may be isolated in large quantities away from protein, RNA, and contaminating chromosomal DNA in one step. The solution containing the plasmid band is removed from the centrifuge tube and may then be concentrated using ethanol precipitation as described above after removal of ethidium bromide.

O3 Restriction enzymes and electrophoresis

Restriction endonucleases

To incorporate fragments of foreign DNA into a plasmid vector, methods for the cutting and rejoining of dsDNA are required. The identification and manipulation of **restriction endonucleases** in the 1960s and early 1970s was the key discovery that allowed the cloning of DNA to become a reality. Restriction–modification systems occur in many bacterial species, and constitute a defense mechanism against the introduction of foreign DNA into the cell. They consist of two components; the first is a restriction endonuclease, which recognizes a short, symmetrical DNA sequence (Figure 1) and cuts (**hydrolyzes**) the DNA backbone in each strand at a specific site within that sequence. Hence, foreign DNA will be degraded to relatively short fragments. The second component of the system is a **methylase**, which adds a methyl group to a C or A base within the same recognition

(a)

(b)

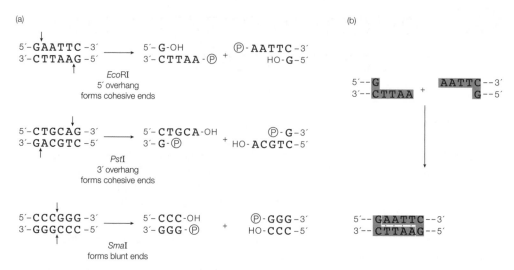

Figure 1. (a) The action of restriction endonucleases at their recognition sequences; (b) the annealing of cohesive ends.

sequences in the cellular DNA (Section C1). This modification renders the host DNA resistant to degradation by the endonuclease.

Recognition sequences

The action of restriction endonucleases (**restriction enzymes** for short) is illustrated in Figure 1a, including the archetypal enzyme EcoRI as an example. This enzyme, which acts as a dimer, will only recognize a 6-bp **palindromic** sequence (i.e. the sequence is the same, reading $5' \to 3'$ on each strand). The products of the cutting reaction at this site on a linear DNA are two double-stranded fragments (**restriction fragments**), each with an identical protruding single-stranded 5'-end with a phosphate group attached. The 3'-ends have free hydroxyl groups. A 6-bp recognition sequence will occur on average every 4^6=4096 bp in random sequence DNA, hence a very large DNA molecule will be cut into specific fragments averaging 4 kb by such an enzyme. Nearly 10 000 restriction enzymes are now known, and many hundreds are commercially available. They recognize sites ranging in size from 4 to 8 bp or more, mostly, but not always, symmetrical, and may give products with protruding 5'- or 3'-tails or blunt ends. Newly formed 5'-ends always retain the phosphate groups. Two further examples are illustrated in Figure 1. The extremely high specificity of restriction enzymes for their sites of action allows large DNA molecules and vectors to be cut reproducibly into defined fragments.

Cohesive ends

Products of restriction enzyme digestion with protruding ends have a further property; these ends are known as **cohesive**, or '**sticky**' ends, since they can bind to any other end with the same overhanging sequence by base-pairing (annealing) of the single-stranded tails. So, for example, any fragment formed by an EcoRI cut can anneal to any other fragment formed in the same way (Figure 1b), and may subsequently be joined covalently by ligation (Section O4). In fact, in some cases, DNA ends formed by enzymes with different recognition sequences may be compatible, provided the single-stranded tails can base-pair together.

Restriction digests

Digestion of plasmid or genomic DNA is carried out with restriction enzymes for analytical or preparative purposes, using commercial enzymes and buffer solutions. All restriction enzymes require Mg^{2+}, usually at a concentration of up to 10 mM, but different enzymes require different pHs, NaCl concentrations or other solution constituents for optimum activity. The buffer solution required for a particular enzyme is supplied with it as a concentrate. The digestion of a sample plasmid with two different restriction enzymes, BamHI and EcoRI, is illustrated in Figure 2. The digestion of a few hundred nanograms (<1 μg) of plasmid DNA is sufficient for analysis by agarose gel electrophoresis; preparative purposes may require a few micrograms. The former amount corresponds to a few percent of a miniprep sample, as described in Section G2. The DNA is incubated with the enzyme and the appropriate buffer at the optimum temperature (usually 37°C) in a volume of perhaps 20 μl. A dye mixture is then added to the solution, and the sample is loaded on to an agarose gel.

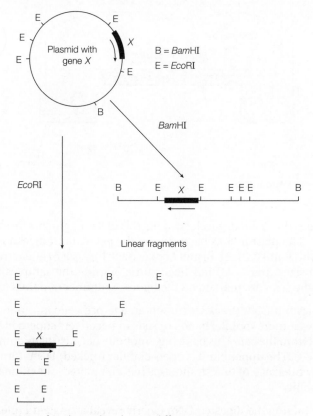

Figure 2. The digestion of a plasmid with two different restriction enzymes.

Agarose gel electrophoresis

Agarose is a long-chain polysaccharide derived from seaweed, which forms a solid gel when dissolved in aqueous solution at concentrations between 0.5 and 2% (w/v). Agarose used for electrophoresis is a more purified form of the agar used to make bacterial culture plates. When an electric field is applied to an agarose gel in the presence of a buffer solution that will conduct electricity, DNA fragments move through the gel towards the positive electrode (DNA is highly negatively charged; Section B1) at a rate that is dependent on their size and shape (Figure 3). Small linear fragments move more quickly than large ones, which are retarded by entanglement with the network of agarose chains forming the gel. Hence, this process of **electrophoresis** may be used to separate mixtures of DNA fragments on the basis of size. Different concentrations of gel [1%, 1.5% (w/v), etc.] will allow the optimal resolution of fragments in different size ranges. The DNA samples are placed in **wells** in the gel surface (Figure 3), the power supply is switched on, and the DNA is allowed to migrate through the gel in separate **lanes** or **tracks**. The added dye also migrates and is used to follow the progress of electrophoresis. The DNA is stained by the inclusion of **ethidium bromide** (Section B3) or another fluorescent dye in the gel, or by soaking the gel in a solution of dye after electrophoresis. The DNA shows up as a fluorescent band on illumination by UV light.

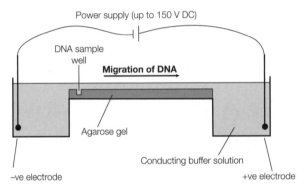

Figure 3. Agarose gel apparatus.

Figure 4a illustrates the result of gel electrophoresis of the fragments formed by the digestions in Figure 2. The plasmid has been run on the gel without digestion (track U), and after digestion with BamHI (track B) and EcoRI (track E). A set of linear marker DNA fragments of known sizes (tracks M) has been run alongside the samples at two different concentrations; the sizes are marked on the figure. A number of points may be noted.

- Undigested plasmid DNA (track U) run on an agarose gel commonly consists of two bands. The lower, more mobile, band consists of negatively supercoiled plasmid DNA isolated intact from the cell. This has a high mobility because of its compact conformation (Section B3). The upper band is open-circular (nicked) DNA, formed from supercoiled DNA by breakage of one strand; this has an opened-out circular conformation and lower mobility.

- The lanes containing the digested DNA clearly reveal a single fragment (track B) and five fragments (track E) whose sizes can be estimated by comparison with the marker tracks (M). The intensities of the bands in track E are proportional to the sizes of the fragments, since a small fragment has less mass of DNA at a given molar concentration. This is also true of the markers, since in this case they are formed by digestion of the 48.5-kb linear DNA of bacteriophage λ (Section P2). More sophisticated sets of markers with more even sizes and intensities are commercially available. The amount of DNA present in tracks U, B, and E is not equal; the quantities have been optimized to show all the fragments clearly.

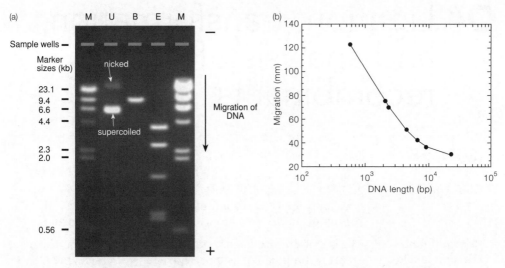

Figure 4. (a) An agarose gel of DNA restriction fragments (see text for details); (b) a calibration curve of migration distance against fragment size.

- A more accurate determination of the sizes of the linear fragments can be made by plotting a calibration curve of the log of the size of the known fragments in track M against the distance migrated by each fragment. This plot (Figure 4b) is a fairly straight line, often with a deviation at large fragment sizes. This may be used to derive the size of an unknown linear fragment on the same gel from its mobility, by reading off the log(size) as shown. It is not possible to derive the sizes of undigested circular plasmids by the same method, since the relative mobility of circular and linear DNA on a gel depends on the conditions (temperature, electric field, etc.).

Isolation of fragments

Agarose gels may also be used preparatively to isolate specific fragments for use in subsequent ligation and other cloning experiments. Fragments are excised from the gel, and treated by one of a number of procedures to purify the DNA away from the contaminating agarose and ethidium bromide stain. If we assume that the EcoRI fragment containing the gene *X* (Figure 2) is the target DNA for a subcloning experiment (Section O1), then the third largest fragment in track E of Figure 4a could be purified from the gel ready for ligation into a new vector (Section O4).

O4 Ligation, transformation, and analysis of recombinants

Key Notes

DNA ligation	T4 DNA ligase repairs breaks in a dsDNA backbone and can covalently rejoin annealed cohesive ends in the reverse of a restriction enzyme reaction, to create new DNA molecules.
Recombinant DNA molecules	The use of a restriction enzyme, followed by DNA ligase, can create recombinant plasmids, with a target DNA fragment inserted into a vector plasmid.
Alkaline phosphatase	Treatment of the linear vector molecule with alkaline phosphatase will remove the 5′-phosphates and render the vector unable to ligate into a circle without an inserted target, so reducing the proportion of recreated vector in the mixture.
Transformation	Transformation is the process of take-up of foreign DNA, normally plasmids, by bacteria. Plasmids are cloned by transfer into strains of *E. coli* with defined genetic properties. The *E. coli* cells can be made competent to take up plasmid DNA by treatment with Ca^{2+} or an electric field. The cells are plated out on agar and grown to yield single colonies, or clones.
Selection	Bacteria that have taken up a plasmid are selected by growth on a plate containing an antibiotic to which the plasmid vector encodes resistance.
Transformation efficiency	The efficiency of the transformation step is given by the number of antibiotic-resistant colonies per microgram of input plasmid DNA.
Screening transformants	In many cases, such as when using DNA libraries, plasmid, and other vectors have been designed to facilitate the screening of transformants for recombinant plasmids. In the case of a simple subcloning experiment, transformants are screened most easily by digesting the DNA from minipreparations of the transformants, followed by analysis on an agarose gel.
Growth and storage of transformants	Single colonies from a transformation plate are grown in liquid medium, maintaining the antibiotic selection for the plasmid, and a portion of the culture is stored for later use as a frozen glycerol stock.

Gel analysis	Recombinant plasmids can be distinguished from vectors by size on an agarose gel and by excising the inserted fragment with the same restriction enzyme(s) used to insert it.
Fragment orientation	The orientation of the insert in the vector may be determined using an agarose gel by digestion of the plasmid with a restriction enzyme known to cut asymmetrically within the insert sequence.
Modern subcloning methods	The classical methods described here have been enhanced by the use of the polymerase chain reaction to modify the ends of sequences, as well as the development of highly efficient subcloning vector systems based on site-specific recombination.
Related topics	(O1) DNA cloning: an overview (Section P) Cloning vectors

DNA ligation

To insert a target DNA fragment into a vector, a method for the covalent joining of DNA molecules is essential. **DNA ligase** enzymes perform just this function; they will repair (**ligate**) a break in one strand of a dsDNA molecule (Section D2), provided the 5′-end has a phosphate group. They require an adenylating agent to activate the phosphate group for attack by the 3′-OH; the *E. coli* enzyme uses NAD⁺, and the more commonly used enzyme from bacteriophage T4 uses ATP. Ligases are efficient at sealing the broken phosphodiester bonds in an annealed pair of cohesive ends (Section O2), essentially the reverse of a restriction enzyme reaction, and T4 ligase can even ligate one blunt end to another, albeit with rather lower efficiency.

Recombinant DNA molecules

We can now envisage an experiment in which a DNA fragment containing a gene (*X*) of interest (the target DNA) is inserted into a plasmid vector (Figure 1). The target DNA may be a single fragment isolated from an agarose gel (Section O3), or a mixture of many fragments from, for example, genomic DNA (Section Q1). If the target has been prepared by digestion with EcoRI, then the fragment can be ligated with vector DNA cut with the same enzyme (Figure 1). In practice, the vector should have only one site for cleavage with the relevant enzyme, since otherwise, the correct product could only be formed by the ligation of three or more fragments, which would be very inefficient. There are many possible products from this ligation reaction, and the outcome will depend on the relative concentrations of the fragments as well as the conditions, but the products of interest will be circular molecules with the target fragment inserted at the EcoRI site of the vector molecule (with either orientation), to form a **recombinant** molecule (Figure 1). The recreation of the original vector plasmid by circularization of the linear vector alone, is a competing side reaction, which can make the identification of recombinant products problematic. One solution is to prepare both the target and the vector using a pair of distinct restriction enzymes, such that they have uncompatible cohesive ends at either end. The likelihood of ligating the vector into a circle is then much reduced.

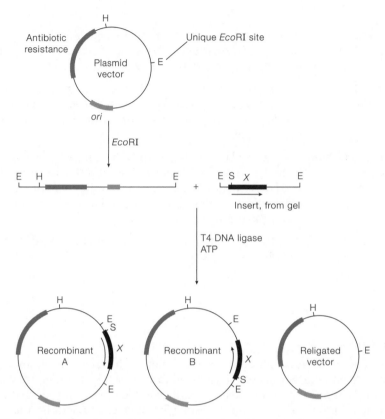

Figure 1. The ligation of vector and target to yield recombinant and nonrecombinant products.

Alkaline phosphatase

If it is inconvenient to use two restriction enzymes, then the linear vector fragment may be treated with the enzyme **alkaline phosphatase** after restriction enzyme digestion. Alkaline phosphatase removes phosphate groups from the 5′-ends of DNA molecules. Hence, the linear vector will be unable to ligate into a circle, as no phosphates are available for the ligation reaction (Figure 2). A ligation with a target DNA insert can still proceed, since one phosphate is present to ligate one strand at each cut site (Figure 2). The remaining nicks in the other strands will be repaired by cellular mechanisms after transformation (see below).

Transformation

The components of the mixture of recombinant and other plasmid molecules formed by ligation (Figure 1) must now be isolated from one another and replicated (**cloned**) by transfer into a **host organism**. By far the most common hosts for simple cloning experiments are strains of *E. coli* with specific genetic properties. One obvious requirement, for example, is that they must not express a restriction–modification system (Section O3).

Escherichia coli cells may be made susceptible to taking up exogenous DNA in a process known as **transformation**, which can be induced chemically or electrically; the latter process is called **electroporation**. In chemical transformation, cells are pre-treated with

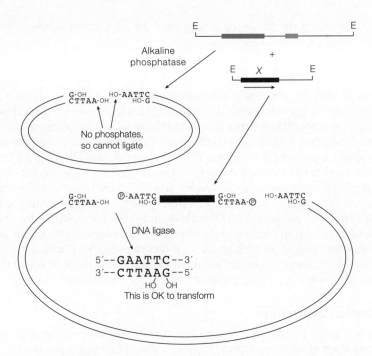

Figure 2. The use of alkaline phosphatase to prevent religation of vector molecules.

Ca²⁺ (and sometimes more exotic metal ions such as Rb⁺ and Mn²⁺), which makes them able to take up DNA, i.e. they become **competent cells**. A solution of a plasmid molecule, or a mixture of molecules formed in a ligation reaction, is then combined with a suspension of competent cells for a period of time to allow the DNA to be taken up. The precise mechanism of the transfer of DNA into the cells is obscure. The mixture is then **heat-shocked** at 42°C for 1–2 min. This induces enzymes involved in the repair of DNA and other cellular components, which allow the cells to recover from the unusual conditions of the transformation process, and increases the efficiency. In electroporation, a suspension of cells and exogenous DNA is treated with a strong electric field, which briefly permeabilizes the cell membrane. In either case, the cells are then incubated in a growth medium and finally spread on an agar plate and incubated until single colonies of bacteria grow (Figure 3). All the cells within a colony originate from division of a single individual. Thus, all the cells will have the same genotype, barring spontaneous mutations

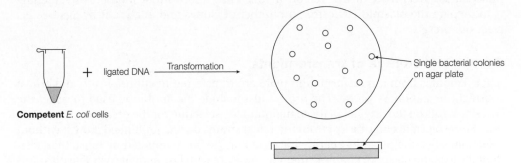

Figure 3. The formation of single colonies after transformation of *E. coli*.

(Section E1), including the presence of any plasmid introduced in the transformation step (in other words, they will be **clones**).

Selection

If all the competent cells present in a transformation reaction were allowed to grow on an agar plate, then many thousands or millions of colonies would result. Furthermore, transformation is an inefficient process; most of the resultant colonies would not contain a plasmid molecule and it would not be obvious which did. A method for the **selection** of clones containing a plasmid is required. This is almost always provided by the presence of an antibiotic resistance gene on the plasmid vector, for example the β-lactamase gene (*amp^r*) conferring resistance to ampicillin (Section O2). If the transformed cells are grown on plates containing ampicillin, only cells that are expressing β-lactamase due to the presence of a transformed plasmid will survive and grow. We can therefore be sure that colonies formed on an ampicillin plate after transformation have grown from single cells that contained a plasmid with an intact β-lactamase gene. If a ligation mixture had been used for the transformation, we would not know at this stage which clones contain recombinant plasmids with a target fragment incorporated (Figure 1).

Transformation efficiency

The quality of a given preparation of competent cells may be measured by determining the **transformation efficiency**, defined as the number of colonies formed (on a selective plate) per microgram of input DNA, where that DNA is a pure plasmid, most commonly the vector to be used in a cloning experiment. Transformation efficiencies can range from 10^3 per µg for crude transformation protocols, which would only be appropriate for transferring an intact plasmid to a new host strain, to more than 10^9 per µg for very carefully prepared competent cells to be used for the generation of libraries (Section Q2). A transformation efficiency of around 10^5 per µg would be adequate for a simple cloning experiment of the kind outlined here.

Screening transformants

Once a set of transformant clones has been produced in a cloning experiment, the first requirement is to know which clones contain a recombinant plasmid with inserted target fragment. Plasmids have been designed to facilitate this process, and are described in Section P1. In many cases, such as the screening of a DNA library (Section Q3), it might be necessary to identify the clone of interest from amongst thousands or even hundreds of thousands of others. In the case of a simple subcloning experiment, the design of the experiment can maximize the production of recombinant clones, for example by alkaline phosphatase treatment of the vector. In this case, the normal method of screening is to prepare the plasmid DNA from a number of clones and analyze it by agarose gel electrophoresis.

Growth and storage of transformants

Single colonies from a transformation plate are transferred to culture broth and grown overnight to stationary phase. The broth must include the antibiotic used to select the transformants on the original plate, to maintain the selection for the presence of the plasmid. Some plasmids may be lost from their host strains during prolonged growth without selection, since the plasmid-bearing bacteria may be out-competed by those that accidentally lose the plasmid, enabling them to replicate with less energy cost. The plasmids are then prepared from the cultures by the miniprep technique (Section O2). It is normal practice to prepare a stock of each culture at this stage by freezing a portion of the culture in the presence of glycerol, which protects the cells from ice crystal formation (a **glycerol stock**). The stock will enable the same strain/plasmid to be grown and prepared again if and when it is required.

Gel analysis

Recombinant plasmids can usually be simply distinguished from recreated vectors by the relative sizes of the plasmids, and further by the pattern of restriction digests. Figure 4 shows a hypothetical gel representing the analysis of the plasmids in Figure 1. Tracks corresponding to the vector plasmid and to recombinants are indicated. The larger size of the recombinant plasmid is seen by comparing the undigested plasmid samples (tracks U) containing super-coiled and nicked bands (Section O3), and the excision of the insert from the recombinant is seen in the EcoRI digest (track E).

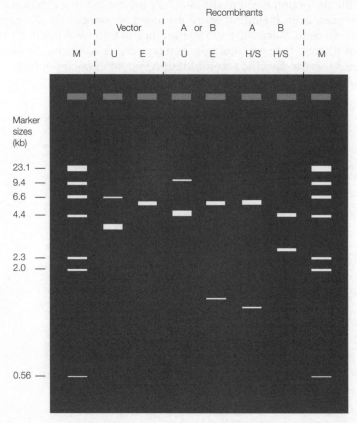

Figure 4. The analysis of recombinant plasmids by agarose gel electrophoresis (see text for details).

Fragment orientation

If a ligation reaction has been carried out using a vector and target prepared with a single restriction enzyme (Section O3), then the insert can be ligated into the vector in either orientation. This might be important if, for example, the target insert contained the coding region of a gene that was to be placed downstream of a promoter in the vector. The orientation of the fragment can be determined using a restriction digest with an enzyme that is known to cut asymmetrically within the insert sequence, together with one which cuts at some specified site in the vector. This is illustrated in Figures 1 and 4, using a **double digest** with the enzyme SalI (S), which cuts in the insert sequence, and HindIII (H), which cuts once in the vector. The patterns expected from the two orientations, A and B, of the inserted fragment (Figure 1) are illustrated in Figure 4 (tracks H/S).

Modern subcloning methods

The preceding discussion of the classical method for subcloning helps to introduce many of the basic concepts of gene cloning. However, it is worth pointing out that some of these procedures have been enhanced and extended by faster and more efficient methods based on more recent technology. For example, the **polymerase chain reaction** (**PCR**; Section R3) may be used to extensively modify the ends of a DNA fragment, to provide specific restriction enzyme sites for very precise cloning (Section Q2), or other specific sequences, even including entire promoter sequences for the expression of a coding region of a gene. Since 2000, the highly efficient Gateway® method of transferring a sequence of interest between multiple cloning vectors has been developed (Section P1), which uses the **site-specific recombination** system of bacteriophage λ integration (Sections E4 and M2).

P1 Design of plasmid vectors

Key Notes

Ligation products	One of the most important steps in a cloning procedure is to distinguish between recreated vector molecules and recombinant plasmids. A number of methods have been developed to facilitate this process.
Blue–white screening	Insertional inactivation of the *lacZ'* gene on a plasmid can be used to screen for recombinants on a plate containing IPTG and X-gal. The X-gal is converted to a blue product if the *lacZ'* gene is intact and induced by IPTG; hence recombinants grow as white colonies.
Multiple cloning sites	A multiple cloning site provides flexibility in choice of restriction enzyme or enzymes for cloning.
Transcription of cloned inserts	A promoter within the vector may be used either *in vivo* or *in vitro* to transcribe an inserted fragment. Some vectors have two specific promoters to allow transcription of either strand of the insert.
Expression vectors	Many vectors have been developed that allow genes within a cloned insert to be expressed by transcription from a strong promoter in the vector. In some cases, for example using T7 expression vectors, a large proportion of the total protein in the *E. coli* cells may consist of the desired product.
Gateway® – subcloning by recombination	The Gateway® subcloning method uses an engineered version of the phage λ integration reaction to enable efficient and precise transfer of target DNA fragments between multiple vectors.
Related topics	(Section O) Gene manipulation (Section Q) Gene libraries and screening	(Section R) Analysis and uses of cloned DNA

Ligation products

When ligating a target fragment into a plasmid vector, the most frequent unwanted product is the recreated vector plasmid formed by circularization of the linear vector fragment (Section O4). Religated vectors may be distinguished from recombinant products by performing minipreparations from a number of transformed colonies, and screening by digestion and agarose gel electrophoresis (Sections O2 and O4) but this is impossibly inconvenient on a large scale, and more efficient methods based on specially developed vectors have been devised.

Blue–white screening

The standard method for screening for the presence of recombinant plasmids or other vectors is called **blue–white** screening. This involves the principle that a recombinant fragment inserted into a gene sequence in the vector will inactivate that gene, or its encoded protein, allowing a test based on the presence of that protein; this is known as **insertional inactivation**. As the name implies, blue–white screening uses the production of a blue compound as an indicator. The gene in this case is **lacZ**, which encodes the enzyme β-**galactosidase**, and is under the control of the *lac* promoter (Section G1). If the host *E. coli* strain used for cloning is expressing the **lac repressor**, then expression of a *lacZ* gene on the vector may be induced using **isopropyl-β-D-thiogalactopyranoside (IPTG)** (Section G1), and the expressed enzyme can utilize the synthetic substrate **5-bromo-4-chloro-3-indolyl-β-D-galactopyranoside (X-gal)** to yield a blue product. Insertional inactivation of *lacZ* in the production of a recombinant plasmid would prevent the development of the blue color. In this method (Figure 1b) the transformed cells are spread on to a plate containing ampicillin (to select for transformants in the usual way), IPTG and X-gal, to yield a mixture of blue and white colonies. The white colonies have no expressed β-galactosidase and so are likely to contain the inserted target fragment. The blue colonies probably contain religated vector. In practice, the vectors used in this method have a shortened derivative of *lacZ*, **lacZ'**, which produces the N-terminal α-**peptide** of β-galactosidase. These vectors must be transformed into a special host strain containing a mutant gene expressing only the C-terminal portion of β-galactosidase that complements the α-peptide to produce active enzyme. This reduces the size of the plasmid-borne gene, but does not alter the basis of the method.

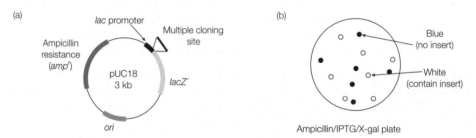

Figure 1. (a) A plasmid vector designed for blue–white screening; (b) the colonies produced by blue–white screening.

Multiple cloning sites

The first vectors to utilize blue-white selection also pioneered the idea of the **multiple cloning site (MCS)**. These plasmids, the pUC series, contain an engineered version of the *lacZ'* gene, which has multiple restriction enzyme sites within the first part of the coding region of the gene (Figures 1a and 2). This region is known as the MCS; insertion of target

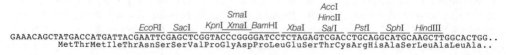

Figure 2. A multiple cloning site at the 5'-end of lacZ'.

DNA in any of these sites or between any pair, inactivates the *lacZ'* gene, to give a white colony on an appropriate plate. An MCS allows flexibility in the choice of a restriction enzyme or enzymes for cloning and is a feature of all cloning vetors.

Transcription of cloned inserts

Since the pUC vectors above have a promoter (*lac*) adjacent to the site of insertion of a cloned fragment, it is easy to imagine that such a promoter could be used to transcribe the inserted DNA, either to produce an RNA transcript *in vitro*, which could be used as a hybridization probe (Section Q3), or to express the protein product of a gene within the insert (see below). Several varieties of transcriptional vector have been constructed that allow the *in vitro* transcription of a cloned fragment. For example, the pGEM series has promoters from **bacteriophages T7** and **SP6** flanking an MCS (which itself is set up for *lacZ'* blue–white screening). The phage promoters are each recognized only by their corresponding bacteriophage RNA polymerases (Section O1, Table 1), either of which may be used *in vitro* to transcribe the desired strand of the inserted fragment.

Expression vectors

The pUC vectors may be used to express cloned genes to yield the **recombinant protein** product in *E. coli*. This requires the positioning of the cloned fragment within the MCS such that the coding region of the target gene is contiguous with and in the same **reading frame** (Section K1) as the *lacZ'* gene. After induction of transcription from *lac*, this results in the production of a **fusion protein** (Section A5) with a few amino acids derived from the N-terminus of *lacZ'* followed directly by the protein sequence encoded by the insert. Innumerable variants of this scheme have been developed, using strong (very active) promoters such as ***lacUV-5*** [a mutant *lac* promoter which is independent of catabolite activator protein (CAP); Section G1], the **phage λ P$_L$** promoter (Section M2) or the **phage T7** promoter. Some vectors rely on the provision of a ribosome binding site (RBS) and translation initiation codon (Section L1) within the cloned fragment, while some are designed to encode a fused sequence at the N-terminus of the expressed protein, which allows the protein to be purified easily by a specific binding step. An example of this is the **His-Tag**, a series of consecutive histidine residues that bind strongly to a chromatography column bearing Ni^{2+} ions (Section A5). Some vectors may even allow the fused N-terminus to be removed from the protein by cleavage with a specific protease.

One example will serve as an illustration: the overexpression of many proteins in *E. coli* is now achieved using a T7 promoter system as in the pET series of vectors (Figure 3). These vectors contain a T7 promoter and an *E. coli* RBS, followed by a start codon, an MCS, and, finally, a transcription terminator (Section F4). Expression is achieved in a special *E. coli* strain (a λ lysogen; Section P2) that produces the T7 RNA polymerase on induction with IPTG. This polymerase is very efficient, and only transcribes from the T7 promoter, usually resulting in the production of high levels of the mRNA encoding the target protein, which after translation may comprise up to 30% of the total *E. coli* protein.

Gateway® – subcloning by recombination

One of the most useful developments in cloning vector methodology in the last ten years has been the **Gateway®** cloning system, which allows extremely easy and efficient transfer of a cloned fragment, often a gene or cDNA, between multiple vectors designed for different purposes – protein expression, expression in alternate hosts (Section P3), etc. The method is based on the site-specific recombination system responsible for the **integration** of bacteriophage λ DNA into the *E. coli* chromosome, forming a λ lysogen (Sections

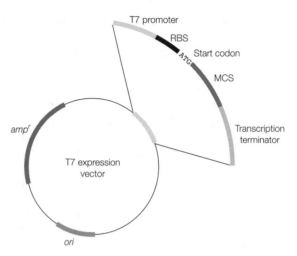

Figure 3. A plasmid designed for expression of a gene using the T7 system.

E4 and M2). The Gateway® system requires two steps (Figure 4). The gene of interest must initially be flanked by two variants of the short ***attB* sequence** (*attB1* and *attB2*), which is the site of insertion of phage λ in the *E. coli* chromosome. This initial construct could either be a plasmid clone produced by classical methods (Section O), or a linear PCR product with the *attB* sites engineered into the terminal primers (Section R3). The *attB*-flanked gene is then recombined *in vitro*, with a **donor plasmid**, which contains two variants of the ***attP* site**, the corresponding recombination site from the phage genome (Section E4). Two recombination reactions occur between *attB1*/*attP1* and *attB2*/*attP2*, analogous to the λ integration reaction carried out by the **Int** protein, to produce a new clone, called the **entry clone**, where the insert gene is flanked by two hybrid sites called *attL*. This plasmid is grown and maintained in the usual way. In the second step, the gene sequence may be transferred to any one of many **destination vectors**, using recombination between the *attL* sites and corresponding *attR* sites on the destination vector. These reactions are analogous to the λ excision reaction, the opposite of integration (Section M2), and recreate the *attB* sites in the new vector containing the gene.

The recombination reactions are highly efficient, and correspond to very precise cutting and religation combined into a single step. As well as a variety of antibiotic resistance genes used to select for the relevant plasmids, an interesting feature of this system is the use of the bacterial toxin gene ***ccdB*** in the destination vectors. This ensures that the other plasmid product of the recombination reaction cannot survive in the host, since the expressed CcdB protein will kill the cell (Figure 4). Destination plasmids may be maintained in a strain expressing the antitoxin, **CcdA**.

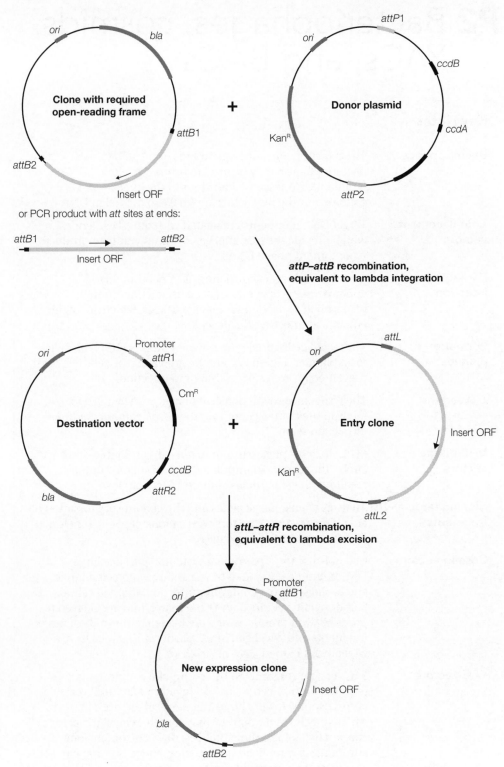

Figure 4. Outline of the Gateway® cloning method.

P2 Bacteriophages, cosmids, YACs, and BACs

Key Notes

Bacteriophage λ

The infection and subsequent lysis of *E. coli* by bacteriophage λ may be used to propagate cloned DNA fragments. Nonessential portions of the linear 48.5-kb λ genome may be replaced by up to 23 kb of foreign DNA.

λ Replacement vectors

Target DNA fragments are ligated with the λ DNA ends, which provide the essential genes for infection, to produce recombinant phage DNAs.

Packaging and infection

A packaging extract consisting of λ coat proteins and processing enzymes may be used to incorporate recombinant λ DNA into phage particles, which are highly efficient at infecting *E. coli* cells.

Formation of plaques

λ-Infected cells spread on a lawn of uninfected *E. coli* cells form plaques, regions where the growth of the cells has been prevented by cycles of cell lysis and infection.

λ Lysogens

The lysogenic growth phase of phage λ can be used to incorporate cloned genes into the *E. coli* genome for long-term expression.

M13 phage vectors

M13 phages replicate inside *E. coli* cells as double-stranded circles that can be manipulated like plasmids, but the resulting phage particles contain ssDNA circles.

Cloning large DNA fragments

Analysis of eukaryotic genes and the genome organization of eukaryotes requires vectors with a larger capacity for cloned DNA than plasmids or phage λ.

Cosmid vectors

Cosmids use the λ packaging system to package large DNA fragments bounded by λ *cos* sites, which circularize and replicate as plasmids after infection of *E. coli* cells. Some cosmid vectors have two *cos* sites, and are cleaved to produce two *cos* ends, which are ligated to the ends of target fragments and packaged into λ particles. Cosmids have a capacity for cloned DNA of 30–45 kb.

YAC vectors

YACs can be constructed by ligating the components required for replication and segregation of natural yeast chromosomes to very large fragments of target DNA, which may be more than 1 Mb in length. Yeast YAC vectors contain two telomeric sequences, one centromere, one autonomously replicating sequence, and genes that can act as selectable markers in yeast.

Selection in yeast	Selection for the presence of YACs or other vectors in yeast is achieved by complementation of a mutant strain unable to produce an essential metabolite, with the correct copy of the mutant gene carried on the vector.	
BAC vectors	BACs are based on the F factor of *E. coli* and can be used to clone up to 350 kb of genomic DNA in a conveniently handled *E. coli* host. They are a more stable and easier to use alternative to YACs.	
Related topics	(Section O) Gene manipulation (P1) Design of plasmid vectors	(Section Q) Gene libraries and screening (R2) Nucleic acid sequencing

Bacteriophage λ

Bacteriophage λ, which infects *E. coli* cells, has classically been used as a cloning vector to accommodate larger DNA fragments than plasmids. The process of infection by phage λ is described in Section M2. In brief, the phage particle injects its **linear DNA** into the cell, where it is ligated into a circle. It may either replicate to form many phage particles, which are released from the cell by lysis and cell death (the **lytic** phase), or the DNA may integrate into the host genome by site-specific recombination (Section E4), where it may remain for a long period (the **lysogenic** phase).

The 48.5-kb λ genome is shown schematically in Figure 1 (also Section M2, Figure 2). At the ends are the *cos* (cohesive) sites, which consist of 12-bp cohesive ends. The *cos* sites are asymmetric, but in other respects are equivalent to very large (16 bp) restriction sites. The *cos* ends allow the DNA to be circularized in the cell. Much of the central region of the genome is dispensable for lytic infection, and may be replaced by unrelated DNA sequence. There are limits to the size of DNA that can be incorporated into a λ phage particle; the DNA must be between 75 and 105% of the natural length, i.e. 37–52 kb. Taking account of the essential regions, DNA fragments of around 20 kb (maximum 23 kb) can be cloned into λ, which is more than can be conveniently incorporated into a plasmid vector.

λ Replacement vectors

A number of so-called **replacement vectors** have been developed from phage λ; examples include **EMBL3** and λ **DASH** (Section Q1). A representative scheme for cloning using such a vector is shown in Figure 2. The vector DNA is cleaved with BamHI and the long (19 kb) and short (9 kb) ends (Figure 1) are purified. The target fragment or fragments are prepared by digestion, also with BamHI or a compatible enzyme (Sections O3 and Q1), and treated with alkaline phosphatase (Section O4) to prevent them ligating to each other. The λ arms and the target fragments are ligated together (Section O4) at relatively high concentration to form long linear products (Figure 2).

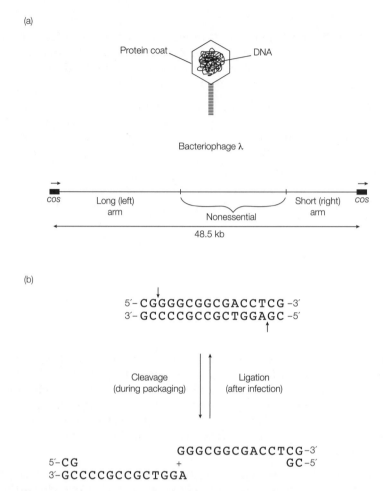

(a)

Protein coat

DNA

Bacteriophage λ

cos — Long (left) arm — Nonessential — Short (right) arm — *cos*

48.5 kb

(b)

5′– C G G G G C G G C G A C C T C G –3′
3′– G C C C C G C C G C T G G A G C –5′

Cleavage
(during packaging)

Ligation
(after infection)

                                    G G G C G G C G A C C T C G –3′
5′– C G   G C –5′
                    +
3′– G C C C C G C C G C T G G A

Figure 1. (a) Phage λ and its genome; (b) the phage λ *cos* ends.

Packaging and infection

Although pure circular λ DNA or derivatives of it can be transformed into competent cells as described for plasmids (Section O4), the infective properties of the phage particle may be used to advantage, particularly in the formation of DNA libraries (Section Q). Replication of phage λ *in vivo* produces long linear molecules with multiple copies of the λ genome. These concatamers are then cleaved at the *cos* sites, to yield individual λ genomes, which are then packaged into the phage particles. A mixture of the phage coat proteins and the phage DNA-processing enzymes (a **packaging extract**) may be used *in vitro* to **package** the ligated linear molecules into phage particles (Figure 2). The packaging extract is prepared from two bacterial strains, each infected with a different **packaging-deficient** (mutant) λ phage, so that packaging proteins are abundant. In combination, the two extracts provide all the necessary proteins for packaging. The phage particles so produced may then be used to infect a culture of normal *E. coli* cells. Ligated λ ends that do not contain an insert (Figure 2), or have one which is much smaller or larger than the

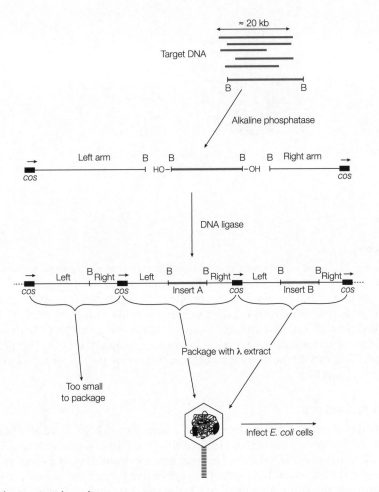

Figure 2. Cloning in a λ replacement vector.

20-kb optimum, are too small or too large to be packaged, and recombinants with two left or right arms are likewise not viable. The infection process is very efficient and can produce up to 10^9 recombinants per microgram of vector DNA.

Formation of plaques

The infected cells from a packaging reaction are spread on an agar plate (Section O4) that has been pre-spread with a high concentration of uninfected cells, which will grow to form a continuous **lawn**. Single infected cells result in clear areas, or **plaques**, within the lawn after incubation, where cycles of lysis and re-infection have prevented the cells from growing (Figure 3). These are the analogs of single bacterial colonies (Section O4). Recombinant λ DNA may be purified for further manipulation from phage particles isolated from plaques or from the supernatant broth of a culture infected with a specific recombinant plaque (Section R1).

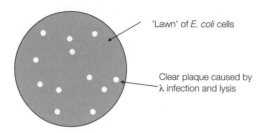

Figure 3. The formation of plaques by λ infection.

λ Lysogens

The lysogenic phase of λ infection is also used in cloning technology. Genes or foreign sequences may be incorporated essentially permanently into the genome of *E. coli* by integration of a λ vector containing the sequence of interest. One example of this method is the strain of *E. coli* commonly used for overexpression of proteins using T7 RNA polymerase. The strain BL21(DE3) and derivatives include the gene for T7 RNA polymerase under control of the *lac* promoter as a λ lysogen, designated DE3. The gene can be induced by IPTG, and the polymerase will then transcribe the target gene in the expression vector (Section P1).

M13 phage vectors

The so-called filamentous phages (Section M2), specifically **M13**, have also been used as *E. coli* vectors. The phage particles contain a 6.4-kb circular single strand of DNA. After infection of a sensitive *E. coli* host, the complementary strand is synthesized, and the DNA replicated as a double-stranded circle, the **replicative form**, with about 100 copies per cell. In contrast to the situation with phage λ, the cells are not lysed by M13, but continue to grow slowly, and single-stranded forms are continuously packaged and released from the cells as new phage particles (up to 1000 per cell generation). The same single strand of the complementary pair is always present in the phage particle. The useful properties of M13 as a vector are that the replicative form can be purified and manipulated exactly like a plasmid, but the same DNA may be isolated in a single-stranded form from phage particles in the medium. ssDNA was originally required for a number of significant applications in cloning, including DNA sequencing (Section R2) and site-directed mutagenesis (Section R5), although these techniques now no longer have that limitation. M13 vectors are now only used for rather specialist applications.

Cloning large DNA fragments

The analysis of genome organization and the identification of genes, particularly in organisms with large genome sizes (human DNA is 3×10^9 bp, for example), is difficult to achieve using plasmid and bacteriophage λ vectors (see Section O and above), as the relatively small capacity of these vectors for cloned DNA means that an enormous number of clones would be required to represent the whole genome in a DNA library (Sections O1 and Q1). In addition, the very large size of some eukaryotic genes due to their large intron sequences means that an entire gene may not fit on a single cloned fragment. Vectors with much larger size capacity have been developed to circumvent these problems. **Pulsed field gel electrophoresis** (**PFGE**) has made it possible to separate, map, and analyze very large DNA fragments. In Section O3 it was seen that the limitation of conventional agarose gel electrophoresis becomes apparent as large DNA fragments above a critical size do not separate, but comigrate instead. This is because nucleic acids alternate between folded (more globular) and extended (more linear) forms as they migrate through a porous matrix such as an agarose gel. However, when the DNA molecules become so large that their globular forms do not fit into the matrix pores even in the lowest percentage agarose gels that can be easily handled (0.1–0.3%), then they comigrate. This limitation can be overcome if the electric field is applied discontinuously (pulsed), and even greater separation can be achieved if the direction of the field is also made to vary. Each time the electric field changes, the DNA molecules reorient their long axes, and this process takes longer for larger molecules. A number of variations of PFGE have been developed that differ in the number of electrodes used and how the field is varied (e.g. **field inversion gel electrophoresis**, **FIGE**, and **contour clamped homogeneous electric field**, **CHEF**). Separations of molecules up to 7 Mb have been achieved and it is now possible to resolve whole chromosome DNA fragments, including artificial chromosomes (see below).

Cosmid vectors

Cosmid vectors are so-called because they utilize the properties of the phage λ *cos* sites in a plasmid molecule. The *in vitro* packaging of DNA into λ particles (see above) requires only the presence of the λ *cos* sites spaced by the correct distance (37–52 kb) on linear DNA. The intervening DNA can have any sequence at all; it need not contain any λ genes, for example. The simplest cosmid vector is a normal small plasmid containing a plasmid origin of replication (*ori*) and a selectable marker, and which also contains a *cos* site and a suitable restriction site for cloning (Figure 4). After cleavage with a restriction enzyme and ligation with target DNA fragments, the DNA is packaged into λ phage particles. The DNA is re-circularized by annealing of the *cos* sites after infection, and propagates as a normal plasmid, under selection by ampicillin. As in the case of phage λ (see above), more sophisticated methods are used in a real cloning situation to ensure that multiple copies of the vector or the target DNA are not included in the recombinant. Libraries of clones prepared with cosmid vectors can be screened as described in Section Q3. The genomes of entire bacteria such as *E. coli* are available as a set of around 100 cosmid clones.

YAC vectors

The realization that the components of a eukaryotic chromosome required for stable replication and segregation (at least in the budding yeast *Saccharomyces cerevisiae*) consist of rather small and well-defined sequences (Section C3) has led to the construction of recombinant chromosomes (**yeast artificial chromosomes**; **YACs**). These can be used as vectors for carrying very large cloned fragments. The centromere, telomere, and replication origin sequences (Sections C3, D1, and D3) have been isolated and combined on plasmids constructed in *E. coli*. The structure of a typical pYAC vector is shown in Figure 5. The method of construction of the YAC clone is similar to that for cosmids in that two end fragments are ligated with target DNA to yield the complete chromosome, which is then introduced (**transfected**) into yeast cells (Section P3). YAC vectors can accommodate genomic DNA fragments of more than 1 Mb, and so can be used to clone entire human genes such as the cystic fibrosis gene, CFTR, which is 250 kb in length. YACs have been invaluable in mapping the large-scale structure of large genomes, for example in the original Human Genome Project.

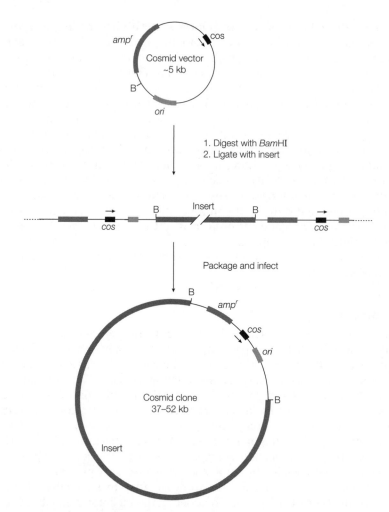

Figure 4. Formation of a cosmid clone.

The yeast sequences that have been included on pYAC3 (Figure 5) are as follows: **TEL** represents a segment of the telomeric DNA sequence, which is extended by the telomerase enzyme inside the yeast cell (Sections C3 and D3). **CEN4** is the centromere sequence from chromosome 4 of *S. cerevisiae*. Despite its name, the centromere will function correctly to segregate the daughter chromosomes even if it is very close to one end of the artificial chromosome. The **ARS** functions as a yeast origin of replication (Section D3). **TRP1** and **URA3** are yeast selectable markers (see below), one for each end, to ensure that only properly reconstituted YACs survive in the yeast cells. **SUP4**, which is insertionally inactivated in recombinants (Figure 3), is a gene that is the basis of a red–white color test analogous to blue–white screening in *E. coli* (Section P1).

Selection in yeast

Saccharomyces cerevisiae selectable markers do not normally confer resistance to toxic substances as in *E. coli* plasmids, but instead enable the growth of yeast on selective media lacking specific nutrients. **Auxotrophic** yeast mutants are unable to make a specific

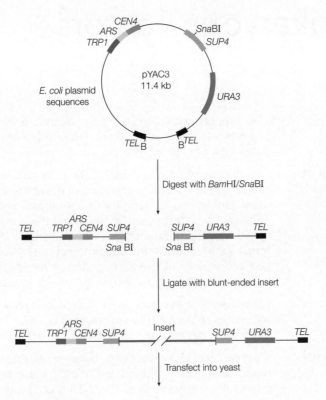

Figure 5. Cloning in a YAC vector.

compound. *TRP1* mutants, for example, cannot make tryptophan, and can only grow on media supplemented with tryptophan. Transformation of the mutant yeast strain with a YAC or other vector containing an intact *TRP1* gene **complements** this deficiency and hence only transfected cells can grow on media lacking tryptophan.

BAC vectors

BAC vectors, or **bacterial artificial chromosomes**, were developed to overcome one or two problems with the use of YACs to clone very large genomic DNA fragments. Although YACs can accommodate very large fragments, quite often these fragments turn out to comprise **noncontiguous** (nonadjacent) segments of the genome and they frequently lose parts of the DNA during propagation (i.e. they are unstable). BACs are able to accommodate up to about 300–350 kb of insert sequence, less than YACs, but with the advantages of stability, ease of transformation and speed of growth of their *E. coli* host. They are also simpler to purify using standard plasmid miniprep techniques (Section O2). The vectors are based on the natural extrachromosomal F factor of *E. coli*, which encodes its own DNA polymerase and is maintained in the cell at a level of one or two copies. A BAC vector incorporates the genes essential for replication and maintenance of the F factor, a selectable marker, and a cloning site flanked by rare-cutting restriction enzyme sites and other specific cleavage sites, which enable the clones to be linearized within the vector region without the possibility of cutting within the very large insert region. BACs are a more user-friendly alternative to YACs and have been used extensively in genomic mapping projects.

P3 Eukaryotic vectors

Key Notes

Cloning in eukaryotes

Many applications of gene cloning require the transfer of genes into eukaryotic cells and their expression, either transiently or permanently.

Transfection of eukaryotic cells

Transfection of DNA into eukaryotic cells such as plants and yeasts may require prior digestion of the cell wall. DNA is usually taken up as a complex with a cationic polymer. Take-up may also be promoted by electroporation, microinjection, liposomes, and bombardment with solid particles. Stable transfection requires the integration of the DNA into the genome, which often requires genetic selection of stably transfected cells using a marker gene such as *neo^r*.

Shuttle vectors

Many eukaryotic vectors also incorporate bacterial plasmid sequences so they can be constructed and checked using *E. coli* hosts. They can shuttle between more than one host.

Yeast episomal plasmids

Yeast vectors (YEps) have been developed using the replication origin of the natural yeast 2μ plasmid, and selectable markers such as *LEU2*. They can replicate as plasmids, but may also integrate into the chromosomal DNA.

***Agrobacterium tumefaciens* Ti plasmid**

The bacterium *A. tumefaciens*, which infects some plants and integrates part of its Ti plasmid into the plant genome, has been used to transfer foreign genes into a number of plant species.

Viral transduction

The transfer of DNA into a cell using a virus vector is called transduction. DNA viruses such as SV40, adenovirus, or vaccinia maintain their genomes episomally in cells. Retroviruses integrate their genome into the host cell genome.

Baculoviruses

Baculoviruses are insect viruses that are used for the overexpression of animal proteins in insect cell culture. Insect cells produce essentially the same post-translational modifications as mammalian cells.

Related topic

(Section R) Analysis and uses of cloned DNA

Cloning in eukaryotes

Many eukaryotic genes and their control sequences have been isolated and analyzed using gene cloning techniques based on *E. coli* as host. However, many applications of genetic engineering, from the large-scale production of eukaryotic proteins to the engineering of new plants and gene therapy, require vectors for the expression of genes in diverse eukaryotic species. Examples of such vectors designed for a variety of hosts are discussed in this section.

Transfection of eukaryotic cells

Introduction of nucleic acids into eukaryotic cells by nonviral means is called **transfection**, as the word 'transformation' when applied to animal cells denotes the conversion of a normal cell to a cancer cell. Transfection can be more problematic than bacterial transformation (Section O4), and the efficiency of the process is much lower. In yeast and plant cells, for example, the cell wall must normally be digested with degradative enzymes to yield fragile **protoplasts**. The cell walls are re-synthesized once the degrading enzymes are removed. Protoplasts and animal cells in culture will take up DNA by **endocytosis** when it is complexed with cationic polymers that neutralize its negative charge. Other methods involve **electroporation** (transient permeabilization by an electric pulse), packaging the DNA into **liposomes** (synthetic membrane vesicles that fuse with the plasma membrane), direct **microinjection** into the nucleus using very fine glass pipettes, and introduction by firing metallic microprojectiles coated with DNA at the target cells, using what has become known as a '**gene gun**'.

The simple transfection of a plasmid (or plasmids) carrying a 'foreign' gene (transgene) into mammalian cells is called transient transfection, since within a few days the plasmid DNA and any resulting gene expression is lost from the cells. For some experiments, stable gene expression may need to be maintained for many cell generations, even permanently. One approach is to select for cells that have integrated the transfected DNA into their chromosomes. This is called stable transfection, but it normally occurs with a low frequency in mammalian cells. It is therefore necessary to use a selectable marker that allows the genetic selection of stably transfected cells (cf. antibiotic selection in Section O4). A common example is the gene for neomycin phosphotransferase (neo^r). The drug geneticin (G418) selectively kills cells *not* expressing this resistance gene. The neo^r gene (and an associated promoter) may be introduced on a separate plasmid or incorporated into the plasmid whose integration into the genome is required. The transgene can be attached to a constitutive promoter so that it is expressed constantly, or it can be placed under the control of an inducible promoter that responds to a synthetic ligand so that it is only expressed when the ligand is added to the cells. Often these approaches lead to the integration of several tandem copies of the plasmids at a single chromosomal site. Another method is to use vectors such as BACs that can either permit independent replication (episomal maintenance) of the DNA in the cells, or integration of these large DNAs into the genome (Section P2).

Shuttle vectors

Most of the vectors for use in eukaryotic cells are constructed as **shuttle vectors**. This means that they incorporate the sequences required for replication and selection in *E. coli* (*oriC*, *amp^r*) as well as in the desired host cells (e.g. SV40 origin, *neo^r*). This enables the vector with its target insert to be constructed and its integrity checked using the highly developed *E. coli* methods before transfer to the appropriate eukaryotic cells. Many shuttle vectors are now set up to use the Gateway® recombinational cloning method (Section P2).

Yeast episomal plasmids

Vectors for the cloning and expression of genes in *Saccharomyces cerevisiae* have been designed based on the natural **2 micron (2μ) plasmid**. The plasmid is named for the length of its DNA, which corresponds to 6 kb of sequence. The 2μ plasmid has an origin of replication and two genes involved in replication, and also encodes a site-specific recombination protein FLP, homologous to the phage λ integrase, Int (Section E4), which can invert part of the 2μ sequence. Vectors based on the 2μ plasmid (Figure 1) called **YEps** normally contain the 2μ replication origin, *E. coli* shuttle sequences, and a yeast gene that can act as a selectable marker (Section P2), for example the *LEU2* gene, involved in leucine biosynthesis. Although they normally replicate as plasmids, YEps may integrate into a yeast chromosome by homologous recombination (Section E4) with the defective genomic copy of the selection gene (Figure 1). *Pichia pastoris* is a useful host cell for the large-scale production of recombinant proteins as it can grow rapidly to high densities on cheap carbon sources such as methanol and the proteins can be engineered to be secreted into the extracellular medium.

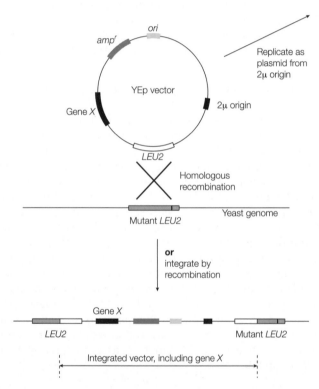

Figure 1. Cloning using a yeast episomal plasmid, based on the 2μ origin.

Agrobacterium tumefaciens Ti plasmid

Plant cells do not contain natural plasmids that can be utilized as cloning vectors. However, the bacterium *A. tumefaciens* which primarily infects dicotyledenous plants (tomato, tobacco, peas, etc.), but which can also infect rice, a monocot, contains the 200-kb **tumor-inducing Ti plasmid**. Upon infection, part of the Ti plasmid, the **T-DNA**, is integrated into the plant chromosomal DNA (Figure 2), resulting in uncontrolled growth of the plant cells directed by genes in the T-DNA, and the development of a **crown gall**, or tumor.

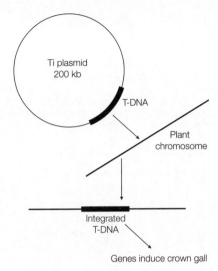

Figure 2. Action of the A. *tumefaciens* Ti plasmid.

Recombinant Ti plasmids with a target gene inserted in the T-DNA region can integrate that gene into the plant DNA, where it may be expressed. In practice, however, several refinements are made to this simple scheme. The size of the Ti plasmid makes it difficult to manipulate, but it has been discovered that if the T-DNA and the remainder of the Ti plasmid are on separate molecules within the same bacterial cell, integration will still take place. The recombinant T-DNA can be constructed in a standard *E. coli* plasmid, then transformed into the *A. tumefaciens* cell carrying a modified Ti plasmid without T-DNA. A further improvement is made by deleting the genes for crown gall formation from the T-DNA. So-called **disarmed** T-DNA shuttle vectors can integrate cloned genes benignly, and, if the host cells are growing in culture, complete recombinant plants can be reconstituted from the transformed cells (Figure 3).

Viral transduction

Several viral vector systems have been developed for the study of gene expression in cells and tissues, and particularly for gene therapy (Section S5). The transfer of DNA into a cell using virus-mediated transfer is called **transduction**. Typical examples of viruses used for transduction are DNA viruses such as SV40, adenoviruses, herpesviruses, and vaccinia virus (Section M3), insect baculoviruses (see below), and retroviruses, which have an RNA genome (Section M4). Recombinant vaccinia and adenoviruses are used for longer-term transient expression studies as they maintain their genomes episomally, while recombinant retroviruses are used for the generation of permanently integrated, transfected cell lines. The generation of recombinant viruses initially requires their construction by DNA transfection of cultured cells using one of the methods described above. The production of a recombinant virus can be time consuming as potential virulence genes need to be deleted, and usually requires strict biological safety and handling procedures.

Baculoviruses

Baculoviruses infect insect cells. One of the major proteins encoded by the virus genome is polyhedrin, which accumulates in very large quantities in the nuclei of infected cells,

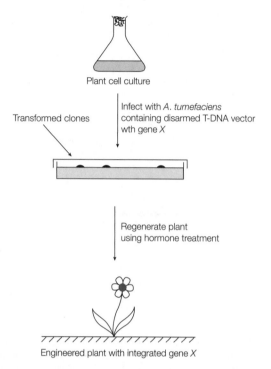

Figure 3. Engineering new plants with *A. tumefaciens*.

as the gene has an extremely active promoter. The same promoter can be used to drive the overexpression of a foreign gene engineered into the baculovirus genome, and large quantities of protein can be produced in infected insect cells in culture. This method is being used increasingly for large-scale culture of proteins of animal origin, since the insect cells can produce many of the post-translational modifications of animal proteins that bacterial and yeast expression systems cannot (Sections P1 and L4). Insect cells also have the advantage of growing at lower temperatures than mammalian cells, which saves energy.

Q1 Genomic libraries

Key Notes

Representative gene libraries	Gene libraries made from genomic DNA are called genomic libraries and those made from DNA complementary to mRNA are known as cDNA libraries. The latter represent only the transcribed sequences from a particular cell. Good gene libraries are representative of the starting material and should not have lost or gained sequences due to cloning artifacts.
Size of library	A gene library must contain a certain number of recombinants for there to be a high probability that it contains any particular sequence of interest. This value can be calculated if the genome size and the average size of the insert in the vector are known.
Genomic DNA	Genomic DNA for constructing libraries is usually prepared by protease digestion and phase extraction then fragmented randomly by physical shearing or restriction enzyme digestion to give a size range appropriate for the chosen vector. Combinations of restriction enzymes are often used to partially digest the DNA.
Vectors	Plasmid, λ phage, cosmid, BAC, or yeast artificial chromosome vectors can be used to construct genomic libraries, the choice depending on the genome size. The upper size limit of these vectors is about 10, 23, 45, 350, and 1000 kb respectively. The genomic DNA fragments are ligated to the prepared vector molecules using T4 DNA ligase.
Related topics	(J3) mRNA processing, hnRNPs, and snRNPs (P1) Design of plasmid vectors (P2) Bacteriophages, cosmids, YACs, and BACs

Representative gene libraries

A gene library is a collection of different DNA sequences from an organism, each of which has been cloned into a vector for ease of purification, storage, and analysis. There are essentially two types of gene library that can be made depending on the source of the DNA used. If the DNA is genomic DNA, the library is called a **genomic library**. If the DNA is a copy of an mRNA population, that is cDNA, then the library is called a **cDNA library** (Section Q2). When producing a gene library, an important consideration is how well it represents the starting material, i.e. does it contain all the original sequences (a **representative library**)? If certain sequences have not been cloned, for example repetitive sequences lacking restriction sites (Section C4), the library is not representative. Likewise, if the library does not contain a sufficient number of clones, then it is probable that some genes will be missing. cDNA libraries that are enriched for certain sequences will obviously lack others, but if correctly prepared and propagated they can be representative of the enriched mRNA starting material (Section Q2).

Size of library

It is possible to calculate the number (N) of recombinants (plaques or colonies) that must be in a gene library to have a particular probability of obtaining a desired sequence. The formula is:

$$N = \frac{\ln (1 - P)}{\ln (1 - f)}$$

where P is the desired probability and f is the fraction of the genome in one insert. For example, for a probability of 0.99 with insert sizes of 20 kb, these values for the 4.6 Mb *E. coli* and 3000 Mb human genomes are:

$$N_{\text{E. coli}} = \frac{\ln (1 - 0.99)}{\ln [1 - (2 \times 10^4 / 4.6 \times 10^6)]} = 1.1 \times 10^3$$

$$N_{\text{human}} = \frac{\ln (1 - 0.99)}{\ln [1 - (2 \times 10^4 / 3 \times 10^9)]} = 6.9 \times 10^5$$

These values explain why it is possible to make good genomic libraries from prokaryotes in plasmids where the insert size is 5–10 kb, as only a few thousand recombinants will be needed. For larger genomes, the larger the insert size the fewer recombinants are needed, which is why cosmid and YAC vectors have been developed (Section P2). However, the methods of cloning in λ and the efficiency of λ packaging still make it a good choice for constructing genomic libraries.

Genomic DNA

To make a representative genomic library, genomic DNA must be purified and then broken randomly into fragments that are the correct size for cloning into the chosen vector. Cell fractionation will reduce contamination from organelle DNA (mitochondria, chloroplasts). Hence, purification of nuclear genomic DNA from eukaryotes is usually carried out by first preparing cell nuclei and then removing proteins, lipids, and other unwanted macromolecules by **protease digestion** and **phase extraction** with something that selectively solubilizes the DNA, such as a phenol–chloroform mixture. Prokaryotic cells can be extracted directly. Genomic DNA prepared in this way is composed of long chromosomal fragments of several hundred kilobase pairs. There are two basic ways of fragmenting this DNA in an approximately random manner – **physical shearing** and **restriction enzyme digestion**. Physical shearing by vigorous pipetting or **sonication** (Section B1) will break the DNA progressively into smaller, random fragments. The choice of method and time of exposure depend on the size requirement of the chosen vector. The ends produced are likely to be blunt ends due to breakage across both DNA strands, however end repair with Klenow polymerase can be performed in case some ends are not blunt. This DNA polymerase will fill in any recessed 3′-ends on DNA molecules if dNTPs are provided (Section O1, Table 1).

Digestion of genomic DNA with restriction enzymes (Section O3) is more prone to nonrandom fragmentation due to the nonrandom distribution of restriction sites. To generate genomic DNA fragments of 15–25 kb or greater (a convenient size for λ and cosmid vectors), it is necessary to perform a **partial digest**, where, by using limiting amounts of restriction enzyme, the DNA is not digested at every recognition sequence that is present, thus producing molecules of lengths greater than in a complete digest. A common enzyme used is Sau3A (recognition sequence 5'-/GATC-3', where '/' denotes the cleavage site) as it cleaves to produce a sticky end that is compatible with a vector that has been cut with BamHI (5'-G/GATCC-3'). The choice of restriction enzyme must take into account the type of ends produced (sticky or blunt), whether they can be ligated directly to the cleaved vector, and whether the enzyme is inhibited by DNA base modifications (such as **CpG methylation** in mammals; Section C3). The time of digestion and ratio of restriction enzyme to DNA are varied to produce fragments spanning the desired size range (i.e. the sizes that efficiently clone into the chosen vector). The correct sizes are then purified using an agarose gel (Section O3) or a sucrose gradient.

Vectors

In the case of organisms with small genomes such as *E. coli*, a genomic library could be constructed in a plasmid vector (Section P1) as only 5000 clones of average insert size 5 kb would give a >99% chance of cloning the entire 4.6-Mb genome. Most libraries from organisms with larger genomes are constructed using **phage λ, cosmid, BAC,** or **YAC** vectors (Section P2). These accept inserts of approximately 23, 45, 350, and 1000 kb respectively, and so fewer recombinants are needed for complete genome coverage than if plasmids were used. The most commonly chosen genomic cloning vectors are λ **replacement vectors** (EMBL3, λDASH, and their derivatives) and cosmids, and a typical λ vector cloning scheme is shown in Section P2, Figure 2. The λ DNA must be digested with restriction enzymes to produce the two λ end fragments, or λ arms, between which the genomic DNA will be ligated. With the original vectors, after digestion the arms would be purified from the central (stuffer) fragment on gradients, but newer λ vectors allow digestion with multiple enzymes, which makes the stuffer fragment unclonable. Following λ arm preparation, the genomic DNA prepared as described above is ligated to them using **T4 DNA ligase** (Sections O1 and O4). Once ligated, the recombinant molecules are ready for **packaging** and **propagation** to create the library (Section P2). With most modern λ-based vectors it is possible to (i) positively select for recombinants that contain the genomic DNA inserts; (ii) avoid having to purify the λ vector arms; and (iii) avoid creating **empty vectors** that contain no insert.

In some circumstances, e.g. limited supplies of starting material, or where there is known sequence information for the target, the PCR (Section R3) could be used to either amplify the starting material prior to genomic library construction, or to directly amplify the target gene and thus avoid having to make and screen a genomic library. However, in many cases the creation of a good quality and representative genomic library is preferable.

Q2 cDNA libraries

Key Notes

Principle

Being derived from mRNA, cDNA libraries represent only the protein-coding genes and are therefore tissue-specific. They are a good source of individual cDNAs that can be used to express the encoded protein in *E. coli*.

mRNA isolation, purification, and fractionation

mRNA can be readily isolated from lysed eukaryotic cells by adding magnetic beads which have oligo(dT) covalently attached. The mRNA binds to the oligo(dT) via its poly(A) tail and so can be isolated from solution. The integrity of an mRNA preparation can be checked by translation *in vitro* and visualization of the translation products by electrophoresis. Integrity can also be studied by electrophoresis of the mRNA itself, which also allows the mRNA to be size-fractionated. Specific sequences can be removed or selected from the total mRNA pool by hybridization.

Synthesis of cDNA

In first strand synthesis, RT is used to make a cDNA copy of the mRNA by extending a primer, usually oligo(dT) or an oligo(dT)-adaptor, by the addition of deoxyribonucleotides to the 3′-end. Oligo(dG) or an oligo(dG)-adaptor can prime second strand synthesis to give duplex cDNA with or without an amplification step. Cutting the duplex with a rare-cutting restriction enzyme prepares it for ligation.

Ligation to vector

The vector is usually dephosphorylated with alkaline phosphatase to prevent self-ligation, and so promote the formation of recombinant molecules. Plasmid or phage vectors can be used to make cDNA libraries, but the phage λgt11, or its derivatives, are preferred for the construction of expression libraries.

Gene synthesis

Chemically synthesized oligonucleotides of up to 250 nt corresponding to overlapping regions of both strands of the desired DNA can be annealed and then ligated together to form larger molecules. In turn these sections can be ligated into longer gene sequences.

Related topics

(J3) mRNA processing, hnRNPs, and snRNPs
(O1) DNA cloning: an overview
(P1) Design of plasmid vectors

(P2) Bacteriophages, cosmids, YACs, and BACs
(Q1) Genomic libraries
(Q3) Screening procedures

Principle

As they are derived from mRNA, eukaryotic cDNAs have no intron sequences and represent only the protein-coding parts of the genome. As each cell or tissue expresses a characteristic set of genes and proteins (which may alter in response to stimuli or during development), cDNA libraries are tissue- and condition-specific. Thus, there is no such thing as a human cDNA library; instead you can have a human normal liver cDNA library, a human liver cancer cDNA library, a human brain cDNA library, etc. Furthermore, as mRNA preparations from particular tissues usually contain some specific sequences at higher abundance (e.g. globin mRNA in erythrocytes), choosing the correct starting tissue can greatly facilitate cloning and isolation of a specific cDNA. This, and the lack of introns, is useful if you wish to express the encoded protein in *E. coli* (Sections A5 and P1). Generally, cDNA libraries are not made from prokaryotic mRNA as it is very unstable and introns in protein-coding genes are very rare in prokaryotes; genomic libraries are easier to make and contain all the genomic sequences.

mRNA isolation, purification, and fractionation

Most eukaryotic mRNAs are polyadenylated (Section J3) and this 3'-tail of about 200 adenine residues provides a useful method for isolating eukaryotic mRNA. Synthetic **oligo(dT)** can be attached to some form of solid support and used to bind the poly(A) tail by hybridization and so recover the mRNA free of other RNAs from a cell lysate. A rapid way to do this that keeps hydrolysis by nucleases to a minimum, is to add oligo(dT) linked to **magnetic beads** to the cell lysate and 'pull out' the mRNA with a magnet. In some circumstances, lysing cells and then preparing mRNA–ribosome complexes (polysomes, Section L1) on sucrose gradients may provide an alternative route for isolating mRNA. Before using an mRNA preparation for cDNA synthesis, it is advisable to check that it is not degraded. This can be done by either translating the mRNA in a **cell-free translation system** (e.g. **wheat germ extract**, or **rabbit reticulocyte lysate**), or by size analysis using agarose or polyacrylamide gel electrophoresis. An intact mRNA preparation usually produces a smear of molecules from about 0.5 kb up to about 10 kb or more. Some sequences derived from other transcripts, e.g. rRNA, can end up in a cDNA library due to nonspecific priming by oligo(dT) or if other primers are used.

Sometimes it can be useful to fractionate or enrich the mRNA prior to cDNA synthesis, especially if the aim is to clone a particular gene or genes rather than make a complete cDNA library. Size fractionation is usually performed on agarose gels and enrichment is usually carried out by **hybridization**. For example, to make a cDNA library of all the mRNA sequences induced in a cell after hormone treatment, mRNA if first prepared from both induced and uninduced cells. First strand cDNA (see below) made from the uninduced cell mRNA is then hybridized to the induced cell mRNA. Sequences common to both preparations form cDNA-mRNA duplexes but the induced mRNAs remain single-stranded. This mRNA can then be isolated (e.g. using magnetic beads) and a cDNA library constructed. Such a library is called a **subtracted cDNA library**.

Synthesis of cDNA

An efficient scheme for making cDNA is shown in Figure 1. During first strand synthesis, the enzyme **RT** (Section O1, Table 1) extends a synthetic primer to make a DNA copy of the mRNA template. This can be oligo(dT), but in the method shown in Figure 1, an oligo(dT)-**adaptor primer** is used. An adaptor primer has an additional sequence at the 5'-end that contains a restriction enzyme cleavage site. Alternatively, to make **random-primed cDNA**, a random mixture of all possible hexanucleotides would be used. All four

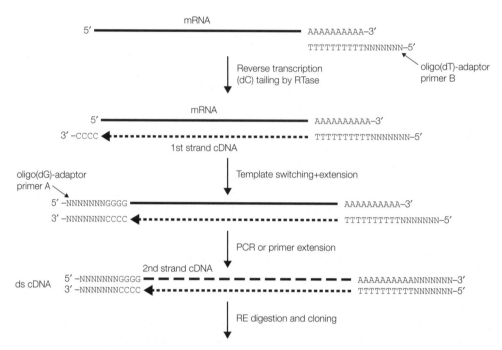

Figure 1. cDNA cloning using adaptor primers. NNNNNNN denotes the adaptor part of the primer sequence that contains the recognition site for a restriction enzyme, usually a rare-cutter.

dNTPs are required. The enzyme copies the mRNA template by adding the complementary nucleotides to the 3′-end of the extending primer.

Second strand synthesis also requires a primer. Although there are a number of variations, perhaps the best way of making full-length cDNA is to '**tail**' the 3′-end of the first strand by the nontemplated addition of nucleotides using **terminal transferase** (Section O1, Table 1) and then use a complementary primer to make the second strand. Originally, this was carried out in separate steps. However, RT adds nucleotides (normally a few C residues) to the 3′-end of the first strand cDNA without the need for a template. Thus, if oligo(dG) (or an oligo(dG)-adaptor primer as shown in Figure 1) is included in the reaction, when the RT reaches the 3′-end of the first strand the oligo(dG)-adaptor primer can anneal to this short run of 3′-C residues. The RT then switches templates from the mRNA to the adaptor primer and extends the first strand cDNA product so it now has the complement of the adaptor primer sequence at its 3′-end. At this point primer extension by a DNA polymerase can be performed to make the second strand, giving duplex cDNA. If starting material is in short supply, a PCR amplification step can be carried out (Section R3). The product of either of these reactions is duplex cDNA, and while the ends of the molecules may not be suitable for direct cloning, the duplex can be digested by a rare-cutting restriction enzyme whose recognition site is present (hopefully) only in the adaptor-derived sequence. Choosing a rare-cutting enzyme site for the adapter increases the chances that the cDNA will not be cut. After digestion, the sticky ended duplex cDNA can be easily cloned into an appropriate vector in a directional manner (Section O3).

Ligation to vector

As cDNAs are relatively short (0.5–10 kb), plasmid vectors are often used; however, for greater numbers of clones and especially for **expression cDNA libraries**, λ phage vectors are preferred (Section P2). After cleaving with the appropriate restriction enzyme(s), the vector is usually **dephosphorylate** with the enzyme **alkaline phosphatase** to prevent the vector fragments (ends) rejoining during ligation. This ensures that only recombinant molecules are produced by joining the vector and the cDNA (Section O4). Expression vectors like λgt11 and its derivatives have restriction site(s) placed within in its *lacZ* gene, enabling expression of the cDNA as part of a β-**galactosidase fusion protein**. This aids screening of the library (Section Q3). Ligation of vector to cDNA is carried out using T4 DNA ligase, and the recombinant molecules are either packaged (Section P2) or transformed (Section O2) to create the cDNA library.

Gene synthesis

Although there have been many developments to both cDNA cloning vectors and methods for cDNA synthesis, significant effort is still required to make a good cDNA library or to obtain a full-length cDNA for a particular gene. For the latter application it may be quicker/cheaper/better to make the required DNA from scratch by chemical synthesis, provided that its DNA sequence is known. Nucleotide precursors with chemically protected groups called **phosphoramidites** are used to build up chains of oligonucleotides up to 250-nt long on a solid support by automated cycles of addition from the 3′-end (i.e. the opposite to DNA synthesis by DNA polymerases) using a **DNA synthesizer**. These oligonucleotides are more than adequate for the production of probes for screening (Section Q3) or primers for sequencing (Section R2), PCR (Section R3), or mutagenesis (Section R5) but, as the efficiency of synthesis drops above 250 nt, full gene synthesis requires additional steps.

Multiple oligonucleotides covering overlapping sections of the entire sequence of both strands are synthesized, phosphorylated, and then annealed and ligated together. Some methods use a form of PCR (Section R3) to enzymatically assemble the complete duplex cDNA, but there is a tendency to introduce errors into the sequence with this approach, which would need to be identified and corrected. Consequently, molecules much larger than 1.5–2 kb are not routinely made this way. However, longer genes can be assembled by ligating several of these 2-kb molecules together. In this way, the entire 1-Mb genome of the small bacterium *Mycoplasma mycoides* was synthesized in the laboratory in 2010 by J. Craig Venter and colleagues and shown to be functional (Section S7). Gene synthesis can be used to compensate for **codon usage bias** (Section K1) when expressing a 'foreign' cDNA (e.g. human) in a host cell (e.g. *E. coli*). The codons used to specify each amino acid can be altered to correspond to the most abundant tRNAs in the host organism, so increasing the efficiency of translation (Section K2).

Q3 Screening procedures

Key Notes

Screening	Screening to isolate one particular clone from a gene library routinely involves a nucleic acid probe for hybridization. The probe binds to its complementary sequence allowing the required clone to be identified.
Colony and plaque hybridization	A copy of the position of colonies or plaques on a petri dish is made on the surface of a membrane, which is then incubated in a solution of labeled probe. After hybridization and washing, the location of the bound label is determined. The group of colonies/plaques to which the label has bound is diluted and re-plated in subsequent rounds of screening until an individual clone is obtained.
Expression screening	Most of the clones in a cDNA expression library will produce their encoded polypeptide (or part of it). Depending on the vector used this may be as a fusion protein with, for example, vector-encoded β-galactosidase. Antibodies to the desired protein can be used to screen the library in a process similar to plaque hybridization to obtain the particular clone.
Functional screening	Batches of expressing clones are assayed for the biological function of the gene of interest, for example an enzyme activity or the restoration of a mutant phenotype by functional complementation. Positive batches are subdivided and re-assayed until a single clone conferring the function is obtained.
Related topics	(P1) Design of plasmid vectors (R1) Characterization of clones (P2) Bacteriophages, cosmids, YACs, and BACs

Screening

The process of identifying one particular clone containing a gene of interest from among the very large number of others in the gene library is called **screening**. Some knowledge of the gene or its product is required, such as a corresponding cDNA fragment or a related sequence that can be used as a **nucleic acid probe**, which is labeled in some way to allow detection. If sufficient of the protein product is available to permit determination of some amino acid sequence (Section A5), this information can be used to derive a mixture of possible DNA sequences that could encode that amino acid sequence, bearing in mind the redundancy of the genetic code (Section K1). This DNA sequence information can be used to make a nucleic acid probe to screen the library by **hybridization**. One of the most common ways to make DNA probes for library screening uses PCR (Section R3) to make **PCR probes**. Short oligonucleotide probes are readily produced by automated chemical synthesis (Section Q2) and these can be used directly for hybridization or to make longer PCR probes for hybridization. If a pair of primers can be designed, PCR can also be used to screen a library since a PCR product will only be detected if the sequence (i.e. clone) is present, and so pools of clones can be successively subdivided until a single positive clone is isolated. If antibodies had been raised to the protein (Section A5) these could be used to detect the presence of a clone that expressed the protein, often as part of a fusion protein. cDNA libraries that express the protein in a functional form can be screened for biological activity.

Colony and plaque hybridization

Colonies are visible groups of cells on an agar plate that are derived from the multiplication of a single cell (e.g. a bacterium containing a plasmid vector), whereas plaques are areas of cell lysis in a lawn of bacteria produced by an infecting phage (e.g. a λ vector) (Section P2). After the initial step, colony, and plaque hybridization of gene libraries in plasmid or phage vectors are essentially the same. The first step involves transferring some of the DNA in the plaque or colony to a **nylon** or **nitrocellulose membrane** as shown in Figure 1. Because plaques are areas of lysed bacteria, the phage DNA is directly available and will bind to the membrane when it is placed on top of the petri dish. Bacterial colonies must be lysed first to release the plasmid DNA. First, a replica of the colonies on the dish is created on the membrane by simply laying it on top, removing it, and allowing the colonies to grow (**replica plating**). The bacteria on the membrane are then lysed by soaking it in a protease solution containing SDS. The original plate is kept to allow subsequent growth and isolation of the required clone (i.e. colony with recombinant plasmid) after identification. Recombinant λ phage can be isolated from the remaining material on the original dish of plaques. In both cases, the DNA on the membrane is denatured

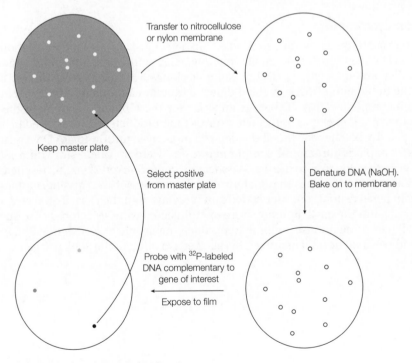

Figure 1. Screening by plaque hybridization.

with alkali to produce single strands, which are then bonded to the membrane by baking or UV irradiation. The membrane is then immersed in a solution containing a labeled nucleic acid probe and incubated to allow the probe to hybridize to its complementary sequence. After hybridization, the membrane is washed extensively to remove unhybridized probe, and regions where the probe has hybridized are then visualized. If the probe was radioactively labeled, this is carried out by **autoradiography** (exposure to X-ray film) or, if the probe was labeled with a modified nucleotide, by using a colorimetric assay that depends on the modification (Section R1). By comparing the membrane with the original dish and lining up the regions of hybridization, the original group of colonies or plaques can be identified. This group is re-plated at a much lower density, and the hybridization process repeated until a single individual clone is isolated.

Expression screening

If a cDNA is cloned into an expression vector, the cDNA will be translated into its gene product. Depending on the vector used the translation product may be a fusion protein such as a *lacZ* or *lacZ'* (β-galactosidase) fusion (Section P1). The fusion protein will contain regions of polypeptide (**epitopes**) that can be recognized by **antibodies** raised to the native protein. These antibodies can therefore be used to screen the **expression library**. The procedure has similarities to the plaque hybridization protocol (above) in that a '**plaque lift**' is taken by placing a membrane on the dish of plaques, though in this case it is the protein encoded by the cDNA rather than the DNA itself that is detected on the membrane. The membrane is treated to covalently attach the protein, and immersed in a solution of the antibody. When the antibody has bound to its epitope, it is detected by standard immunochemical procedures. In this way, the location of the expressing plaque can be narrowed down. Repeat cycles of screening are again required to isolate pure plaques.

Functional screening

While expression screening with antibodies can detect just a part of the translation product of the cDNA (epitope), functional screening relies on the detection of the function of the encoded protein in either cDNA or genomic clones. This usually requires the expression of the full-length cDNA but could detect a functional domain of the protein made from an incomplete cDNA. If there is an assay for the biological activity of the target protein, then pools of clones induced to express their products can be assayed in batches and any positive batches can be subdivided and re-assayed until a single cDNA is isolated. Another example is **functional complementation**. Here an organism with a lethal or mutant gene (e.g. the mouse deafness-associated gene, **Shaker-2**) can be rescued/cured by providing pools of clones that include one that expresses the missing/mutant function. Again positive pools are subdivided and re-assayed. In the case of Shaker-2, mouse eggs carrying the Shaker-2 mutation were microinjected with BAC clones corresponding to a small part of the mouse genome to produce transgenic mice (Section S5), some of which had regained normal hearing. This allowed isolation of the Shaker-2 gene.

R1 Characterization of clones

Key Notes

Characterization
Clone characterization involves determining properties of a recombinant DNA molecule such as size, restriction map, orientation of any transcribed sequence present, and nucleotide sequence. It requires a purified preparation of the cloned DNA.

Restriction mapping
Digesting recombinant DNA molecules with single or multiple restriction enzymes allows the construction of a restriction map of the molecule indicating the cleavage positions and fragment sizes.

Labeling nucleic acid
DNA and RNA can be end-labeled using polynucleotide kinase or terminal transferase. Uniform labeling requires DNA or RNA polymerases to synthesize a complete labeled strand.

Southern and northern blotting
Nucleic acid species separated in a gel are transferred (blotted) on to a membrane and then hybridized with a labeled nucleic acid probe. After washing to remove unhybridized probe, the membrane is processed to reveal the bands of interest. When genomic DNA or parts of cloned genes are blotted, this is called a Southern blot; if RNA species, it is known as a northern blot.

Related topics

(O1) DNA cloning: an overview	(O3) Restriction enzymes and electrophoresis
(O2) Preparation of plasmid DNA	(Q1) Genomic libraries
	(Q3) Screening procedures
	(R2) Nucleic acid sequencing

Characterization

The characterization of a genomic or cDNA clone begins with the preparation of pure DNA. Subsequently some or all of the following may be determined: clone size, DNA insert size, features of the insert such as its pattern of cleavage by restriction enzymes (**restriction map**), the position and polarity of any transcribed sequence that it may contain (Section P1), and some or all of the insert sequence. Before DNA sequencing became simple and routine (Section R2), these characterizations were performed roughly in the order listed, but nowadays sequencing all or part of a clone is generally the first step, and this sequence will provide the size, restriction map, and the orientation of any predicted genes (**ORFs**; Section K1). Nevertheless, experiments must still be performed to verify that the predicted gene is actually transcribed into RNA in some cells of the organism and that, for protein-coding genes, a protein of the size predicted is made *in vivo*. The preparation of plasmid DNA from bacterial colonies has been described in Section O2. DNA can readily be made from bacteriophage clones by isolating the phage particles from infected bacterial cell cultures after phage-induced cell lysis and then extraction of the DNA with phenol–chloroform (Section O2).

Restriction mapping

The sizes of linear DNA molecules can be determined by agarose gel electrophoresis using marker fragments of known sizes (Section O3). To determine the size of the cloned genomic or cDNA insert in a plasmid or phage clone, it is best to digest the DNA with a restriction enzyme that separates the insert and vector sequences. For a cDNA constructed using EcoRI adaptors (Section Q2), digestion with EcoRI would achieve this. Running the digested sample on an agarose gel with size markers gives the size of the insert fragment(s) as well as the size of the vector (already known). This information is part of what is needed to draw a map of the recombinant DNA molecule, but the **orientation of the insert** relative to the vector (Section P2), and possibly the order of multiple fragments, is not known. This can be determined by performing digests with different restriction enzymes, particularly in combinations. Figure 1 illustrates the process for a lEMBL3 genomic clone. SalI cuts adjacent to the EcoRI or BamHI cloning sites releasing the insert from the vector. As there are no internal SalI sites in the insert, there are only three fragments of approximately 9, 15, and 19 kb. HindIII cuts only once in the short arm of the vector, about 4 kb from the end. In this example, HindIII cuts twice in the insert as well, giving fragments of 4, 7, 11, and 21 kb. Although the 4 kb is part of the short arm and the 21 kb must be the long arm plus 2 kb (19+2=21), the order of the other two fragments is unknown at present. The SalI+HindIII **double digest** gives fragments of 2, 4, 5, 6, 7, and 19 kb, of which the 19 kb and the 4 and 5 kb together are the two vector arms. Since SalI cuts at the cloning sites, it cuts the 21 kb into 19+2 kb and the 11 kb into 6+5 kb, of which the 6 kb is part of the insert. The 7-kb HindIII fragment has not been cut by SalI, and this confirms that it is in the central region of the insert, giving the map as shown.

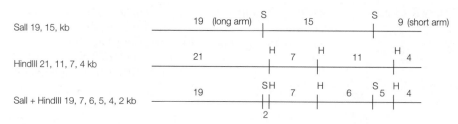

Figure 1. Restriction mapping using single and double digestion to completion.

Labeling nucleic acid

DNA and RNA molecules can either be labeled at their ends (**end labeling**) or throughout their length (**uniform labeling**). **5′-End labeling** is performed using **polynucleotide kinase** to add a radioactive phosphate (^{32}P or ^{33}P) to nucleic acids with a free 5′-OH group, which can be created by **dephosphorylation** using **alkaline phosphatase**. The external γ-phosphate of ATP is the source of the label (Section O1, Table 1). **3′-End labeling** can be performed using the enzyme **terminal transferase** to add one or more nucleotides to the 3′-end of nucleic acids (Section O1, Table 1, and Section Q2). These can be radiolabeled, e.g. [^{32}P]dNTP, or preferably tagged with a nonradioactive moiety such as the B vitamin biotin, e.g. biotinylated-dUTP (a dTTP analog). The biotin is detected through its high affinity for the protein **streptavidin** which, when used as a streptavidin–enzyme conjugate, can catalyze a visible colorimetric reaction. Other tags can be bound by specific antibodies. Nucleotides tagged with fluorescent dyes are used to label DNA for automated sequencing (Section R2).

DNA polymerases can be used to make uniformly labeled, **high specific activity** DNA for use as probes, etc. In **nick translation**, the duplex DNA is treated with a tiny amount of **DNase I**, an **endonuclease** that introduces random nicks along both strands. **DNA polymerase I** binds to these nicks and removes nucleotides using its **5′→3′ exonuclease** activity. As it removes a 5′-residue at the site of the nick, it adds a nucleotide to the 3′-end, incorporating labeled nucleotides using its polymerizing activity in a reaction identical to primer removal and gap filling during DNA replication (Section D2). DNA polymerases can also be used to make probes by first denaturing the duplex DNA template in the presence of all possible hexanucleotides. These anneal at positions of complementarity along the single strands and act as **primers** for the polymerases to synthesize a complementary, labeled strand (see also PCR, Section R3).

Being double-stranded, DNA can also be labeled **strand-specifically**. Strand-specific DNA probes can be made by using single-stranded DNA obtained after cloning in an M13 phage vector (Sections M2 and P2), or duplex DNA that has had part of one strand removed by a **3′→5′ exonuclease**. DNA polymerases can re-synthesize the missing strand incorporating labeled nucleotides in the process. Strand-specific RNA probes are generated by *in vitro* **transcription** using **SP6**, **T7**, or **T3 phage RNA polymerases** (Section P1). If the desired sequences are cloned into an *in vitro* transcription vector (e.g. pGEM, pBluescript, or their derivatives) which has one of these polymerase promoter sites at each end of the cloning site, then a **sense** RNA transcript (the same strand as the natural RNA transcript) can be made using one polymerase, and an **antisense** one (complementary to the natural transcript) using the other. Labeled NTPs are incorporated. Strand-specific probes are useful for northern blots (see below) as the antisense strand will hybridize to cellular RNA while the sense probe will act as a control.

Southern and northern blotting

These are used to detect specific DNA or RNA molecules in the many that separate from a mixture on an agarose gel. **Southern blots**, named after their inventor Ed Southern, are used for DNA while **northern blots,** named by analogy, are used for RNA. Both employ specific hybridization probes. After separation, the nucleic acid molecules are transferred to a nylon or nitrocellulose membrane (Figure 2). This is often done by capillary action, but can be achieved by electrotransfer, vacuum transfer, or centrifugation. In Southern blotting, the DNA is usually denatured with alkali before transfer, (Section B1), so it is single stranded and ready for hybridization. This is not necessary for RNA and would in any case hydrolyze the molecules (Section B1). Once transferred, the procedure for Southern and northern blotting is identical. The nucleic acid must be bonded to the membrane, hybridized to the labeled probe, washed extensively and the hybridized probe then detected by autoradiography, if radiolabeled, or by some other method if labeled with tagged nucleotides (see above). The **stringency** of the hybridization and washing conditions is critical. Stringency is determined by the hybridization temperature and the salt concentration in the hybridization buffer (high temperature plus low salt is more stringent as only perfectly matched hybrids will be stable). If the probe and target are 100% identical in sequence, then high stringency hybridization can be carried out, giving high specificity of detection. For probes that do not match the target completely, the stringency must be reduced to a level that allows imperfect hybrids to form. If the

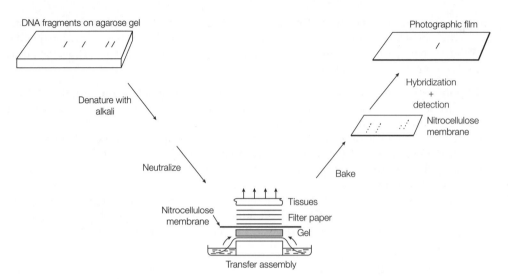

Figure 2. Southern blotting.

stringency of washing or hybridization is too low, then the probe may bind to too many sequences to be useful. Formamide (Section B1) can be included in the hybridization buffer to reduce the hybridization temperature by about 25°C, from the usual 68°C to the more convenient 43°C. The washing step should be carried out at 12°C below the theoretical melting temperature (T_m) of the probe and target sequences, using the formula: $T_m = 69.3° + 0.41\,[\%(G+C)] - 650/l$, where l is the length of the probe molecule.

Northern blots give information about the size of an mRNA and any precursors, and can be used to determine whether a cDNA clone used as a probe is full-length and whether it is one of a family of related (perhaps alternatively processed) transcripts. Northern blots can help to identify whether a genomic clone has regions that are transcribed and, if the RNA on the blot is made from different tissues, where these transcripts are made. Southern blots of cloned genomic DNA fragments can be probed with cDNA molecules to find which parts of the genomic clone correspond to the cDNA fragment. If the Southern blot contains genomic DNA fragments from the whole genome, the probe will give information about the size of the fragment that contains the gene and how many copies of the gene are present in the genome. Blots with DNA or RNA samples from different organisms (**zoo blots**) can show how conserved a gene is between species.

R2 Nucleic acid sequencing

Key Notes

Sanger DNA sequencing	Sanger's original method of DNA sequencing uses one of the four dideoxynucleotide triphosphates as a chain terminator to produce a mixture of different-length molecules generated by polymerase extension of a primer, all ending in the same specific nucleotide. These are separated by PAGE, adjacent to mixtures derived from use of the other three dideoxynucleotides, allowing the original sequence to be read up the gel. Improvements to the original method include the use of fluorescent chain terminators, capillary electrophoresis, laser detection and automation, ultimately reading up to 300 000 nt of sequence per run.
Shotgun sequencing	The sequencing of large stretches of DNA, or whole genomes, by the sequencing of random overlapping clones or amplified fragments, followed by assembly of the complete sequence by searching for overlaps of sequence between fragments by computer.
Emulsion PCR	A method for the separate clonal amplification of many millions of individual random DNA fragments, where PCR reactions take place on microscopic beads in separate aqueous droplets of an oil–water emulsion. Arrays of the beads can then form the basis of high-throughput parallel sequencing methods yielding anything up to 10s of billions of nucleotides of sequence per run.
Fluorescent reversible terminator sequencing	A high-throughput sequencing method related to Sanger sequencing, in which a mixture of four fluorescent terminator nucleotides is added enzymatically to a large array of primer-template pairs, leading to detection by imaging of the array. The terminators are then unblocked and fluorophores removed chemically to allow a new cycle of addition to read the next nucleotide in the template.
Pyrosequencing	A high-throughput sequencing method in which DNA polymerase and each of the four dNTPs in turn is added to an array of primer-template pairs. Addition of the correct nucleotide at a particular location in the array is detected by an enzymatic light-producing reaction driven by the release of pyrophosphate in the polymerase reaction.
Sequencing by ligation and other methods	The sequence of a template can also be detected by sequential specific ligation of one of a set of fluorescently labeled oligonucleotides to a primer annealed to the template. Other methods in development include those based on the sequence of single DNA molecules.

RNA sequencing	Four different RNases that cleave on the 3′ side of specific nucleotides are used to produce a ladder of fragments from end-labeled RNA. PAGE analysis allows the sequence to be read.	
Related topics	(O1) DNA cloning: an overview (P2) Bacteriophages, cosmids, YACs, and BACs	(Section S) Functional genomics and the new technologies

Sanger DNA sequencing

Consideration of the process of DNA replication (Section D1) means that we can appreciate how a DNA polymerase 'reads' the sequence of a template strand, in the process of synthesizing a new complementary strand from the 3′-end of a primer. We could determine the sequence of the template DNA if we could detect which nucleotide had been added by the enzyme at each addition step. The classical **Sanger** (**dideoxy**, or **chain terminator**) method of sequencing DNA was the first method that allowed this stepwise detection of the addition of nucleotides by a DNA polymerase. It has now largely been supplanted by newer, high-throughput methods (see below), but an understanding of its principles may be applied to the newer methods. The method involves the enzymatic synthesis of a set of DNA molecules with one common end but differing in length by one nucleotide. These are separated according to length by denaturing PAGE, using urea to denature the DNA (Section B1), to allow the sequence to be read. It uses a single-stranded DNA (ssDNA) template, DNA polymerase, dNTPs, and small amounts of each of the four **dideoxynucleotides** (**ddNTPs**) to synthesize complementary copies of the template, some of which terminate at each position (Figure 1). The ddNTPs act as **chain terminators** since they lack the deoxyribose 3′-OH group needed by the polymerase to extend the growing chain. A **sequencing primer** about 15–17-nt long is annealed to the template DNA, dNTPs added, and the reaction mixture split into four. A different ddNTP is added to each sample (e.g. one tube would contain dATP, dCTP, dGTP, and dTTP, with a small amount of ddATP, etc.). DNA polymerase (usually **Klenow** or **T7** DNA polymerase) is then used to extend the primer and copy the template. In each reaction, copying stops when the less abundant ddNTP (about 1% of the dNTP) is used by chance instead of the dNTP, leading to limited chain termination; statistically, this will occur at every position in the pool of DNA molecules. If one dNTP is labeled (e.g. $[\alpha\text{-}^{32}P]dATP$ or $[\alpha\text{-}^{35}S]dATP$, in which one oxygen atom on the α-phosphate is replaced by sulfur) or if the primer is end-labeled (Section R1) then all the products will be detectable by autoradiography after the four samples are separated side-by-side by denaturing PAGE. The sequence can then be read directly from the gel (Figure 1). This is of course the complementary product sequence, from which the original template sequence can be inferred. The original method required a ssDNA template with which to synthesize the complementary copy, which meant that genomic or cDNA DNA fragments had to be subcloned into the phage vector M13 (Sections M2 and P2) before sequencing.

Many improvements have been made to Sanger's original method, e.g. the use of double-stranded templates, which removes the need to subclone to produce ssDNA, or PCR products (Section R3), which eliminates the need to subclone at all. Thermostable DNA polymerases are commonly used with temperature cycling (**cycle sequencing**) to increase

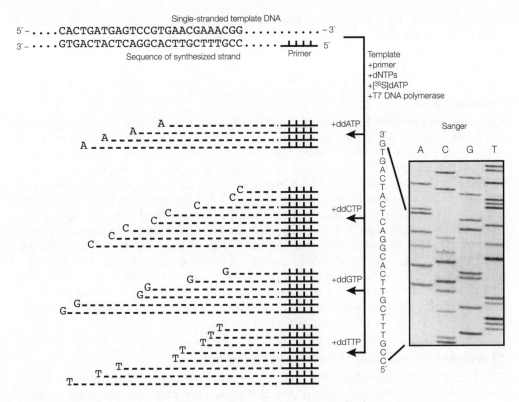

Figure 1. DNA sequencing by the Sanger chain terminator method.

the yield of sequencing products (Section R3). Fluorescent dye-tagged ddNTPs (**dye terminator sequencing**) are safer than radiolabeling, and, if each ddNTP is labeled with a dye that fluoresces at a different wavelength, then all four reactions can be run in a single gel lane with each color being distinguished separately by a laser scanner as each product band runs past a location at the foot of the gel. A further elaboration is to replace the slab gel with a single, narrow capillary of gel for each sample, which greatly reduces the separation time. These developments have allowed automated multi-capillary **DNA sequencers** based on the principle of chain termination to analyze as many as 384 samples simultaneously and produce up to 800 nt of sequence per sample in under 3 h.

Shotgun sequencing

Classically, most DNA sequencing up to and including much of the human genome project during the 1990s was based on the sequencing of a series of well-defined clones from cDNA or genomic DNA libraries (Sections Q1 and Q2) in a largely structured way. However, as more efficient methodologies were developed, the idea arose that it would be possible to sequence large stretches of DNA, including entire genomes, by sequencing in parallel a large array of random overlapping DNA segments, derived from a DNA library or later simply from fragments of genomic DNA, and then assembling the complete sequence computationally, by identifying the overlaps between the individual sequenced fragments. A number of next generation sequencing technologies have been developed that effectively rely on the use of this 'shotgun sequencing' approach, and which increase the throughput of DNA sequencing from the 300 000 nt per machine run of the capillary gel method, to 400–1200 million nt per run. These high throughput sequencing machines in some cases use elaborations of the Sanger method, although some use entirely different methods (see below).

Emulsion PCR

One important advance has been the use of emulsion PCR to clonally amplify single DNA molecules (Section R3) to provide sufficient identical template molecules to allow detection of the sequenced products. Essentially, random DNA fragments are modified by ligation of DNA adapters on to each end to provide common primer binding sites. These template molecules, primers and PCR reagents are mixed and emulsified in oil to form tiny aqueous droplets (microreactors) of around 10 mm diameter within the oil phase in which the PCR reaction takes place. One of the primers is attached to microscopic beads, many copies to each bead. If the DNA concentrations are controlled appropriately, each microreactor should contain only one template DNA molecule, and the reaction will then yield a mixture of many thousands or millions of beads, each with multiple copies of a single amplified DNA sequence attached. An array of these beads can then form the basis of the sequencing of many DNA fragments in parallel, using the high-throughput methods described below. A related method can directly produce tiny clusters of amplified random DNA sequences on the flat surface of a chip.

Fluorescent reversible terminator sequencing

The high-throughput method that it closest in principle to the Sanger method makes use of fluorescent reversible terminators. Several million small clusters, each containing around 1000 amplified DNA molecules, are arrayed on a treated surface (Figure 2). DNA polymerase plus primer plus a mixture of four fluorescent terminators are washed across the surface. A single nucleotide is added to each DNA and hence each DNA spot fluoresces a different color depending on the nucleotide added, and the color is recorded in an image of the surface. However, the chemistry of the terminators allows the fluorescent tag to then be removed, and the terminator to be unblocked, so repeated cycles of addition can be carried out. Comparison of successive images of each spot allows the sequence to be read as the color changes at each cycle. Over one billion spots can be analyzed in a single 4–11-day run, with 75 nt read from each spot. This corresponds to 100 billion nt of sequence per run.

Pyrosequencing

Pyrosequencing is a method that also relies on polymerase addition to a primer-template. In this case, the addition of a normal dNTP is detected through the production of a flash of light when pyrophosphate (PPi) is released as the dNTP is joined to the elongating DNA chain. Since only the correct (i.e. complementary) dNTP will be added to the growing DNA strand by the DNA polymerase, each dNTP is provided on its own in four stepwise reactions that include other light-generating reagents where the reagents are washed away between each step. When the correct dNTP is provided and incorporated as dNMP, the released pyrophosphate causes light emission. The light emission is proportional to the amount of PPi released so that runs of the same base in the sequence will give light signals that are 2, 3, 4, etc. times the strength of a single base. Pyrosequencing is also carried out, in parallel, with a million beads from emulsion PCR, each with a different random sequence, arrayed on a silicon plate. The DNA is then sequenced by washing the plate with the first dNTP, polymerase and lumigenic reagents, producing a light flash

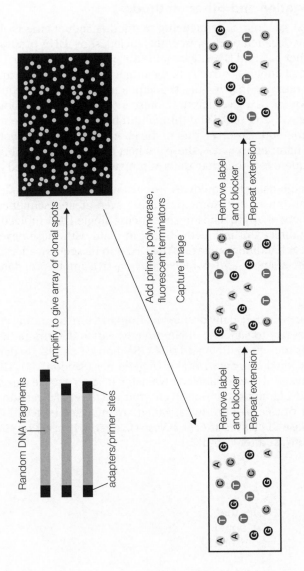

Figure 2. DNA sequencing by the fluorescent reversible terminator method. The sequences of many spots in parallel is revealed by the changing fluorescent color of each spot in successive cycles.

only at those locations where that dNTP is the next to be added. A repeating cycle of the four dNTPs is used and the light emission at each step is captured by successive images of the plate. Reads of about 400 nt are possible from each well, per 8-h run, around 400 million nucleotides in all.

Sequencing by ligation and other methods

The third major next generation sequencing method is **sequencing by ligation** (SBL) whereby a mixture of degenerate, fluorescently labeled, short DNA probes are annealed to a target DNA/primer pair and DNA ligase joins the probe to the primer. DNA ligase will only ligate a probe that anneals correctly to the target adjacent to the primer. The actual methodology is complex but, in principle, the sequence is read by the sequential ligation of short annealed sequences with different fluorescent labels. Reads of about 50 nt can be obtained per sample with run times and throughput being similar to the reversible fluorescent terminator method. However, due to the short read lengths of both the ligation and reversible terminator approaches, these are best used for determining variations in known reference sequences rather than determining whole unsequenced genomes.

Several other DNA sequencing technologies are currently being developed, some at more advanced stages than others, and some promising even greater sequence throughput. These include a variety of approaches to sequencing single DNA molecules by detecting individual bases directly or indirectly. Going in hand with these developments are improved methods for handling/preparing the DNAs to be sequenced as well as for the computer acquisition, analysis, and storage of the massive amounts of data produced.

RNA sequencing

Although RNA sequences can usually be inferred from the corresponding gene sequence, it is sometimes necessary to sequence RNA directly, e.g. to determine the positions of modified nucleotides present in tRNA and rRNA (Sections J1 and J2), to detect RNA editing (Section J4), or when dealing with viral RNA genomes (Section M4). This is achieved by base-specific cleavage of 5′-end-labeled RNA using **RNases** that cleave on the 3′ side of a particular nucleotide. Limiting amounts of enzyme and times of digestion are employed to generate a ladder of cleavage products, which are analyzed by PAGE. The following RNases are used: **RNase T1** cleaves after G, **RNase U2** after A, **RNase Phy M** after A and U, and ***Bacillus cereus* RNase** after U and C.

R3 Polymerase chain reaction

Key Notes

PCR

The PCR is used to amplify a sequence of DNA using a pair of oligonucleotide primers each complementary to one end of the DNA target sequence. These are extended towards each other by a thermostable DNA polymerase in a reaction cycle of three steps: denaturation, primer annealing, and polymerization.

The PCR cycle

The reaction cycle comprises a 95°C step to denature the duplex DNA, an annealing step of around 55°C to allow the primers to bind, and a 72°C polymerization step. Mg^{2+}, and dNTPs are required in addition to template, primers, buffer, and enzyme.

Template and primers

In theory, almost any source that contains one or more intact target DNA molecule can be amplified by PCR, providing appropriate primers can be designed. A pair of oligonucleotides of about 18–30 nt with similar G+C content will serve as PCR primers as long as they direct DNA synthesis towards one another. Primers with some degeneracy can also be used if the target DNA sequence is not completely known. Multiplex PCR uses multiple primer pairs to amplify more than one sequence simultaneously while inverse PCR allows regions outside that flanked by the primers to be amplified.

Enzymes

Thermostable DNA polymerases (e.g. Taq polymerase) are used in PCR as they survive the hot denaturation step. Some are more error-prone than others.

PCR optimization

It may be necessary to vary the annealing temperature and/or the Mg^{2+} concentration to obtain faithful amplification. With complex mixtures, a second pair of nested primers can improve specificity.

RT-PCR and RACE

Performing PCR after carrying out a reverse transcription reaction on mRNA is known as RT-PCR. This is useful for cDNA cloning, and for mRNA detection and quantification. If the resulting cDNA is not full-length, the missing ends can be generated by rapid amplification of cDNA ends (RACE).

Real-time and quantitative PCR

Real-time PCR monitors the progress of the PCR reaction by detecting the amount of product after each cycle. This can allow quantitation of the amount of template in a reaction by determining the number of PCR cycles required to produce a threshold level of product during the exponential phase of the reaction.

Related topics	(O1) DNA cloning: an overview	(R1) Characterization of clones
		(R2) Nucleic acid sequencing
	(Q1) Genomic libraries	(R5) Mutagenesis of cloned
	(Q3) Screening procedures	genes

PCR

If a pair of **oligonucleotide primers** complementary to the two strands of a duplex target DNA molecule are designed such that they can be extended by a DNA polymerase towards each other, then the region of the template between the primers can be copied over and over again by carrying out cycles of **denaturation**, **primer annealing**, and **polymerization**. This basic amplification process, for which Kary Mullis received a Nobel prize in 1993, is known as the **PCR** and it and its many variants have become essential tools in molecular biology as aids to cloning and gene analysis. The use of **thermostable DNA polymerases** has allowed automation of the cycling process. PCR can be used to make cDNAs, to generate labeled hybridization probes, to measure gene expression, and to introduce mutations, and has revolutionized DNA sequencing, genetic fingerprinting, and disease diagnosis.

The PCR cycle

Figure 1 shows how PCR works. The reaction mixture contains **target** (**template**) **DNA**, **PCR primers**, a **thermostable DNA polymerase**, **dNTPs**, and **Mg^{2+} ions**. In the first cycle, the DNA is denatured by heating to **95°C**, typically for around **60 s** (Section B2). The temperature is then reduced to around **55°C** for about **30 s** to allow the primers to anneal to the separated DNA strands. The actual temperature depends on the primer lengths and sequences (see below). After annealing, the temperature is increased to **72°C** for **60–90 s** for optimal polymerization, which requires **dNTPs** and **Mg^{2+}**. In the first polymerization step, the target strands are copied from the primer sites for various distances on each target molecule until the beginning of cycle 2, when the reaction is heated to 95°C again which denatures the newly synthesized molecules. In the second annealing step, the other primer can bind to the newly synthesized strand and during polymerization can only copy till it reaches the end of the first primer. Thus, at the end of cycle 2, some newly synthesized molecules of the correct length exist, though these are base paired to variable length molecules. In subsequent cycles, these soon outnumber the variable length molecules and increase two-fold with each cycle. If PCR were 100% efficient, one target molecule would become 2^n after n cycles. In practice, 20–40 cycles are commonly used.

Template and primers

Because of the extreme amplification achievable, PCR can amplify as little as one molecule of target DNA. Therefore, any source of DNA that provides one or more target molecules can in principle be used as a template for PCR. This includes DNA prepared from blood, sperm, or any other tissue, from older forensic specimens, from ancient biological samples, or in the laboratory from bacterial colonies or phage plaques as well as purified DNA. Whatever the source of template DNA, PCR can only be applied if some sequence information is known so that primers can be designed. However, when the amount of available target is extremely low, contamination by other DNAs, e.g. from the

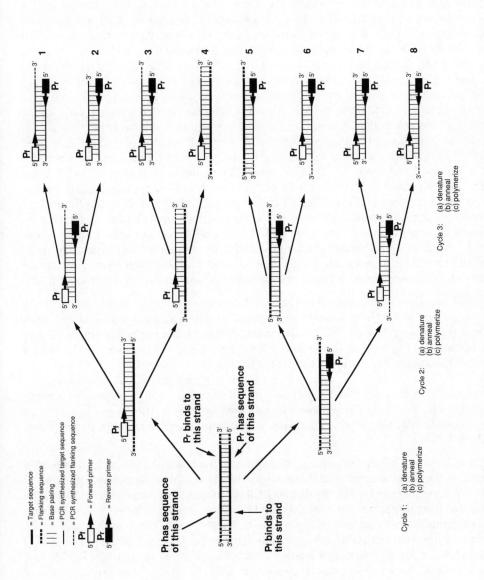

Figure 1. The first three cycles of a polymerase chain reaction. Only after cycle 3 are there any duplex molecules which are the exact length of the region to be amplified (molecules 2 and 7). After a few more cycles these become the major product.

experimenter, can be a major problem. Not all the DNA in a sample will be amplified, only the region between the two primer-binding sites.

Correct primer design is important but relatively easy if the target DNA sequence is known. The target region should be inspected for two sequences of about 18–30 nt with similar and suitable **G+C contents**, one on either side of the sequence to be amplified, and primers synthesized corresponding to these. The primers need to have a similar G+C content so that they anneal to their complementary sequences at similar temperatures (Section B2). They should not form secondary structure with themselves or with each other. For short oligonucleotides (<25 nt), the annealing temperature in °C can be calculated using the formula: $T_m=2(A+T)+4(G+C)$, where T_m is the melting temperature; the annealing temperature is chosen to be approximately 3–5°C lower. The **forward** and **reverse** primers are designed to anneal on opposite strands of the target sequence so that they will be extended towards each other by addition of nucleotides to their 3′-ends. They should preferably have a G or C at the 3′-end as either will form a more stable GC base pair with the target for initial elongation. As they will become part of the amplified product they are added in great excess over the target. Short target sequences amplify more easily, so this distance is often less than 500 nt, but with optimization PCR can amplify fragments over 10 kb in length. Since the primers are incorporated into the product, any novel features that are required in the product can be included in the primer sequences. Commonly, if the PCR product is to be inserted into a cloning vector, restriction enzyme recognition sequences are added to 5′-ends of the primers. Initially, these will form unpaired 5′ overhangs as only the 3′-end of the primer is complementary to the target, but the PCR process itself makes them double stranded. Mutations can also be deliberately introduced into the primers so that they appear in the product (Section R5).

If the DNA sequence of the target is not known, for example when trying to make a cDNA for a protein for which there is only some limited amino acid sequence available, then primer design is more difficult. For this, **degenerate primers** are designed using the genetic code (Section K1) to work out the various DNA sequences that could encode the known amino acid sequence. For example HisPheProPheMetLys is encoded by the DNA sequence 5′-CAYTTYCCNTTYATGAAR-3′, where Y=pyrimidine, R=purine, and N=any base. This sequence is 2×2×4×2×2=64-fold degenerate. Thus, if a mixture of all 64 sequences is made and used as a primer, then one of these sequences will be correct. The second primer must be made in a similar way. PCR using **degenerate oligonucleotide primers** is sometimes called **DOP-PCR**.

If multiple pairs of primers are added, PCR can be used to amplify more than one DNA fragment in the same reaction and these fragments can easily be distinguished on gels if they are of different lengths. This **multiplex PCR** is often used as a quick test to detect the presence of microorganisms in contaminated food or water or infected tissue. **Inverse PCR** is a modification that makes it possible to amplify (and hence clone) sequences upstream or downstream of the region amplified by the basic primer pair. For example, if genomic DNA is first digested by a restriction enzyme and then circularized by ligation, a pair of back-to-back primers can be used to amplify round the circle from a region of known sequence to obtain the 5′- and 3′-flanking regions up to the joined restriction sites.

Enzymes

The use of thermostable DNA polymerases isolated from thermophilic bacteria allows multiple PCR cycles to be performed without the need to add fresh enzyme after each cycle. For example, **Taq polymerase** from *Thermus aquaticus* has a half-life of 1.6 h at

95°C and so easily survives the 1–2-min denaturation step. Because it has no associated 3′→5′ proofreading exonuclease activity (Sections E2 and O1, Table 1), Taq polymerase is liable to introduce errors when it copies DNA – roughly one per 250 nt polymerized. For this reason, high fidelity proofreading polymerases, such as **Pfu** from *Pyrococcus furiosus*, are better for certain applications.

PCR optimization

PCR reactions are not usually 100% efficient, even when using pure DNA and primers of defined sequence. Usually the reaction conditions must be varied to improve the efficiency, or even to get it to work at all. This is very important when trying to amplify a particular target from a population of other sequences, for example one gene from genomic DNA or one cDNA from a cDNA library. If the reaction is not optimal, PCR often generates a smear of products on a gel rather than a defined band. The usual parameters to vary include the **annealing temperature** and the **Mg^{2+} concentration**. Too low an annealing temperature favors mispairing between primers and more distantly related sequences in the sample. The optimal Mg^{2+} concentration varies with each new sequence, but is usually between 1 and 4 mM. The specificity of the reaction can be improved by carrying out **nested PCR**, where, in a second round of PCR, a new set of primers is used that anneal within the fragment amplified by the first pair, giving a shorter PCR product. If on the first round of PCR some nonspecific products have been produced, giving a smear or a number of bands, nested PCR ensures that only the desired product is amplified from this mixture as it should be the only sequence present containing both sets of primer-binding sites.

RT-PCR and RACE

RT-PCR involves making first strand cDNA from an mRNA preparation, as in cDNA library construction (Section Q2), then amplifying one cDNA from this population by PCR using **gene specific primers**. It is generally the quickest way to produce a specific double-stranded cDNA for cloning (e.g. for recombinant protein production) and avoids the need to make a full cDNA library. It can also be used for mRNA quantitation. When only a fragment of cDNA has been produced by RT-PCR, or isolated from a cDNA library, it is possible to amplify the 5′-flanking sequence by first using terminal transferase to add a tail, e.g. oligo(dC), to the first strand cDNA (Figure 1, Section Q2). This allows a gene specific primer to be combined with an oligo(dG) primer to amplify the 5′-region. This technique is called **rapid amplification of cDNA ends** (**RACE**). 3′-RACE to amplify the 3′-flanking sequence of eukaryotic mRNAs uses a gene specific primer and an oligo(dT) primer which will anneal to the poly(A) tail at the 3′-end of the mRNA.

Real-time and quantitative PCR

Although in principle a PCR reaction results in exponential amplification of the product DNA, a standard reaction that is usually intended to provide enough product to be visible on a gel (generally 20–40 cycles) has generally progressed well beyond the exponential phase. The amount of product finally formed has been limited by the availability of primers or nucleotides, or possibly by the declining activity of the enzyme. This means that the amount of product detected bears almost no relation to the amount of template originally present, and the reaction is not at all quantitative in this mode. **Real-time** PCR can overcome this limitation of the method. In real-time PCR, the amount of product formed is monitored continuously as the PCR cycles progress. In the simplest version, use of a fluorescent dye such as ethidium bromide (Section B3) will produce increasing fluorescence as the amount of double-stranded product increases. Alternatively, fluorescent single-stranded probes complementary to the expected product can be used in a number of ways to report on the accumulation of that product specifically; in fact, with different fluorescent probes, **multiplexed** real-time PCR can be carried out to measure multiple products simultaneously.

In order to use real-time PCR to quantitate template DNAs (**qPCR**), the exponential phase of the reaction must be identified. Figure 2 shows an idealized pair of real-time PCR traces for different starting amounts of template, where the fluorescence, indicating amount of product, is plotted against the number of PCR cycles. As discussed above, the two reactions ultimately give a similar amount of product, even though the amount of starting template was very different. The exponential phase of the reaction is very close to the bottom axis, where detection of product is only just possible – this can be confirmed by plotting the log of the fluorescence against cycle number, which will give a linear plot if the reaction is exponential. A threshold level of product is chosen within the exponential phase, and the **threshold cycle** (C_t) at which a particular reaction exceeds the threshold is determined (Figure 2). A calibration curve for a particular reaction can then be determined by plotting the log of the template concentration (or number of molecules – **copy number**) against C_t for known template samples, and the amount of template in unknown samples can then be calculated. When combined with a reverse transcription step, qPCR has proved particularly valuable for the quantitation of individual gene transcripts, and provides a sensitive and potentially high throughput alternative to northern blotting (Section R1) for this purpose.

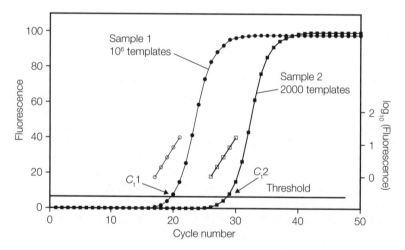

Figure 2 Real-time-PCR and the quantitation of template (qPCR) by determining the threshold cycle (C_t) for different amounts of template. A linear plot of log (fluorescence) against cycle number (open symbols) indicates that the reaction is exponential in that range.

R4 Analysis of cloned genes

Key Notes

Sequence organization | The polarity of oligo(dT)-primed cDNA clones is often apparent from the location of the poly(A), and so the coding region can be deduced. The presence and polarity of any gene in a genomic clone may be obvious from sequencing, but can be determined and confirmed by mapping and probing experiments.

S1 nuclease mapping | The 5′- or 3′-end of a transcript can be identified by hybridizing a longer, end-labeled antisense fragment to the RNA. The hybrid is treated with S1 nuclease to remove single-stranded regions, and the size of the remaining fragment is measured on a gel.

Primer extension | A primer is extended by a polymerase until the end of the template is reached and the polymerase dissociates. The length of the extended product indicates the 5′-end of the template.

Gel retardation | Mixing a protein extract with a labeled DNA fragment and running the mixture on a native gel will show the presence of DNA–protein complexes as retarded bands on the gel.

DNase I footprinting | The 'footprint' of a protein bound specifically to a DNA sequence can be visualized by treating the mixture of end-labeled DNA plus protein with small amounts of DNase I prior to running the mixture on a gel. The footprint is a protected region with few bands in a ladder of cleavage products.

Reporter genes | To verify the function of a promoter, it can be joined to the coding region of an easily detected gene (reporter gene) and the protein product assayed under conditions when the promoter should be active.

Related topics | (O1) DNA cloning: an overview   (Q1) Genomic libraries

Sequence organization

cDNA clones have a defined sequence organization, especially those synthesized using oligo(dT) as primer. Usually a run of A residues is present at one end of the clone, and this clearly defines its 3′-end. At some variable distance upstream of this there will be an ORF ending in a stop codon (Section K1). If the 5′-end of the cDNA clone is complete, it will have an ATG start codon preceded usually by only 20–100 nt. As genomic clones from eukaryotes are larger and may contain intron sequences as well as nontranscribed sequences, they provide a greater challenge to understanding their organization. After isolating a cDNA clone, it is common to subsequently obtain a genomic clone for the gene under study or vice versa. The problem is then to find which parts of the two clones correspond to each other. This means establishing which genomic sequences are present in the mature mRNA transcript. The genomic sequences absent from the cDNA clones are usually introns as well as sequences upstream of the transcription start site and downstream of the 3′-processing site (Sections H4 and J3). Most, but not all, of these questions can be answered by simply comparing the sequences of the clones. However, other important features that cannot be identified in this way are the start and stop sites for transcription and the sequences that regulate transcription (Sections F1 and H4), for which the techniques described below are necessary.

S1 nuclease mapping

This technique determines the precise 5′- and 3′-ends of RNA transcripts, although different end-labeled probes (Section R1) are required in each case. As shown in Figure 1, for 5′-end mapping, a labeled **antisense DNA** molecule is hybridized (Section B2) to the **RNA** preparation. If duplex DNA is used (still with only the antisense strand labeled), 80% formamide is required in the hybridization buffer to favor RNA–DNA hybrids rather than DNA duplex formation. The hybrids are then treated with the single strand-specific **S1 nuclease**, which will remove the single strand protrusions at each end. The remaining material is analyzed by PAGE next to size markers or a sequencing ladder. The size of the nuclease-resistant band, usually revealed by autoradiography, allows the end of the RNA molecule to be deduced.

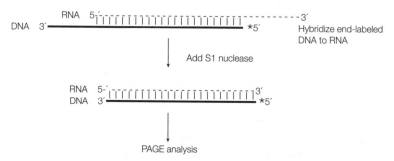

Figure 1. S1 nuclease mapping the 5′-end of an RNA. *=position of end label.

Primer extension

The 5′-end of an RNA transcript can also be determined using **RT** to extend an antisense DNA **primer** in the 5′ to 3′ direction, from the site where it base-pairs on the target to where the polymerase dissociates at the end of the template (Figure 2). The primer extension product is run on a gel next to size markers and/or a sequence ladder from which its length can be established.

Gel retardation

When the 5′-end of a gene transcript has been determined (e.g. by S1 mapping), the corresponding position in the genomic clone is the transcription start site. The DNA sequence upstream usually contains the regulatory sequences controlling when and where the

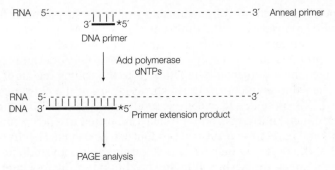

Figure 2. Primer extension. *=position of end label.

gene is transcribed. Transcription factors (Sections H5 and I1) bind to these sequences to activate or repress transcription. The technique of **gel retardation** (**gel shift analysis**) shows the effect of **protein binding** to a labeled nucleic acid and can be used to detect transcription factors binding to regulatory sequences. A short, labeled nucleic acid, such as the region of a genomic clone upstream of the transcription start site, is mixed with a cell or nuclear extract expected to contain the binding protein. Then, samples of labeled nucleic acid with and without extract are run on a nondenaturing gel of agarose or poly-acrylamide. If a large excess of unlabeled nucleic acid of different sequence is also present to bind proteins that interact nonspecifically, then the specific binding of a factor to the labeled molecule to form one or more DNA-protein complexes is shown by the presence of slowly migrating (**retarded**) **bands** on the gel by autoradiography, or other detection methods.

DNase I footprinting

Although gel retardation shows that a protein is binding to a DNA molecule, it does not provide the sequence of the binding site, which could be anywhere within the DNA fragment used. DNase I footprinting shows the actual region where the protein binds. Again, an end-labeled DNA fragment is required, and this is mixed with the protein preparation (e.g. a nuclear extract). After binding, the complex is very gently digested with the endonuclease **DNase I** to produce on average one cleavage per molecule. In the region of protein binding, the nuclease cannot easily gain access to the DNA backbone, and so fewer cuts take place there. When the partially digested DNA is analyzed by PAGE, a ladder of bands is seen showing all the random nuclease cleavage positions in control DNA. In the lane where protein was added, the ladder will have a gap, or region of reduced cleavage, corresponding to the **protein-binding site** ('**footprint**') where the protein has protected the DNA from nuclease digestion. Other DNA cleaving reagents may also be used in footprinting experiments, e.g. dimethyl sulfate or a mixture of Fe^{2+} ions and H_2O_2, which generates hydroxyl radicals, ($\bullet OH$) (Section E1). While gel retardation and DNase I footprinting allow visualization of protein-binding to the regulatory region(s) of one particular cloned gene, other methods, such as ChIP-seq (Section S2), permit genome-wide identification of all the sequences that bind a specific transcription factor.

Reporter genes

When a transcriptional promoter (Section F1) has been identified by sequencing, S1 mapping and DNA–protein binding experiments, it is common to attach the promoter region to a **reporter gene** to study its action and verify that the promoter has the properties being ascribed to it, e.g. level and timing of expression. A reporter gene is one whose product, usually a protein, is not normally expressed in the cell of interest and is easily detected (most proteins have no easily detectable property). For example, the promoter of the *HSP70* heat-shock gene could be attached to the coding region of the β-galactosidase gene. When this gene construct is expressed, and if the chromogenic substrate (X-gal, Section P1) is present, a blue color is produced. If the *HSP70* promoter–reporter construct is introduced into a cell, and the cell is subjected to a heat shock, β-galactosidase transcripts are made and the protein product can be detected by the blue color. This would show that the normally inactive *HSP70* promoter is activated after a heat shock because the **heat shock transcription factor** has bound to the regulatory sequence in the promoter and activated the reporter gene. Reporter genes can be attached directly to a promoter as described or can be fused to the end of the downstream coding region, in which case a **fusion protein** (Section A5) is produced containing the protein of interest (e.g. HSP70) fused to the reporter protein (e.g. β-Gal). Other examples of reporter genes are those for **chloramphenicol acetyltransferase** (which requires a **radioassay**), **green fluorescent protein (GFP)**, and the light-emitting **firefly luciferase**. The last two are very easy to detect and are particularly useful as tools for biological imaging (Section S4).

R5 Mutagenesis of cloned genes

Key Notes

Types of mutagenesis	Mutating DNA is very useful for defining the importance of particular sequences. Three major types of DNA alteration are site-specific mutations, insertions or deletions of longer sequences, and random mutations along the entire sequence.
Site-directed mutagenesis	Changing one or a few nucleotides at a particular site can most easily be achieved using PCR by annealing a pair of mutagenic primers to each strand of a plasmid containing the cloned gene. After PCR, the original unmutated template is digested using DpnI and the mutated plasmids are transformed into *E. coli*.
Insertion/deletion mutagenesis	Using forward and reverse mutagenic primers and other primers that anneal to common vector sequences, two PCR reactions are carried out to amplify the 5′- and 3′-portions of the DNA to be mutated. The two PCR products are mixed and used for another PCR using the outer primers only. Part of this product is then subcloned to replace the region to be mutated in the starting molecule.
Random mutagenesis by PCR	Error-prone PCR can be used to produce random mutations throughout a target sequence. It relies on the misincorporation of dNTPs by a DNA polymerase as it amplifies the target molecules. The population of mutant molecules can be cloned as a library.
Related topics	(O1) DNA cloning: an overview (O4) Ligation, transformation, and analysis of recombinants	(P2) Bacteriophages, cosmids, YACs, and BACs (R3) Polymerase chain reaction

---

Types of mutagenesis

One of the major reasons for cloning genes is to establish how they function, whether one is investigating regulatory sequences in DNA or amino acids in an enzyme. This often involves changing the sequence by mutation (Section E2) and observing the outcome. Once cloned genes have been isolated, it is easy to mutate them *in vitro* and then assay for the effects by expressing the mutant gene and/or its product *in vitro* or *in vivo*. The three most common kinds of mutation made in cloned genes are: (i) changes to one or a few nucleotides at a specific point (**site-specific**, or **site-directed mutagenesis**); (ii) insertion or deletion of longer sections of the gene (**insertion** or **deletion mutagenesis**); and

(iii) random mutations along the entire length of the gene or a section of it (**random mutagenesis**). One efficient example of each type from the many available is described below.

Site-directed mutagenesis

It is very useful to be able to change just one, or a few specific nucleotides in a sequence to test a hypothesis. The importance of every base in a transcription factor-binding site can be examined by changing each one in turn. The role of a specific amino acid in the structure and function of a protein can be assessed by expressing it from a site-mutated cDNA that causes a single substitution to be made. One of the most widely used methods for site-directed mutagenesis is the **Quikchange™** procedure (Figure 1). First, the sequence to be mutated is cloned in an appropriate plasmid vector (Section P1). A pair of complementary **mutagenic primers** is then synthesized (Section Q2) that span the region where the mutation is to be created. These contain the required base differences compared to the template sequence and anneal to opposite strands of the template. In a PCR (Section R3), these primers direct the synthesis of a copy of the entire DNA (vector plus insert), now with the required mutation and with staggered nicks on each strand. After multiple cycles of PCR, the mutant sequence predominates as the mutant primers are of course part of the product DNA. However, copies of the original sequence persist and need to be destroyed. Like most bacterial DNAs, the original plasmid vector was methylated on both strands (Section A2). However, as the PCR is carried out *in vitro*, the bulk of the PCR products are completely unmethylated. As DNA synthesis is semi-conservative (Section D1), the original template strands exist as **hemimethylated** dsDNA molecules. These can be degraded by incubating the mixture with the restriction endonuclease DpnI, which, unusually for a restriction enzyme, only degrades methylated or hemimethylated DNA (its recognition site is 5′-GmATC-3′, Section O3). The unmethylated mutant PCR product is resistant. This is then transformed into a host *E. coli*, which repairs the staggered nicks and produces closed circular DNA.

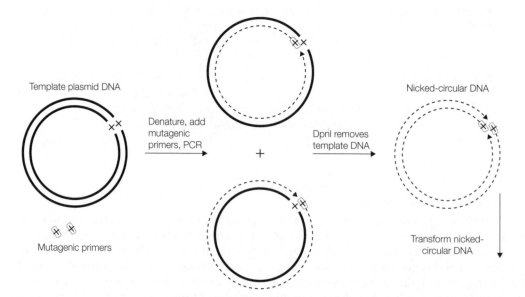

Figure 1. Site-directed mutagenesis. ✕ is the mutated site in the mutagenic primers and the PCR products.

Insertion/deletion mutagenesis

Sometimes it is useful to make larger changes to DNA sequences than is possible by site-directed mutagenesis, such as the insertion or deletion of longer stretches of nucleotides. The method of **overlap extension PCR (OE-PCR)** is an efficient way of achieving this goal. Its use to produce a specific deletion is shown in Figure 2. The two internal primers are **chimeric** in that they contain sequences that anneal at either side of the sequence to be deleted. By carrying out two separate **primary reactions**, the 5′- and 3′- portions of the target are produced, each lacking the sequences to be deleted. When these PCR products are combined in a **secondary PCR** using just the outer, nonchimeric, primers, the final product is the complete target DNA lacking the precise deletion. Not only has OE-PCR been developed to allow the precise insertion of long DNA sequences, it can now be used to clone an insert into a vector without the use of DNA ligase (cf. Section O4), i.e. **OE-PCR cloning**.

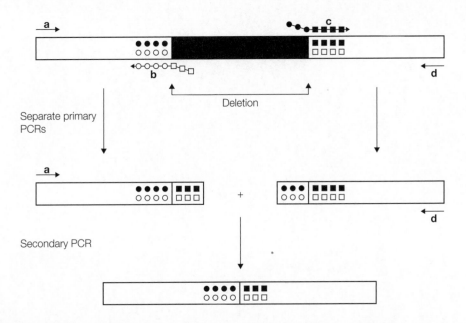

Figure 2. Deletion mutagenesis by overlap extension PCR. Filled symbols represent one strand of the DNA around the deletion site, and open symbols the complement. Primers **a** and **b** amplify upstream of the deletion, and primers **c** and **d** amplify downstream. After purification, the PCR products are mixed and amplified in a secondary PCR using primers **a** and **d**.

Random mutagenesis by PCR

Producing a collection, or library, of clones containing random mutations throughout a target sequence has many applications, such as in **directed evolution** experiments *in vitro* that aim to improve the activity of a specific protein by testing many different amino acid substitutions at once. One of the best methods is a form of **error-prone PCR** in which a modified low-fidelity DNA polymerase is used to PCR amplify the target sequence in a plasmid using reaction conditions that favor misincorporation of dNTPs. The mutated target molecules are produced in a primary PCR and then purified to remove the error-prone polymerase. They are then mixed with more of the original plasmid, denatured, and used as large primers in a secondary PCR that uses a high-fidelity DNA polymerase. The original plasmid can be destroyed using DpnI (see above) and the amplified mutant molecules used to transform competent cells to produce a library of mutant sequences that can be screened, for example, for improved/altered protein activities in **protein engineering** experiments.

S1 Introduction to the 'omics

Key Notes

Genomics

Genomics involves the sequencing of the complete genome, including structural genes, regulatory sequences, and noncoding DNA segments, in the chromosomes of an organism, and the interpretation of all the structural and functional implications of these sequences and of the many transcripts and proteins that the genome encodes. Genomic information offers new therapies and diagnostic methods for the treatment of many diseases.

Transcriptomics

Transcriptomics is the systematic and quantitative analysis of all the transcripts present in a cell or a tissue under a defined set of conditions (the transcriptome). The major focus of interest in transcriptomics is the mRNA population, although there is increasing interest in noncoding RNAs. The composition of the transcriptome varies markedly depending on cell type, growth, or developmental stage and on environmental signals and conditions.

Proteomics

The proteome is the total set of proteins expressed from the genome of a cell via the transcriptome. Proteomics is the quantitative study of the proteome using techniques of high-resolution protein separation and identification. Like the transcriptome, the proteome varies in composition depending on conditions. Post-translational modifications to proteins make the proteome highly complex.

Metabolomics

Metabolomics is the study of all the small molecules, including metabolic intermediates (amino acids, nucleotides, sugars, etc.) that exist within a cell. The metabolome provides a sensitive indicator of the physiological status of a cell and has potential uses in monitoring disease and its management.

Other 'omics

The suffixes '-ome' and '-omics' have been applied to many other molecular sets and subsets. Examples include the kinome (protein kinases) and the phosphoproteome (phosphoproteins) and their study, such as glycomics (carbohydrates) and lipidomics (lipids).

Related topics

(A3) Protein structure and function
(R2) Nucleic acid sequencing

(S7) Systems and synthetic biology

Genomics

The **genome** of an organism can be defined as its total DNA complement. It contains all the RNA- and protein-coding genes required to generate a functional organism as well as the noncoding DNA. It may comprise a single chromosome or be spread across multiple chromosomes. In eukaryotes, it can be subdivided into nuclear, mitochondrial, and chloroplast genomes. The aim of **genomics** is to determine and understand the complete DNA sequence of an organism's genome. From this, the number of potential proteins encoded in the genome can be estimated by searching for **ORFs** (Section K1), and functions for many of these proteins can be predicted from sequence similarities to known proteins (Section S6). In this way, a considerable amount can be inferred about the biology of that organism simply by analyzing the DNA sequence of its genome. However, such analyses have great limitations because the genome is only a source of information; in order to generate cellular structure and function, it must be expressed. The ambitious goal of **functional genomics** is to determine the functions of all the genes and gene products that are expressed in the various cells and tissues of an organism under all sets of conditions that may apply to that organism (Figure 1). This requires large-scale, high-throughput technologies to analyze the transcription of the complete genome (**transcriptomics**) and the eventual expression of the full cellular protein complement, including all isoforms arising from alternative splicing (Section J4), post-translational modification, etc. (**proteomics**). Similarly, **structural genomics** uses techniques such as X-ray crystallography, and NMR spectrometry (Section A5) with the aim of producing a complete structural description of all the proteins and macromolecular complexes within a cell. Ultimately, it is the interactions between molecules, both large and small, that define cellular and biological function. The vast amount of information produced by these technologies poses challenges for data storage and retrieval, and requires publicly accessible databases that use agreed standards to describe the data, allowing meaningful data comparison and integration (Section S6).

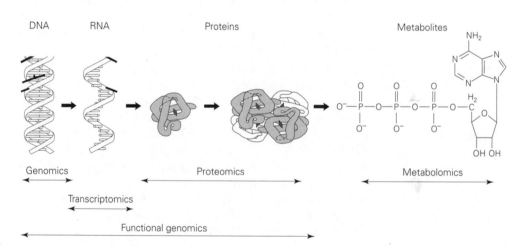

Figure 1. Relationships between the '-omics.'

Transcriptomics

The **transcriptome** is the full set of RNA transcripts produced from the genome at any given time. It includes the multiple transcripts that may arise from a single gene through the use of alternative promoters and alternative processing (Section J4) as well as the many regulatory noncoding RNAs (Section A2). With few exceptions, the genome of an organism is identical in all cell types under all circumstances. However, patterns of gene expression vary markedly depending on cell type (e.g. brain vs. liver), developmental, or cell cycle stage, presence of extracellular effectors (e.g. hormones, growth factors) and other environmental factors (e.g. temperature, nutrient availability). **Transcriptomics** provides a global and quantitative analysis of transcription under a defined set of conditions. The main techniques used are based on nucleic acid hybridization (Section B2) and PCR (Sections R3 and S2). In addition to providing a wealth of information about how cells function, **transcription** (or **expression**) **profiling** can have important medical uses. For example, there are significant differences between the transcriptome of normal breast tissue and breast cancer cells, and even between different types of breast cancer that require different treatments. Using **DNA microarrays** (Section S2) to study cancer tissue can aid early and accurate diagnosis and so determine the most appropriate therapy.

Proteomics

The term proteome is used both to describe the total set of proteins encoded in the genome of a cell and the various subsets that are expressed from the transcriptome at any one time. The proteome includes all the various products encoded by a single gene that may result from multiple transcripts (Section J4), from the use of alternative translation start sites on these mRNAs, and from different post-translational modifications of individual translation products (Section L4). Like the transcriptome, the proteome is continually changing in response to internal and external stimuli but is much more complex than the transcriptome. Thus, while there may be as few as 20 000–25 000 genes in the human genome, these may encode >10^5 transcripts and, on average, as many as $5×10^5$–10^6 distinct proteins. Figures like these may require us to reconsider the definition of a gene.

Proteomics is the quantitative study of the proteome using techniques of high-resolution protein separation and identification (Sections A5 and S3). More broadly, proteomics research also considers protein modifications, functions, subcellular localization, and the interactions of proteins in complexes. Such analyses are of crucial importance to our understanding of how cells function and of how function changes during disease. A protein found only in the diseased state or whose level is altered in disease may represent a useful diagnostic marker or drug target. Since most of what happens in a cell is carried out by large macromolecular complexes rather than by individual proteins, scientists are now trying to detect all the protein–protein interactions within the cell and so produce the **interactome**, an integrated map of all such interactions. The association of one protein with another can reveal functions for unknown proteins and novel roles for known proteins (Section S7).

Metabolomics

Since many proteins are enzymes acting upon small molecules, it follows that alterations to the proteome arising from environmental change, disease, etc. will be reflected in the **metabolome**, the entire set of small molecules – amino acids, nucleotides, sugars, etc. – and all the intermediates that exist within a cell during their synthesis and degradation. **Metabolomics**, or **metabolic profiling**, is the quantitative analysis of all such cellular metabolites at any one time under defined conditions. Because of the very different chemical nature of small cellular metabolites, a variety of methods is required to measure these, including gas chromatography, high performance liquid chromatography, and capillary electrophoresis, coupled to NMR and mass spectrometry. The output from these methods can generate fingerprints of groups of related compounds, or, when combined, the whole metabolome. Given the much lower number of small metabolites in a cell compared with RNA transcripts and proteins – around 600 have been detected in yeast – metabolomics is likely to provide a simpler, yet detailed and sensitive indicator of the physiological state of a cell and its response to drugs, environmental change, etc. The effect on the metabolome of inactivating genes by mutation can also link genes of previously unknown function to specific metabolic pathways.

Other 'omics

Scientists love jargon. Hence, by analogy with genomics and proteomics, we now have **glycomics**, the study of carbohydrates (mostly extracellular polysaccharides, glycoproteins, and proteoglycans) and their involvement in cell–cell and cell–tissue interactions, and **lipidomics**, the analogous study of cellular lipids. We also have the **kinome**, the full cellular complement of protein kinases, and the **degradome**, the repertoire of proteases and protease substrates, as well as sub-'omes like the **phosphoproteome** and the **pseudogenome**. Like the other 'omes and 'omics, these relate to the global picture; using appropriate technologies they describe the structures and functions of the full set of species within the 'ome rather than individual members. However, no 'ome is alone and it must not be forgotten that cell function is dictated by tightly controlled interactions between the 'omes. Thus, in its widest sense, functional genomics encompasses all of the above and describes our attempts to explain the molecular networks that translate the static information of the genome into the dynamic phenotype of the cell, tissue, and organism.

S2 Global gene expression analysis

Key Notes

Genome-wide analysis

Traditional methods for analyzing gene expression, or the phenotypic effect of gene inactivation, can only study small numbers of predetermined genes. The techniques of genome-wide analysis permit the study of the expression or systematic disruption of all genes in a cell or organism. These approaches can therefore identify previously unknown responses to environmental signals or gene loss.

DNA microarrays

DNA microarrays are small, solid supports (e.g. glass slides) on to which are spotted individual DNA samples, corresponding to every gene in an organism. When hybridized to labeled cDNA representing the total mRNA population of a cell, each cDNA (mRNA) binds to its corresponding gene DNA and so can be separately quantified. Commonly, a mixture of cDNAs labeled with two different fluorescent tags and representing a control and experimental condition are hybridized to a single array to provide a detailed profile of differential gene expression.

Chromatin immuno-precipitation

Protein binding sites on DNA can be identified by a technique called chromatin immunoprecipitation (ChIP). This involves fixing the protein–DNA complexes, shearing the DNA, immunoprecipitation with an antibody to the protein of interest, and purification of the coprecipitated DNA fragments. Real-time PCR analysis can then be used to quantify the target DNA that is bound by the protein of interest (ChIP). The genome wide analysis of protein–DNA binding sites can be achieved by hybridization of the DNA fragments to a genomic DNA microarray chip (ChIP-chip), or the pulled down DNA fragments can be sequenced directly (ChIP-Seq).

Gene knockouts

Chromosomal genes can be deleted and replaced with a selectable marker gene by the process of homologous recombination. Yeast strains are available in which every gene has been individually deleted, allowing the corresponding phenotypes to be assessed. Using embryonic stem cells, a similar process is used to create knockout mice. These can be useful models for human genetic disease.

RNA knockdown

The suppression of gene function using siRNA can also be applied on a genome-wide scale as an alternative to gene

	knockout. A collection of strains of the nematode *C. elegans* has been prepared in which the activity of each individual gene is suppressed by RNAi.	
Related topics	(C4) Genome complexity	(R3) Polymerase chain reaction
	(L4) Translational control and post-translational events	(R5) Mutagenesis of cloned genes
	(R1) Characterization of clones	(S1) Introduction to the 'omics
		(S5) Transgenics and stem cell technology

Genome-wide analysis

Several techniques are available for the analysis of the differential gene expression of cells. They compare the patterns of gene expression (the **transcriptome**) between different cell types or between cells exposed to different stimuli. Traditionally, these techniques have focused on experiments to measure just one gene, or a relatively small number of genes. In **northern blotting** (Section R1), total RNA extracted from cells is fractionated on an agarose gel, then transferred to a membrane and hybridized to a solution of a radiolabeled cDNA probe corresponding to the mRNA of interest. The amount of labeled probe hybridized gives a measure of the level of that particular mRNA in the sample. The **ribonuclease protection assay** is a more sensitive method for the detection and quantitation of specific RNAs in a complex mixture. Total cellular RNA is hybridized in solution to the appropriate radiolabeled cDNA probe; then any unhybridized single-stranded RNA and probe are degraded by a single-strand-specific **nuclease** (cf. S1 mapping, Section R4). The remaining hybridized (double-stranded) probe:target material is separated on a polyacrylamide gel, then visualized and quantified by autoradiography. **RT-PCR** (Section R3) is a sensitive method that allows the detection of RNA transcripts of very low abundance. In RT-PCR, the RNA is copied into a cDNA by reverse transcriptase. The cDNA of interest is then amplified exponentially using PCR and specific primers. When detecting and quantifying the PCR products, it is critical to do this while the reactions are still in the exponential phase. **Real-time PCR** (Section R3) ensures this.

The above methods assume that it is known which genes are required for study so that the correct probes and primers can be synthesized. The techniques of **genome-wide analysis** require no prior knowledge of the system under investigation and allow examination of the whole transcriptome (potentially thousands of RNAs) at once. One of the most widely used techniques for studying whole transcriptomes is **DNA microarray analysis**.

DNA microarrays

DNA microarray analysis follows the principles of Southern and northern blot analysis (Section R1), but in reverse, with the sample in solution and the gene probes immobilized. DNA microarrays (DNA chips) are small, solid supports on to which DNA samples corresponding to thousands of different genes are attached at known locations in a regular pattern of rows and columns. The supports themselves may be made of glass, plastic, or nylon and are typically the size of a microscope slide. The DNA samples, which may be gene-specific synthetic oligonucleotides or cDNAs, are spotted, printed, or synthesized directly onto the support. Thus each dot on the array contains a DNA sequence that is unique to a given gene. Each dot will therefore hybridize specifically to mRNA corresponding to that gene. Consider an example of its use – to study the differences in the transcriptomes of a normal and a diseased tissue. Total RNA is extracted from samples of the two tissues and separately reverse transcribed to produce cDNA copies that precisely reflect the two mRNA populations. One of the four dNTP substrates used for cDNA synthesis is tagged with a fluorescent dye, a green dye for the normal cDNA and a red one for the diseased-state cDNA. The two cDNA samples are then mixed together and hybridized to the microarray (Figure 1). The red- and green-labeled cDNAs compete for binding to the gene-specific probes on each dot of the microarray. When excited by a laser, a dot will fluoresce red if it has bound more red than green cDNA. This would occur if that particular gene is expressed more strongly (up-regulated) in the diseased tissue compared with the normal one. Conversely, a spot will fluoresce green if that particular gene is down-regulated in the diseased tissue compared with the normal one. Yellow fluorescence indicates that equal amounts of red and green cDNA have bound to a spot and, therefore, that the level of expression of that gene is the same in both tissues. A fluorescence detector comprising a microscope and camera produces a digital image of the microarray from which a computer quantifies the red to green fluorescence ratio of each spot and, from this, the precise degree of difference in the expression of each gene between the normal and diseased states. In order to ensure the validity of the results, adequate numbers of replicates and controls are used and a complex statistical analysis of the data is required. Since a full human genome microarray could have as many as 30 000 spots, the amount of data generated is enormous. However, by clustering together genes that respond in a similar way to a particular condition (disease, stress, drug, etc.), physiologically meaningful patterns can be observed. In terms of pure research, the clustering of genes of unknown function with those involved in known metabolic pathways can help define functions for such orphan genes.

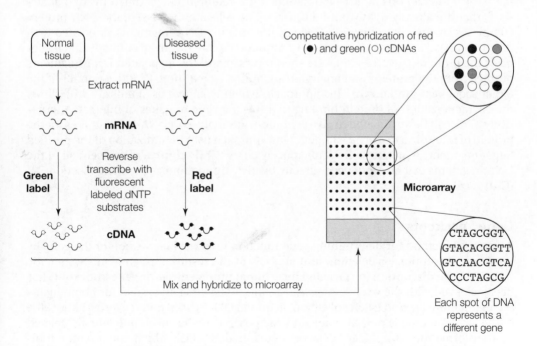

Figure 1. Microarray analysis of differential gene expression in normal and diseased tissue.

The type of microarray described above is sometimes called an **expression microarray** as it is used to measure gene expression. A **comparative genomic microarray**, in which the arrayed spots and the sample are both genomic DNA, can be used to detect loss or gain of genomic DNA that may be associated with certain genetic disorders. A **mutation microarray** detects single nucleotide polymorphisms (SNPs, Section C4) in the sample DNA of an individual. The array usually consists of many different versions of a single gene containing known SNPs associated with a particular disease, while the sample is genomic DNA from an individual. Hybridization of the sample DNA to one particular spot under stringent conditions identifies the SNP in the patients' DNA and, hence, their disease susceptibility. In theory, this type of array could be expanded to cover hundreds of known disease-associated genes and so provide a global disease susceptibility fingerprint for an individual.

An alternative to the use of DNA microarrays that is becoming more popular is **RNA-Seq**, which involves the sequencing and quantification of all of the RNA molecules in a cell population. This approach is powerful and quantitative and is likely to make a considerable impact in the near future.

Chromatin immunoprecipitation

Chromatin immunoprecipitation (**ChIP**) is used to investigate the specific interactions between proteins and DNA (chromatin) in the cell. This method uses immunoprecipitation (Sections A5 and S3) to show the specific genomic regions where proteins such as transcription factors bind. ChIP can also be used to determine the genomic locations of specifically modified histones.

The basic ChIP method (Figure 2) involves reversibly cross-linking the DNA to any closely associated proteins by treating the cells with formaldehyde. Cross-linking preserves the *in vivo* interactions after isolation and is not required for modified histone analysis as they are already tightly bound to DNA. The cells are then lysed, the DNA-protein complexes sheared by sonication (Section B1) and any DNA fragments associated with the protein(s) of interest are selectively immunoprecipitated with an antibody that recognizes the protein. The cross-links are then reversed and the associated DNA fragments purified from the proteins and analyzed/quantified by real-time PCR (Section R3) if the target sequences are known. This approach, which is known as standard ChIP, allows the characterization of protein binding to a specific gene or genes of interest. Alternatively, ChIP can be formulated to give genome-wide analysis of DNA-binding sites for the protein of interest to reveal new targets. One approach uses hybridization of the purified target fragments to genomic DNA microarray chips (**ChIP-chip**); alternatively, all of the DNA fragments can be sequenced directly by high-throughput sequencing (Section R2) (**ChIP-Seq**).

Gene knockouts

A powerful method for determining gene function involves inactivation of the gene by mutation, disruption or deletion, and analysis of the resulting phenotype. Systematic **targeted gene disruption** (or **targeted insertional mutagenesis**, or **gene knockout**) has been achieved with the yeast *S. cerevisiae* thanks to its efficient system for homologous recombination (Section E4). A collection of around 6000 yeast strains covering about 96% of the yeast genome is now available in which each gene has been individually deleted by transformation with linear DNA fragments made by PCR which contain an antibiotic selection marker gene (e.g. kanamycin resistance, Kan^r) flanked by short sequences

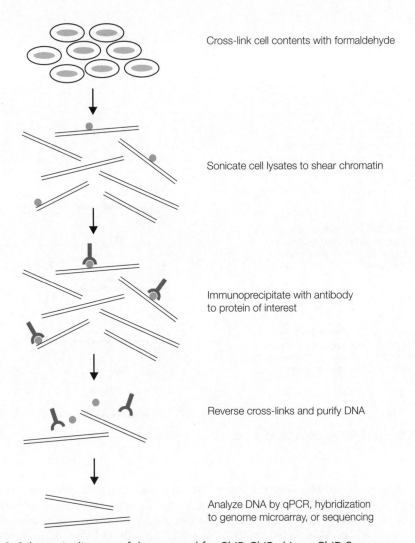

Cross-link cell contents with formaldehyde

Sonicate cell lysates to shear chromatin

Immunoprecipitate with antibody
to protein of interest

Reverse cross-links and purify DNA

Analyze DNA by qPCR, hybridization
to genome microarray, or sequencing

Figure 2. Schematic diagram of the protocol for ChIP, ChIP-chip or ChIP-Seq.

corresponding to the ends of the target gene. The target gene is swapped for the marker gene by recombination (Figure 3). Even strains with essential genes deleted can be maintained by rescuing them with a plasmid carrying the wild-type gene expressed under certain conditions. The effect of the deletion can then be studied by eliminating expression from the plasmid. Around 12–15% of yeast genes may be essential. The deletion strategy also results in the incorporation of short, unique sequence tags ('**molecular barcodes**') into each strain. These permit identification of individual strains in genome-wide experiments where mixed pools of deletion mutants are used.

As many yeast genes have human homologs, such analyses can shed light on human gene function. Since the functions of many genes in animals are associated with their multicellular nature, **knockout mice** have been prepared in which individual genes have been disrupted in the whole animal so that the developmental and physiological consequences can be studied. First, pluripotent **embryonic stem (ES) cells** (Section S5) are

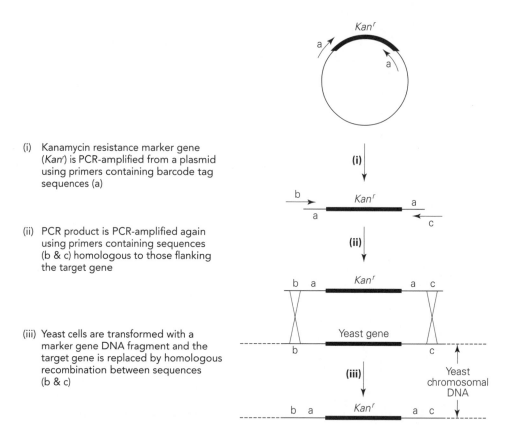

(i) Kanamycin resistance marker gene
(*Kan^r*) is PCR-amplified from a plasmid
using primers containing barcode tag
sequences (a)

(ii) PCR product is PCR-amplified again
using primers containing sequences
(b & c) homologous to those flanking
the target gene

(iii) Yeast cells are transformed with a
marker gene DNA fragment and the
target gene is replaced by homologous
recombination between sequences
(b & c)

Figure 3. Strategy for deleting yeast genes.

removed from a donor **blastocyst,** cultured *in vitro,* and a specific gene replaced with
a marker by recombination as above. The engineered ES cells are inserted into a recipient blastocyst, which is then implanted into a foster mother. The resulting offspring are
mosaic, with some cells derived from the engineered ES cells and some from the original
ES cells of the recipient blastocyst. Since the germ line is also mosaic, a pure knockout
strain can be bred from these offspring. Not only are such mice valuable research tools for
understanding gene function, they can also provide models for human genetic disease
where no natural animal model exists, such as cystic fibrosis. **Conditional gene knockout,** in which a gene is deleted from only one or a few specific tissues rather than from
the whole animal, is an even more useful and powerful tool, and relies mainly on the use
of the **Cre-loxP** system for site-specific recombination (Section E4). Currently, an international consortium of laboratories is aiming to provide the research community with a
library of all 22 000 or so mouse gene knockouts, each strain available as a frozen embryo
or sperm. At the time of writing, about 17 000 such strains had been created, albeit in a
variety of ways.

A complementary approach to studying genes through deletion phenotypes (**loss-of-
function**) is to look at the effect of overexpressing a gene (**gain-of-function**). Collections
of yeast strains are also available in which each gene is individually overexpressed from a
plasmid under the control of an inducible promoter. These strains can also be used as a
source of easily purified recombinant protein.

RNA knockdown

Introduction of small, double-stranded **siRNAs** into cells represses the translation of homologous mRNA transcripts through **RNAi**, part of the wider **gene silencing** (**RNA silencing**) phenomenon, a natural group of mechanisms for down-regulating gene expression (Sections J2 and L4). This provides a simple yet powerful alternative experimental approach to gene knockout to eliminate the function of a gene, and is called **RNA knockdown**. Transient knockdown is achieved by transfecting cells with the appropriate siRNA, while permanent knockdown requires expression of the siRNA from a vector, often as a **hairpin**, which is then processed by the cell (Section J2). Translation of the target mRNA is rarely eliminated completely but a reduction of 80–90% is possible and usually sufficient. A global loss-of-function analysis has been achieved by this means in the nematode *Caenorhabditis elegans*. In *C. elegans*, RNA knockdown can be achieved by feeding the worms with *E. coli* engineered to express a specific dsRNA. Remarkably, this dsRNA can find its way into all cells of the worm resulting in specific inhibition of gene expression. A feeding library of about 17 000 *E. coli* strains each expressing a specific dsRNA designed to target a different gene has been created. This library covers about 85% of the *C. elegans* genome. As many *C. elegans* genes have human homologs, a phenotypic analysis of worms fed on these *E. coli* strains can help define human gene function. Once a potentially interesting gene has been found in *C. elegans*, the human ortholog can be individually knocked down by siRNA in cultured cells and the consequences determined. RNAi is now widely used to investigate individual genes in many cell types. Global studies are also possible in mammalian cells and tissues using a library of viral vectors each expressing a different, gene-specific siRNA. Andrew Fire and Craig Mello received a Nobel prize in 2006 for their work on RNAi in *C. elegans*.

S3 Proteomics

Key Notes

Proteomics	Proteomics is the quantitative study of the proteome using techniques of high-resolution protein separation and identification, such as high performance liquid chromatography followed by mass spectrometry. The proteome is the total set of proteins expressed from the genome of a cell. Like the transcriptome, the proteome varies in composition depending on conditions. Post-translational modifications to proteins make the proteome highly complex.
Protein–protein interactions	Protein–protein interactions are essential for the operation of intracellular signaling systems and for the maintenance of multi-subunit complexes. They can be detected by various techniques including immunoprecipitation, pull-down assays, and two-hybrid analysis.
Two-hybrid analysis	This method allows the detection *in vivo* of weak protein–protein interactions that may not survive cell disruption and extraction. It depends on the activation of a reporter gene by the reconstitution of the two domains of a transcriptional activator, each of which is expressed in cells as a fusion protein with one of the two interacting proteins.
Protein arrays	Protein arrays consist of proteins, protein fragments, peptides or antibodies immobilized in a grid-like pattern on a miniaturized solid surface. They are used to detect interactions between individual proteins and other molecules.
Related topics	(A3) Protein structure and function (A4) Macromolecular assemblies (A5) Analysis of proteins	(L4) Translational control and post-translational events (S4) Cell and molecular imaging

---

Proteomics

Proteomics is the study of the entire protein complement (the **proteome**) of a cell or biological sample, e.g. blood serum, and encompasses both identification and quantitation. Proteomics is of crucial importance to our understanding of how cells function and of how function changes during disease, and is of great interest to pharmaceutical companies in their quest for new drug targets. However, it is highly challenging, because of the sheer complexity of cellular protein content and the huge variation in abundance. For example, a typical cell of the budding yeast *S. cerevisiae* contains about 5 pg of protein comprising, in total, about 7×10^7 protein molecules. Some, such as the ribosomal subunits (of which there about 2×10^5 in the yeast cell, Section J1) and glycolytic enzymes are very abundant, and account for over one-third of the 7×10^7 molecules. Others, such as some transcription factors (Section I1), are present at perhaps less than 20 copies per cell.

A second complexity arises from post-translational modifications, such as phosphorylation, acetylation, ubiquitinylation, and glycosylation, resulting in many proteins existing in several different chemically distinct forms (Section L4). It has been predicted that the combined effects of alternative splicing and post-translational modification mean that the 20–25 000 genes of a typical mammalian cell may encode as many as 10^6 different mature proteins. A true definition of a proteome should therefore include precise quantitative measurements of the partition of each protein into all of its post-transcriptional and post-translational variants.

To identify and quantify a single protein in a cell is relatively straightforward (e.g. by western blotting, Section A5), but enormously complex with entire proteomes. Usually, proteomes are studied by mass spectrometry (Section A5). Until recently, protein mixtures were commonly separated by **two-dimensional gel electrophoresis** and individual spots representing single proteins cut out from the gel and analyzed. Now, however, most proteome studies make use of high-resolution liquid chromatography coupled to very high performance mass spectrometry to analyze complex mixtures directly. A sample containing many proteins is first digested with a proteolytic enzyme, usually trypsin, which specifically cleaves peptide bonds to the C-terminal side of lysine and arginine residues. This breaks every protein into peptides of around 10–20 amino acids in length, and of a size 1000 Da to 2500 Da. The set of tryptic peptides generated from any given protein is characteristic of that protein and is called its **peptide mass fingerprint** (**PMF**). Thus a sample containing, say, 5000 different proteins would yield about 300 000 peptides after digestion. Why go through this enormous increase in complexity instead of analyzing the proteins directly? The main reasons are that the short peptides generated can be separated with very high resolution by **reversed phase chromatography** and because the low masses of tryptic peptides are perfectly suited to the m/z range of modern mass spectrometers (Section A5). As the mixture of peptides is streamed from the chromatography system into the mass spectrometer, the m/z values of all the resulting component ions are measured. Then, under the control of the data system, the most abundant peptide ions are individually drawn into the central 'collision' region of the instrument, where they are fragmented further. (The isolation of one peptide from the first mass analyzer to be fragmented and studied in a second analyzer is known as **tandem mass spectrometry**, or **MS/ MS**). If the energy of collision is carefully controlled, a peptide will fragment randomly at each peptide bond, and by measuring the masses of the products of this fragmentation, it is possible to reconstruct the amino acid sequence of that peptide. Since it is simple to predict the tryptic digestion pattern of known proteins from knowledge of the positions of the lysines and arginines in their primary structures, the experimental fragmentation data can be matched against databases of predicted fragmentation patterns of known proteins to identify all the proteins in the sample. Since the databases also include information on the masses of known post-translational modifications, the nature and positions of these can also be readily identified.

Finally, because mass spectrometers are able to measure masses very accurately, techniques based on stable heavy isotopes such as ^{13}C or ^{15}N can be used to quantify the levels of proteins in a cell or sample. For example, **SILAC** (**s**table **i**sotope **l**abeling with **a**mino acids in **c**ell culture) can be used to measure changes in the relative amounts of (theoretically) all proteins between, say, an unstimulated and hormone-stimulated cell or a normal liver cell and a liver cancer cell. One population of cells (the 'experimental' sample) is labeled in culture with a heavy (^{13}C or ^{15}N-labeled) amino acid while the other (the 'control' sample) is grown with the 'normal' ^{12}C or ^{14}N amino acid. Proteins are extracted from both, mixed, digested with trypsin, and the proteins identified from their PMFs. The corresponding peptides derived from each pair of control and experimental proteins in the mixture are easily identified because they differ in mass by a precisely known amount due to the heavy label. A comparison of the signals generated by the peptides from each pair gives an accurate measure of the relative amounts of that protein in the two cells. Absolute quantification (i.e. how many molecules) can also be achieved by preparing a heavy-labeled recombinant version of a protein and adding a known amount of the purified protein to an unlabeled sample before digestion. This technique can even be extended to multiple proteins.

Protein–protein interactions

Many proteins are involved in multiprotein complexes, and transient protein–protein interactions underlie many intracellular signaling systems. Thus, characterization of these interactions (the **interactome**, Section S1) is crucial to our understanding of cell function. Such interactions may be stable, and survive extraction procedures, or they may be weak and only detectable inside cells. Stable interactions can be detected by **immunoprecipitation** (**IP**). Here, an antibody (Section A5) raised against a particular protein antigen is added to a cell extract containing the antigen to form an **immune complex**. Next, insoluble agarose beads covalently linked to **protein A**, a protein with a high affinity for immunoglobulins, is added. The resulting **immunoprecipitate** is isolated by centrifugation and, after washing, the component proteins are analyzed by SDS-PAGE. These will consist of the antigen, the antibody, and any other protein in the extract that interacted stably with the antigen. These proteins can be identified by mass spectrometry. The **pull-down assay** is a similar procedure (Figure 1). Here, the '**bait**' protein (the one for which interacting partners are sought) is added to the cell extract in the form of a recombinant fusion protein (Section A5), where the fusion partner acts as an **affinity tag**. A commonly used fusion partner is the enzyme **glutathione-S-transferase** (**GST**). Next, agarose beads

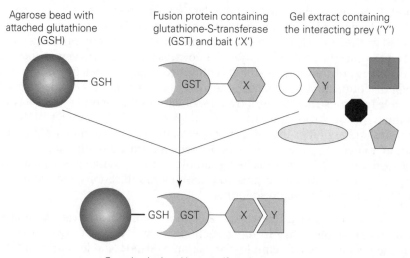

Figure 1. GST pull-down assay for isolating proteins that interact with the bait protein 'X'.

containing immobilized **glutathione**, a tripeptide for which GST has a high affinity, are added. The GST-bait fusion protein binds to the beads, along with any other protein that interacts with the bait. The beads are then processed as for IP. This technique can be used for proteome-wide analysis. A collection of 6000 yeast clones is available with each clone expressing a different yeast protein as a GST-fusion expressed from a plasmid with a controllable promoter. The yeast **TAP-fusion** library is another collection where each yeast protein is tagged with an affinity tag called TAP and expressed, not from a plasmid, but from its normal chromosomal location following homologous recombination (Section E4) between the normal and the tagged version of the gene. The TAP tag is also an **epitope tag** (an amino acid sequence recognized by an antibody, Section A5) so it not only allows purification of proteins that interact with each bait protein fused to the tag in an *in vivo* environment, but, since each tagged protein is expressed from its own chromosomal promoter, it also allows quantification of the natural abundance of every cellular protein by **immunofluorescence** (Section S4).

Two-hybrid analysis

Two-hybrid analysis reveals protein–protein interactions that may not survive the rigors of protein extraction, by detecting them inside living cells. Yeast has commonly been used to provide the *in vivo* environment. This procedure relies on the fact that many gene-specific transcription factors (transcriptional activators) are modular in nature and consist of two distinct domains – a DNA-binding domain (BD) that binds to a regulatory sequence upstream of a gene and an activation domain (AD) that activates transcription by interacting with the basal transcription complex and/or other proteins, including RNA polymerase II (Sections H5 and I1). Although normally covalently linked together, these domains will still activate transcription if they are brought into close proximity in some other way. Two hypothetically interacting proteins X and Y are expressed from plasmids as fusion proteins to each of the separate domains, that is BD–X and AD–Y. They are transfected and expressed in a yeast strain that carries the regulatory sequence recognized by the BD fused to a suitable **reporter gene** (Section R4), for example β-galactosidase (Section P1). If X and Y interact *in vivo*, then the AD and BD are brought together and activate transcription of the reporter gene (Figure 2). Cells or colonies expressing the

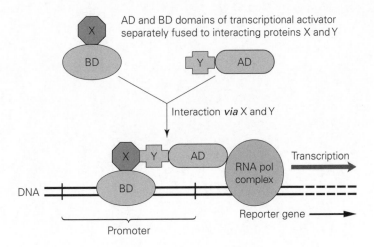

Figure 2. Principles of yeast two-hybrid analysis: activation of a reporter gene by reconstitution of active transcription factor from its two separated domains, activation domain (AD), and binding domain (BD).

reporter gene are easily detected as they turn blue in the presence of X-gal, a chromogenic β-galactosidase substrate. The real power of this system lies in the detection of new interactions on a proteome-wide scale. Thus, if protein X is expressed as a BD–X fusion (the **bait**) and introduced into a yeast culture previously transfected with a cDNA library of AD–N fusions (the **prey**), where N represents the hundreds or thousands of proteins encoded by the cDNAs, any blue colonies arising after plating out the culture to separate the clones represent colonies expressing a protein that interacts with X. If the plasmids are isolated from these colonies and the specific AD–N cDNAs sequenced, the interacting proteins (N) can be identified. Although yeast is often used as the environment, the bait and prey can be from any organism. Bacteria and mammalian cells are also used to provide the *in vivo* setting for the interactions.

Protein arrays

By analogy with DNA microarrays (Section S2), a **protein array** (or **protein chip**) consists of large numbers of individual proteins, protein fragments, peptides, or antibodies immobilized in a gridlike pattern on a miniaturized solid surface. The arrayed molecules are then used to screen for interactions in complex samples applied to the array. Antibody arrays are likely to find future use in medicine to probe specific protein levels in blood or tissue samples, e.g. from cancer patients. Protein arrays can also be screened with substrates to detect enzyme activities, or with DNA, drugs, or other proteins to detect binding.

S4 Cell and molecular imaging

Key Notes

Cell imaging

The process of visualizing cells, subcellular structures, or molecules in cells is called cell imaging. This requires the use of an easily detectable label that allows visualization of a biological molecule or process by eye, film, or electronic detector. Detection often involves the use of radioactive, colored, luminescent, or, most commonly, fluorescent labels. Fluorescence involves the excitation of a fluorophore with a photon of light at the excitation wavelength. The fluorophore then emits a photon of light of lower energy at the emission wavelength.

Imaging of biological molecules in fixed cells

Detection of biological molecules in cells and tissues allows measurement of cell heterogeneity and subcellular localization. Cell fixation is often required for entry of nucleic acid probes or antibodies into cells to allow efficient labeling of the molecule(s) of interest.

Detection of molecules in living cells and tissues

Analysis of biological processes in intact cells requires maintenance of cell viability. The expression of labeled proteins from transfected vectors has been useful for the study of processes in cells, since cell viability is retained.

Fluorescent proteins and reporter genes

Fluorescent proteins have been isolated from various marine organisms. GFP produces green fluorescence emission when excited by blue light. GFP is often used as an easily visualized reporter, for example for the real-time imaging of protein localization and translocation in living cells. These proteins have been engineered to enhance expression and improve their spectral and physical properties. The luminescence resulting from expression of the firefly luciferase reporter gene can be visualized in living cells by imaging with sensitive cameras.

Related topics

(A5) Analysis of proteins
(P1) Design of plasmid vectors
(P3) Eukaryotic vectors

(Q3) Screening procedures
(S5) Transgenics and stem cell technology
(S7) Systems and synthetic biology

Cell imaging

Cell imaging is the visualization of cells, subcellular structures, or molecules within them to follow events or processes. The detection and quantification of molecules and molecular processes requires specific **assays**. Common assays employ radioactive, colored, luminescent, or fluorescent **labels** and are based on the use of labeled enzyme substrates, labeled antibodies or easily detectable proteins from luminescent organisms that can be fused to a protein of interest. A favored technique for the detection of molecules in cells is **fluorescence**, which involves a molecule called a **fluorophore** that absorbs a high-energy photon of a suitable wavelength within its **excitation spectrum**. This excites an electron in the fluorophore, which then decays, leading to emission of a photon at the **emission wavelength**, which is always of a lower energy (and at a longer, red-shifted wavelength) compared with the excitation wavelength. The fluorescence may then be detected in a **fluorescence microscope** and recorded using photographic film or an appropriate digital electronic light detector for computer display, depending on the magnification desired, and whether quantification or simple visual resolution of the molecule is required.

Imaging of biological molecules in fixed cells

The spatial detection of proteins and nucleic acids in cells and tissues is important for distinguishing cell-to-cell differences (heterogeneity) and for measurement of intracellular spatial distribution. It is often sufficient to be able to detect biological molecules in dead cells, and killing and fixing cells can allow easy detection of the molecules of interest. Chemical cell fixation breaks down permeability barriers and allows access of labeled antibody and nucleic acid probes to intracellular and subcellular compartments. Labeling of fixed cells with nucleic acid probes in order to detect complementary nucleic acids (for example specific genes in the nucleus) is called *in situ* hybridization (**ISH**) and employs either radioactive or fluorescent (**FISH**) probes.

Labeling with antibodies to detect proteins is usually called **immunocytochemistry** (**ICC**) for cells and **immunohistochemistry** (**IHC**) for tissues. Detection usually involves two stages. First, a **primary antibody** that recognizes a particular protein (protein X) is used in an unlabeled form. This antibody may have been generated using protein X as the antigen, in which case endogenous protein X can be detected (Figure 1a). Alternatively, a primary antibody that recognizes an **epitope tag** may be used (Section A5). In this case, cells are transfected with an expression vector (Section P3) that encodes protein X fused

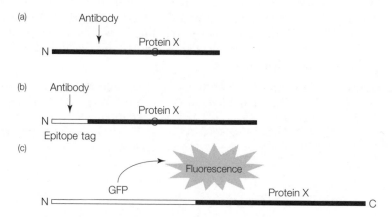

Figure 1 (a–c). Detection methods for locating a protein in a cell. See text for details.

to an additional, short sequence of amino acids, the epitope tag, which is specifically recognized by the antibody (Figure 1b). Therefore, only recombinant protein X is actually detected but it is assumed that it will occupy the same subcellular location as endogenous protein X. The advantage of this approach is that the same primary antibody can be used to detect many proteins, as long as they are expressed fused to the same epitope tag, thus saving time, money, and animals. In the next stage, a secondary antibody that recognizes all antibodies from the species in which the primary antibody was raised is then used in a labeled form for detection of the primary antibody. The secondary antibody is usually labeled with a fluorescent label (**immunofluorescence**) (Figure 2a). IHC generally involves colorimetric rather than fluorescent detection due to the high level of background fluorescence in many tissues.

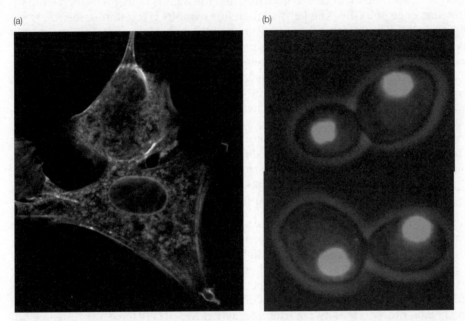

(a)   (b)

Figure 2. (a) Localization of an epitope-tagged recombinant protein to the endoplasmic reticulum of a transfected mammalian cell by immunofluorescence. (b) Localization of a GFP-DNA binding protein fusion protein to the nuclei of yeast cells.

In the past few years, there have been considerable developments in microscopy for the detection of fluorescence in cells. **Confocal microscopy** allows three-dimensional sectioning of cells and tissues but its resolution is limited by the diffraction limit of light (approximately 200 nm). Recently, a set of new, super-resolution techniques has increased the achievable spatial resolution of light microscopy beyond this optical diffraction limit.

Detection of molecules in living cells and tissues

The detection of biological molecules and processes in living cells has many advantages. For example, it allows one to follow the dynamics of biological processes in real-time, e.g. the translocation of a protein from one subcellular compartment to another, or the rapid transcriptional activation and subsequent repression of a specific gene in response to a signal. This requires that any detection method must be noninvasive (i.e. it must not kill the cells nor damage or perturb them), so any procedures for the entry of reagents into cells must be designed so that they maintain cellular viability and function. Detector proteins such as those encoded by **reporter genes** (Section R4) may be expressed genetically. DNA expressing these proteins is usually transferred into the cell by gentle transfection (Section P3). Natural genes and proteins from bioluminescent organisms have been particularly useful as reporters for live cell imaging. These include **fluorescent proteins** from marine organisms, such as *Aequoria victoria*, and **luciferases**, such as firefly luciferase from *Photinus pyralis*. Their uses are described below.

Fluorescent proteins and reporter genes

The jellyfish *A. victoria* produces flashes of blue light in **response** to a release of calcium, which interacts with the luciferase **aequorin**. The blue light is converted to green light due to the presence of a **GFP**. The jellyfish therefore exhibits green luminescence. Both aequorin and GFP have become important tools in biological research. *Aequoria* GFP consists of 238 amino acids and has a central α-helix surrounded by 11 antiparallel β-sheets in a cylinder of about 3-nm diameter and 4-nm long. The fluorophore in the center of the molecule is formed from three amino acids, Ser65-Tyr66-Gly67, and forms via a two-step process involving oxidation and cyclization. This process requires no additional substrates and can therefore occur whenever GFP is expressed in any type of cell. The wild-type GFP from *A. victoria* has a major excitation peak at a wavelength of 395 nm and a minor peak at 475 nm, but the emission peak is at 509 nm. The major application of GFP has been as a marker for protein localization in living cells. The GFP is expressed as a **fusion protein** (Section A5) attached to either the C- or N-terminus of the protein of interest (Figure 1c). The resulting fusion protein can then be used to locate the protein within a cell or follow its translocation in response to a signal or stimulus (Figure 2b).

In addition to *A. victoria* GFP, related fluorescent proteins have been isolated from a variety of marine organisms including fluorescent corals. These proteins have a variety of different spectral properties and span the visible color range, e.g. **red fluorescent protein** (**RFP**) from the coral *Discosoma*. Versions of fluorescent proteins with improved properties have also been engineered by site-directed mutagenesis (Section R5) for a range of applications. These have an optimized level of expression, improved spectral characteristics and greater protein stability. For example, derivatives of GFP have been engineered to report on pH, calcium, and protein interactions in living cells.

Fluorescent proteins are ideally suited for localization studies as they are stable, and different-colored proteins can be used together to locate more than one protein with high resolution. However, their stability is a disadvantage for transcriptional analysis, where a reporter protein with a short half-life is required in order to follow the rapid on-off dynamics of this process. Although GFP with a reduced half-life has been engineered, a better option is **firefly luciferase**. When its substrate **luciferin** is added to intact, luciferase-expressing cells, light is emitted as a result of a chemical reaction (**luminescence**). This reaction also requires oxygen and ATP, both of which are normally present in living cells. The resulting luminescence can be detected with extremely high sensitivity using a camera attached to a microscope. This permits the time-lapse imaging of expression of luciferase in live cells and is most commonly used to report on transcription where the luciferase is expressed under the control of a specific promoter of interest (Section R4).

S5 Transgenics and stem cell technology

Key Notes

Genetically modified and transgenic organisms	Genetically modified organisms have had their genome altered in some way. More specifically, transgenic organisms contain a gene(s) from another species (a transgene) in some or all cells. Microinjection or transfection techniques can be used to create transgenic animals and analogous procedures have been used to develop transgenic plants. Transgenic animals and plants have many potential applications in human medicine and agriculture but ethical issues have restrained their widespread use.
Stem cells	Stem cells are undifferentiated cells that can divide indefinitely and develop into specialized cell types. Somatic stem cells are found in many adult tissues. They have great potential for cell-based treatment of disease. Transfected mouse embryonic stem cells can be introduced into an embryo, and can be used to derive a transgenic mouse. This technology is also used to generate knockout mice for characterization of gene function in whole animals.
Induced pluripotent stem cells	Induced pluripotent stem cells are formed when differentiated adult cells are reprogrammed through expression of a defined set of four genes for transcription factors, or treated with the proteins themselves.
Gene and cell therapy	Gene therapy is the use of gene transfer for the treatment of disease. Most gene therapy protocols use viral transduction. Typically, episomal DNA virus vectors based on adenovirus or vaccinia, or integrating RNA retrovirus vectors, are used. Stem cell therapy involves the use of stem cells to treat diseased or damaged tissues, where the stem cells may repopulate the target tissue and differentiate into the normal cell types of the tissue of interest.
Related topics	(I1) Eukaryotic transcription factors (P3) Eukaryotic vectors (M3) DNA viruses (S4) Cell and molecular imaging (M4) RNA viruses

Genetically modified and transgenic organisms

Genetically modified organisms (**GMOs**) are created when the genome of a prokaryote or eukaryote has been modified in some way. This could be as simple as a point mutation, but GMO usually refers to an organism with a more substantial alteration, such as the deletion or addition of one or more genes or other DNA sequences. More specifically, organisms expressing a gene from another species are called **transgenic organisms** and the transferred gene is called a **transgene**. The design of effective transfection methods has been critical for the development of transgenic technologies. For many scientific and medical experiments, it is important to engineer an intact organism rather than just manipulate cells in culture. For long-term inheritance, the transgene must be stably integrated into the germ cells of the organism. This was originally achieved by microinjection of the DNA into one of the two pronuclei of a fertilized mouse ovum; however, current methods generally employ transfected embryonic stem cells (see below and Section S2), or, for large animals, transfection of somatic cells (e.g. fibroblasts) in culture followed by **somatic cell nuclear transfer** (**SCNT**), where the nucleus of an egg is replaced by the somatic cell nucleus by micromanipulation. After transplantation, the egg develops into the transgenic organism. In 1996, SCNT was used in the **reproductive cloning** of Dolly the sheep, the first mammal to be cloned from an adult cell, using an *unmodified* somatic cell nucleus.

Procedures have now been developed for the generation of many transgenic organisms. These include farm animals such as cows, pigs, and sheep, where the animals can be engineered to put on weight more rapidly, or to be more resistant to common diseases. After their initial creation, the animals can be propagated by normal breeding. Transgenic sheep and cows have also been created to produce useful therapeutic human proteins in their milk, e.g. clotting factors and the bactericidal protein lysozyme; this is known as **pharming**. The desired gene is placed under the control of a milk gene promoter, e.g. casein or lactalbumin, which is only expressed in the mammary gland. Large-scale purification of the protein from milk is much easier than from cultured cells or blood. It has also been suggested that transgenic organisms such as pigs could be used to provide 'humanized' organs for transplantation that would not be rejected (**xenotransplantation**). This process might fill the significant shortfall in availability of human organs. However, as with all genetic engineering, there are significant ethical issues, and concerns have arisen regarding the potential for viral diseases to jump between species following xenotransplantation.

The creation of transgenic plants requires rather different transfection techniques (Section P3). Plants have typically been engineered to provide herbicide or pathogen resistance, or to improve crop characteristics such as climate tolerance, growth characteristics, flowering, or fruit ripening. One that is in widespread use is herbicide-resistant soybean but many others have failed to catch on, such as the FlavrSavr tomato that had the expression of a ripening gene repressed by stable transfection with a siRNA (Section S2). This strain was intended to have a longer shelf life and other improved qualities but was not a commercial success. Despite significant scientific progress, public opinion relating to their potential environmental impact continues to prevent widespread adoption of transgenic plants.

Stem cells

Stem cells are unspecialized or **undifferentiated** cells that have the ability to develop into many different cell types in the body. **Adult stem cells** or **somatic stem cells** exist in many tissues and can continue to divide without limit (unlike other cell types). They are thought to form a reservoir available to replenish specialized tissue cells by differentiation into the appropriate type as required. The ultimate stem cells are those that occur in the **inner cell mass** of early embryos (**embryonic stem (ES) cells**), and which subsequently give rise to all of the specialized cell types in the developing embryo; they are said to be **pluripotent**. In contrast, stem cells from adult tissues are restricted in the range of cell types they can produce. A great deal of interest has been generated in the characterization of stem cells, since they have the potential to be used for cell-based treatment of disease. If they can be induced to develop into specific cell types they could be used for conditions where those cell types are missing or defective, such as diabetes, Parkinson's disease and in the treatment of spinal injury.

One strategy for the creation of transgenic organisms (see above) is based on the manipulation of ES cells. DNA transfected into ES cells may integrate randomly into the genome by **illegitimate recombination** (Section E4). However, if the vector is constructed to contain sequences flanking the transgene that are homologous to a pre-determined genomic target site (**homology boxes**), then it can integrate at this chosen location by **site-specific recombination** (Section E4). Having a selectable marker such as the neo^r gene (Section P3) adjacent to the transgene and *between* the homology boxes allows recombinant ES cells to be selected and enriched in culture. If the vector also carries one or two herpes simplex virus **thymidine kinase** (**HSVTK**) genes *outside* the homology boxes, cells guilty of illegitimate recombination are likely to incorporate at least one of these and can be killed by **ganciclovir**, an anti-herpes drug that requires phosphorylation by HSVTK to be active. When reintroduced into an early embryo, the transgenic ES cells will be incorporated into the developing animal and give rise to parts of all the adult tissues (**mosaicism**). In the male germ line, some sperm will carry the transgene, and after breeding these will produce a subsequent generation of mice with the transgene in every cell. This technique has also been applied to the construction of **knockout mice**, where the gene of interest is permanently inactivated (Section S2).

Induced pluripotent stem cells

An alternative to the use of embryonic or somatic stem cells is to reprogram differentiated adult cells to become stem cells. Work on reprogramming somatic cell nuclei was pioneered in the 1960s by Sir John Gurdon, who first transferred adult frog cell nuclei into enucleated eggs; this ultimately led to the development of animal cloning. In 2007, it was found that genetic over-expression of a set of four defined transcription factors was sufficient to generate human and mouse **induced pluripotent stem** (**iPS**) **cells** from fibroblasts. Direct delivery of the proteins to the cell nuclei showed that these proteins only need to be present in cells up to a key commitment point. Therefore, this procedure can now be used without gene transfer. The key question regarding iPS technology has been whether these cells have the characteristics of true stem cells. The capability of iPS cells to yield transgenic animals in the same way as ES cells has provided considerable evidence of their fundamental stem cell properties. Nevertheless, minor differences may still exist. If iPS cells can be shown to fully replace ES cells, this technology would overcome the ethical problems surrounding the use of embryonic material.

Gene and cell therapy

The **OMIM®** (**Online Mendelian Inheritance in Man**) database lists several thousand disorders caused by heritable mutations and the sequencing of the human genome has now allocated many of these to specific genes. Many of these conditions have the potential to be treated by gene therapy. Somatic cell gene therapy is the transfer of new genetic material to the cells or tissues of an individual in order to provide therapeutic benefit to that individual without affecting any offspring. Not all cells in a tissue necessarily have to be modified in order to alleviate the symptoms of many of these diseases. In theory, stable transfection could lead to a lifetime cure or alleviation of symptoms, although in other cases repeated treatment might be required. Gene therapy also has the potential to treat somatic disease, such as the replacement of a lost tumor suppressor gene (e.g. p53) in cancer patients (Section N3). The most commonly used protocols for gene therapy involve viral transduction (Section P3) and include the use of retroviral vectors (Section M4). They have the advantage that the integrated retroviral DNA can be stably maintained in the cells. The retroviruses used for gene therapy have been engineered to disable their replication. Alternative gene therapy vectors include those based on DNA viruses such as adenoviruses (Section M3) that can maintain their genomes extra-chromosomally in infected cells for a significant period.

Gene therapy may be applied *ex vivo* by removing cells from the body and transducing them with the recombinant vector *in vitro* before replacing them in the body. This approach is most easily achieved with blood cells and has been trialled as an alternative to bone marrow transplantation for the treatment of **severe combined immunodeficiencies (SCIDs)**, caused by a malfunction of B cells and T cells. An alternative is *in vivo* gene therapy where the vector is applied topically to the tissue of interest. An example is the treatment of cystic fibrosis, a relatively common and serious genetic disease caused by a defect in chloride transport that leads to abnormally thick mucus in the lungs and other tissues. It is caused by any one of several mutations in the **cystic fibrosis transmembrane conductance regulator (CFTR)** gene. In one form of treatment, an aerosol containing an adenovirus-based vector including the wild type CFTR gene is inhaled into the lungs. Despite successful examples of gene therapy, there have been notable setbacks. Although the vectors are designed to promote site-specific recombination of the transgene (see above), the use of retroviruses is associated with a slight risk of random integration into the genome at a site that could lead to the development of cancer, while adenovirus infection has been associated with tissue inflammation.

An alternative approach to gene therapy is the introduction of one of the different types of genetically modified stem cells into damaged or diseased tissue. The aim is that the stem cells could self-renew and differentiate to repopulate the target tissue. A great deal of work is being performed to characterize different types of stem cells for this purpose for the treatment of many diseases.

S6 Bioinformatics

Key Notes

Definition and scope

Bioinformatics is the interface between biology and computer science. It comprises the organization of many kinds of large-scale biological data, particularly that based on DNA and protein sequence, into databases (normally Web-accessible), and the methods required to analyze these data.

Applications of bioinformatics

Applications of bioinformatics include: manipulation of DNA and protein sequence, maintenance of sequence and other databases, analytical methods such as sequence similarity searching, genome annotation, multiple sequence alignment, sequence phylogenetics, protein secondary, and tertiary structure prediction, statistical analysis of genomic and proteomic data.

Sequence similarity searching

One of the primary tools of bioinformatics, sequence similarity searching, involves the pairwise alignment of nucleic acid or protein sequences, normally to identify the closest matches from a sequence database to a test sequence. Tools such as BLAST can be used to help determine the possible function of unknown sequences derived from genome sequencing, transcriptomic, or proteomic experiments.

Multiple sequence alignment

The alignment of multiple nucleic acid or protein sequences using tools such as Clustal makes it possible to identify domains, motifs or individual residues from their evolutionary conservation across many species. This has led to the definition of many protein families on the basis of the similarity of specific motifs, or specific sequences.

Phylogenetic trees

We can derive information about evolutionary relationships between proteins or nucleic acids, and potentially the species containing them, by using the differences between sequences to group them into phylogenetic trees. The accumulated differences between sequences are taken to represent time since the evolutionary divergence of species, the so-called 'molecular clock.'

Structural bioinformatics

The mismatch between the number of known protein sequences and the number of determined three-dimensional structures has led to the development of methods to derive structure direct from sequence. These include secondary structure prediction, comparative or homology modeling, which uses sequence similarity between an unknown and a known structure to derive a plausible structure for the new protein, domain or sequence, and *ab initio* modeling, which attempts to derive structure direct from sequence.

Related topics	(A5) Analysis of proteins	(S2) Global gene expression
	(R2) Nucleic acid sequencing	analysis
	(S1) Introduction to the 'omics	(S3) Proteomics

Definition and scope

Bioinformatics is the interface between **biology** and **computer science**. It may be defined as the use of computers to store, organize and analyze biological information. Its origins may perhaps be traced to the use of computers to manipulate raw data and three-dimensional structural information from X-ray crystallography experiments (Section A5). However, bioinformatics as we now know it is really coincident with the recent increase in the amount of **DNA** and related **protein sequence** information. The amount of sequence data rapidly outstripped the ability of anything other than computer databases to handle, and has now grown to encompass many other kinds of data, as outlined below. The Internet and the Web have also been crucial in making this wealth of biological information available to users throughout the world. This means that almost anyone can be a user of **biological databases** and **bioinformatic tools** that are provided online at expert centers throughout the world. Bioinformatics is a complex and rapidly growing field, and this survey is necessarily brief. Readers are urged to consult the companion volume *Instant Notes in Bioinformatics* for a much more comprehensive discussion.

A large number of different kinds of data have now been incorporated into biological databases and can be said to be part of the domain of bioinformatics:

- **DNA sequence**, originally from sequencing of small cloned fragments and genes, but now including the large-scale high throughput sequencing of whole genomes (Section R2), and latterly comparative information about sequence variation between strains, including mutations and human polymorphisms (Section C4) from resequencing multiple genomes.

- **RNA sequence**, most often derived from the sequence of, for example, ribosomal RNA genes (Section J1) to establish evolutionary relationships.

- **Protein sequence**, originally determined directly by Edman degradation but more recently by mass spectrometry (Section A5). However, the vast majority of protein sequence is now deduced by the **theoretical translation** (Section K1) of protein coding regions identified in DNA sequences, usually without the protein itself ever having been identified or purified.

- **Expressed sequence tags** (**ESTs**), short sequences of random cDNAs derived from the mRNA of particular species, tissues, etc., and more recently sequence derived from high-throughput sequencing of entire cDNA libraries or pools.

- Structural data from **X-ray crystallography** and NMR, consisting of three-dimensional coordinates of the atoms in a protein, DNA, or RNA structure, or of a complex, such as a DNA–protein or enzyme–substrate complex (Sections A4 and A5).

- Data from the analysis of transcription across whole genomes (**transcriptomics**; Sections S1 and S2), and analysis of the expressed protein complement of cells or tissues under particular circumstances (**proteomics**; Sections S1 and S3).

- Data from other 'omics experiments, including **metabolomics**, **phospho-proteomics**, and methods for studying interactions between molecules on a large scale, such as **two-hybrid analysis** (Section S3).

- Data on biological networks derived from **systems biology** (Section S7).

Applications of bioinformatics

The scope of bioinformatics could be said to range across:

- The manipulation and analysis of individual nucleic acid and protein sequences.

- The maintenance of sequence data and the other data types above in **databases** designed to allow their easy manipulation and inspection, often in the context of access on the Internet. The best-known sequence databases include **EMBL** and **GenBank**, both of which contain all known DNA sequence data, and **Swiss-Prot** and **TrEMBL** (now combined into the overarching **UniProt**), which contain protein sequences, including those derived from translation of putative protein-coding genes in DNA sequences. The **Protein Data Bank** (**PDB**) contains all the three-dimensional structural information derived from X-ray crystallography and NMR of proteins, as well as nucleic acids and macromolecular complexes.

- Analytical methods, particularly for the comparison of DNA and protein sequences (**sequence similarity searching**, **sequence alignment**), used to identify unknown sequences and assign functions; identification of **structural** or **functional motifs**, **protein domains** (see below). In the context of whole-genome sequencing, this includes tools for the **assembly** of contiguous sequence from short sequenced fragments, a computationally intensive process, and methods for the automated **annotation** of sequence data, i.e. the identification of gene sequences and assignment of likely gene functions based on sequence similarity.

- Analysis of **phylogenetic** relationships by **multiple sequence alignment**, most often of protein or rDNA sequences; these methods help to identify evolutionary relationships between organisms at the sequence level, and aid the identification of functionally important regions of DNA, RNA, and proteins (see below).

- **Structural analysis**. Prediction of **secondary structure** from protein sequence; prediction of protein **tertiary structures** from the known structures of homologous proteins, a process known as **comparative** or **homology modeling** (see below); prediction of protein structure directly from protein sequence – **ab initio** structural analysis.

- **Statistical analysis** of transcription or protein expression patterns. The identification of, for example, clusters of genes whose transcription varies in a similar way in response to a particular disease state, is a crucial step in analyzing large quantities of transcriptomic or proteomic data (Section S2).

- Development of large database systems (**Entrez, SRS**) to encompass many types of bioinformatic information, allowing cross-comparisons to be made. Whilst the generation and maintenance of many databases is being increasingly automated, sequence data are also being combined with many other types of biological data in large manually curated databases, such as **Wormbase**, dealing with the biology of *C. elegans*, and **Flybase**, for *Drosophila* biology.

Sequence similarity searching

Possibly the most basic bioinformatics question is: 'What is this sequence I have just acquired?' A new protein with an interesting expression pattern may have been identified in a **proteomics** experiment and had its sequence determined by **mass spectrometry** (Section S3). Alternatively, it may be one of many putative gene-coding sequences from an **EST** or **genome sequencing project**, an rDNA sequence, or a noncoding region of a genome. In itself, the sequence may give very few clues to a function, but if related sequences can be found about which more information is available, then a possible function can be deduced, important features such as individual protein domains or putative active site residues can be identified, and investigative approaches suggested. The normal approach is to compare the unknown sequence with a database of previously determined sequences to identify the most closely related sequence or sequences from the database. This method relies on the fact that sequences of DNA and proteins both within and between species are related by **homology**, i.e. through having common evolutionary ancestors, and so proteins (and RNA and DNA motifs) tend to occur in families with related sequences, structures, and functions.

Two DNA or two protein sequences will always show some **similarity**, if only by chance. The principle of searching for meaningful sequence similarity is to **align** two sequences, i.e. array them one above the other with instances of the same base or amino acid at the same position (Figure 1). It is then possible to quantify the level of similarity between them and relate this to the probability that the similarity may merely be due to chance. In order to maximize the quality of the **alignment** between two sequences, there will most likely turn out to be some **mismatches** between the two sequences, and the overall alignment may well be improved if **gaps** are introduced in one or both of the sequences. **Computer algorithms** exist that will generate the best alignment between two input sequences, the best known of which is the **Smith–Waterman** algorithm. The algorithms attempt to maximize the number of identically matching letters, corresponding either to DNA bases or to amino acids. However, there is a problem with the introduction of gaps; it is possible to make a perfect alignment of any two sequences as long as many gaps can be introduced, but this is obviously unrealistic. So, the algorithms assign a positive **score** to matching letters (or, in the case of proteins, varying scores to the substitution of more or less related amino acids; Figure 1), but impose a **gap penalty** (negative score) to the introduction or lengthening of a gap. In this way, the final alignment is a trade-off between maximizing the matching letters, and minimizing the gaps. In Figure 1, the introduction of a gap in the top sequence improves the alignment, by pairing the basic (K, R), acidic (D, E) and aromatic (F, Y) residues, even though this incurs a gap penalty. The simplest way of quantifying the similarity between the two sequences is to quote the percentage of the residues that are identical (or perhaps similar in the case of a protein sequence; see below). However, an **identity score** of, for example, 50% will be more meaningful in a

```
Sequence 1    MILVKP-VVLKGDFG
Sequence 2    MILLKPAIIIRAEY-
Score         544357033220230

Total alignment score =
(sum of scores) - (gap penalty)
= 43 - 11 = 32
```

Figure 1. Principles of sequence alignment. An illustration of the principles of assigning scores and gap penalties in the alignment of two protein sequences. The sequences are written using the single-letter amino acid code (Section A3). From C. Hodgman, A. French, and D. Westhead (2009) *Instant Notes in Bioinformatics*, 2nd Edn. Taylor & Francis, Oxford.

longer sequence, since in a short sequence this is quite likely to happen by chance. So, in practice, more complex statistical measures of similarity are used (see below).

Although algorithms such as Smith–Waterman can give the theoretically best alignment between two sequences, faster, marginally less accurate algorithms are used in practice, including **FASTA (Fast-All)** and perhaps the most widely used, **BLAST (Basic Local Alignment Search Tool)**. Using these tools, particularly BLAST, it is possible to compare a test (or query) sequence against Web-based databases containing all known nucleic acid or protein sequences (EMBL, SwissProt, Uniprot), in a few seconds. The programs perform an alignment of the query sequence to each database sequence in turn (many millions for protein sequences) and return a list of the closest matching entries, with the alignment for each sequence and related information. A typical BLAST result is shown in Figure 2. In this case, the query sequence was the first 400 amino acids of the *E. coli* GyrB protein, one of the subunits of DNA gyrase (Sections B3 and D2). The result shown was the 23rd most similar sequence in the Swiss-Prot database at that time, as judged by the sequence alignment score, and the matched sequence corresponds to the GyrB protein from a rather distantly related bacterium, *Staphylococcus aureus*, of total length 643 amino acids. From the header at the top of the entry, we can see that the **percentage identity** in the alignment is 53% (214/403 amino acids; three gaps have been introduced into the query sequence so it now has a total length of 403). This is increased to 69% if we include matches of **similar amino acids** (those with chemically related side chains), and a total of four gaps have been introduced to improve the alignment between the two sequences. The **E (Expect) value** (10^{-115} in this case) indicates the probability that such an alignment would be found in a database of this size by chance. This extremely low value effectively guarantees that this alignment corresponds to a real evolutionary relationship between these two proteins, and indeed there is independent genetic and biochemical evidence for their equivalent role and activity. The alignment itself is then shown in Figure 2, with the **identities** and

```
>SW:GYRB_STAAU P0A0K8 DNA gyrase subunit B (EC 5.99.1.3).
          Length = 643

 Score = 413 bits (1062), Expect = e-115
 Identities = 214/403 (53%), Positives = 279/403 (69%), Gaps = 4/403 (0%)

Query: 1    SNSYDSSSIKVLKGLDAVRKRPGMYIGDTDDGTGLHHMVFEVVDNAIDEALAGHCKEIIV 60
            +++Y +   I+VL+GL+AVRKRPGMYIG T +   GLHH+V+E+VDN+IDEALAG+  +I V
Sbjct: 10   TDNYGAGQIQVLEGLEAVRKRPGMYIGSTSE-RGLHHLVWEIVDNSIDEALAGYANQIEV 68

Query: 61   TIHADNSVSVQDDGRGIPTGIHPEEGVSAAEVIMTVLHAGGKFDDNSYKVSGGLHGVGVS 120
            I  DN + V D+GRGIP  I  + G  A EVI+TVLHAGGKF   YKVSGGLHGVG S
Sbjct: 69   VIEKDNWIKVTDNGRGIPVDIQEKMGRPAVEVILTVLHAGGKFGGGGYKVSGGLHGVGSS 128

Query: 121  VVNALSQKLELVIQREGKIHRQIYEHGVPQAPLAVTGETEKTGTMVRFWPSLETFTNVTE 180
            VVNALSQ LE+ + R   I+ Q Y+ GVPQ  L   G T+KTGT++RF     E FT  T
Sbjct: 129  VVNALSQDLEVYVHRNETIYHQAYKKGVPQFDLKEVGTTDKTGTVIRFKADGEIFTETTV 188

Query: 181  FEYEILAKRLRELSFLNSGVSIRLRDKRDG---KEDHFHYEGGIKAFVEYLNKNKTPIHP 237
            + YE L +R+REL+FLN G+ I LRD+RD    +ED +HYEGGIK++VE LN+NK PIH
Sbjct: 189  YNYETLQQRIRELAFLNKGIQITLRDERDEENVREDSYHYEGGIKSYVELLNENKEPIHD 248

Query: 238  NIFYFSTEKDGIGVEVALQWNDGFQENIYCFTNNIPQRDGGTHLAGFRAAMTRTLNAYMD 297
              Y    KD I VE+A+Q+N G+  N+  + NNI   +GGTH  GF+ A+TR LN+Y
Sbjct: 249  EPIYIHQSKDDIEVEIAIQYNSGYATNLLTYANNIHTYEGGTHEDGFKRALTRVLNSYGL 308

Query: 298  KEGYSKKAKVSATGDDAREGLIAVVSVKVPDPKFSSQTKDKLVSSEVKSAVEQQMNELLA 357
            K+ K   +G+D REG+ A++S+K  DP+F   QTK KL +SEV+  V++  +E
Sbjct: 309  SSKIMKEEKDRLSGEDTREGMTAIISIKHGDPQFEGQTKTKLGNSEVRQVVDKLFSEHFE 368

Query: 358  EYLLENPTDAKIVVGKIIDAARAREAARRAREMTRRKGALDLA 400
            +L ENP  A+ VV K I AARAR AA++ARE+TRRK ALD+A
Sbjct: 369  RFLYENPQVARTVVEKGIMAARARVAAKKAREVTRRKSALDVA 411
```

Figure 2. A typical sequence alignment produced by BLAST. The alignment of one a series of matches derived from a BLAST search of the Swiss-Prot database using the first 400 amino acids of the *E. coli* GyrB protein as the query sequence.

similarities (+) being shown in the central line between the query and the subject lines, and gaps introduced in the sequences indicated as (–).

BLAST searches of this kind have become an important tool in the interpretation of data from many genomic and post-genomic technologies. The assignment of possible relationships and functions to 'new' sequences in genome sequencing is normally done this way, and the process is known as **annotation**. For putative protein-coding genes, it makes most sense to carry out similarity searching at the protein sequence level, since information about substitution of similar amino acids (which may be evolutionarily selected over a random substitution) can be used. In addition, the DNA sequence encoding a protein will frequently contain silent changes that do not affect the derived amino acid sequence. In transcriptomics and proteomics experiments, transcripts and protein sequences identified as having 'significant' expression patterns in the context of the particular study may potentially be identified by a BLAST search to find related sequences for which the function may already be known, if the genome for the organism in question has not been fully sequenced.

Multiple sequence alignment

As the name suggests, **multiple sequence alignments** are related to the alignments discussed above, but involve the alignment of more than two sequences. The advantage of aligning multiple DNA or protein sequences is that **evolutionary relationships** between individual genes or sequences become clearer, and it becomes possible to identify important residues or **motifs** (short segments of sequence) from their conservation between genes or proteins from different organisms (**orthologs**), or proteins of similar, but not identical function (**homologs**, **paralogs**, Section A3). The most commonly used multiple sequence alignment tool is called **Clustal**, and, as with the other bioinformatics tools, is available from many centers as a Web-based application. The method works by initially carrying out pairwise alignments between all the sequences, as in the BLAST method described above, but then uses this information to gradually combine the sequences, beginning with most similar, and finishing with the most distant, optimizing the overall match of the sequences as it goes. Submitting a series of protein sequences to Clustal results in an alignment such as that in Figure 3. This shows the best alignment between all the sequences, and indicates the degree of conservation at each position in the conservation line at the bottom of each block. A '*' indicates that a single amino acid residue is fully conserved at this position, and ':' and '.' indicate respectively that one of a number of strongly and more weakly similar groups of amino acids is conserved at that position. This particular example aligns a series of related type II topoisomerase sequences from bacteria and eukaryotic organisms. The GyrA and ParC subunits of DNA gyrase and topoisomerase IV respectively are paralogs that occur together in many bacterial species, whereas Top2 (topoisomerase II) proteins are orthologs from eukaryotes (Section C4). Despite the large evolutionary distance between the organisms, there is very significant conservation of sequence and a number of entirely conserved amino acids, including the tyrosine (Y) indicated by a '•', which is known to be an active site residue conserved in all enzymes of this type.

Techniques such as protein sequence alignment have helped to identify families of related proteins in terms of conserved amino acids or groups of amino acids that serve to identify members of a group. The identification and analysis of such protein families has become a bioinformatic specialty in its own right, resulting in the development of databases of sequence patterns identifying protein families, specific functions, and post-translational modifications. The best known such databases are **PROSITE** and **PFAM**.

```
CLUSTAL W (1.82) multiple sequence alignment

PARC_ECOLI    SAKFKKSARTVGDVLGKYHPHGDSACYEAMVLMAQPFSYRYPLVDGQGNWGAP-DDPKSF
PARC_BACSU    DKNFRKAAKTVGNVIGNYHPHGDSSVYEAMVRMSQDWKVRNVLIEMHGNNGS--IDGDPP
PARC_STRPN    DKSYRKSAKSVGNIMGHFHPHGDSSIYDAMVRMSQNWKNREILVEMHGNNGS--MDGDPP
GYRA_BACSU    DKPYKKSARIVGEVIGKYHPHGDSAVYESMVRMAQDFNYRYMLVDGHGNFGS--VDGDSA
GYRA_STRPN    DKPHKKSARITGDVMGKYHPHGDSSIYEAMVRMAQWWSYRYMLVDGHGNFGS--MDGDSA
GYRA_ECOLI    NKAYKKSARVVGDVIGKYHPHGDSAVYDTIVRMAQPFSLRYMLVDGQGNFGS--IDGDSA
TOP2A_CHICK   EVKGAQLAGSVAEMSSYHHGEASLMMTIINLAQNFVGSNNLNLLQPIGQFGTRLHGGKDS
TOP2A_HUMAN   EVKVAQLAGSVAEMSSYHHGEMSLMMTIINLAQNFVGSNNLNLLQPIGQFGTRLHGGKDS
TOP2_DROME    EVKVAQLSGSVAEMSAYHHGEVSLQMTIVNLAQNFVGANNINLLEPRGQFGTRLSGGKDC
              .  :  : ..::..*  .          :         . *:: *: *:   . .

PARC_ECOLI    AAMRYTESRLSKYSELLLSELGQGTADWVPNFDGTLQEPKMLPARLPNILLNGTTGIAVG
PARC_BACSU    AAMRYTEARLSPIASELLRDIDKNTVEFVPNFDDTSKEPVVLPAMFPNLLVNGSTGISAG
PARC_STRPN    AAMRYTEARLSEIAGYLLQDIEKKTVPFAWNFDDTEKEPTVLPAAFPNLLVNGSTGISAG
GYRA_BACSU    AAMRYTEARMSKISMEILRDITKDTIDYQDNYDGSEREPVVMPSRFPNLLVNGAAGIAVG
GYRA_STRPN    AAQRYTEARMSKIALEMLRDINKNTVDFVDNYDANEREPLVLPARFPNLLVNGATGIAVG
GYRA_ECOLI    AAMRYTEIRLAKIAHELMADLEKETVDFVDNYDGTEKIPDVMPTKIPNLLVNGSSGIAVG
TOP2A_CHICK   ASPRYIFTMLSPLARLLFFPVVDDNVLRFLYD-DNQRVEPEWYMPIIPMVLINGAEGIGTG
TOP2A_HUMAN   ASPRYIFTMLSSLARLLFPPKDDHTLKFLYD-DNQRVEPEWYIPIIPMVLINGAEGIGTG
TOP2_DROME    ASARYIFTIMSPLTRLIYHPLDDPLLDYQVD-DGQKIEPLWYLPIIPMVLVNGAEGIGTG
              *: ** :: :::     .   : *    *      .:* :*:**: **..*

PARC_ECOLI    MATDIPPHNLREVAQAAAIALID----QPKTTLDQLLDIVQGPDYPTEAEIITSRAEIRKI
PARC_BACSU    YATDIPPHHLGEVIDAVIKRIQ----MPSCSVDELMELIKGPDFPTGGIIQG-VDGIRKA
PARC_STRPN    YATDIPPHNLAEVIDAAVYMID----HPTAKIDKLMEFLPGPDFPTGAIIQG-RDEIKKA
GYRA_BACSU    MATNIPPHQLGEIIDGVLAVSE----NPDITIPELMEVIPGPDFPTAGQILG-RSGIRKA
GYRA_STRPN    MATNIPPHNLGETIDAVKLVMD----NPEVTTKDLMEVLPGPDFPTGALVMG-KSGIHKA
GYRA_ECOLI    MATNIPPHNLTEVINGCLAYID----DEDISIEGLMEHIPGPDFPTAAIING-RRGIEEA
TOP2A_CHICK   WSCKIPNFDIRETVNNIRCLLDGKEPLPMLPSYKNFKGTIDELGPNQYVISG-EVSILDS
TOP2A_HUMAN   WSCKIPNFDVREIVNNIRRLMDGEEPLPMLPSYKNFKGTIEELAPNQYVISG-EVAILNS
TOP2_DROME    WSTKISNYNPREIMKNLRKMINGQEPSVMHPWYKNFLGRMEYVSDGRYIQTG-NIQILSG
              :  .*. ..  *   .       :            :          *  .
```

Figure 3. Multiple sequence alignment produced by Clustal. Part of an alignment of type II topoisomerase proteins (three GyrA, three ParC and three Top2 sequences) produced by the Clustal program.

Phylogenetic trees

The usefulness of sequence similarity searching is based on the fact that similar sequences are related through evolutionary descent. Hence, we can use the similarity between sequences to infer something about these evolutionary relationships. Since individual species have separated from each other at different times during evolution, their DNA and protein sequences should have diverged from each other and show differences that are broadly proportional to the time since divergence, in other words the sequences should form a **molecular clock**. The amount of sequence similarity can therefore be used to generate **phylogenetic** relationships, and **phylogenetic trees** such as that shown in Figure 4. In practice however, such a straightforward assumption is very dangerous for a variety of reasons, including the fact that many sequences may not be changing randomly but in response to some form of selection, so that the rate of sequence change may not be easy to determine. Hence, the details of the methods involved in building and analyzing evolutionary relationships from sequence are extremely complex, and largely beyond the scope of this book. We can outline the basic principles, however. The starting point is a multiple sequence alignment of the type in Figure 3. This is converted to a table of percentage matches between pairs of sequences, or of differences between sequences (these are essentially equivalent). The most closely similar sequences are combined to form the most closely branched pair in the tree. In Figure 4, the closest pair comprises the GyrA sequences from *E. coli* and *Aeromonas salmonicida*. Progressively, more and more distant sequences are included in the tree to give a result like that in Figure 4. The horizontal distances in the tree represent the differences between the individual sequences, with the scale bar indicating a length corresponding to 0.1 substitutions per position, or a 10% difference in sequence. The type of tree in Figure 4 represents the simplest case, a so-called unrooted tree showing only relative differences between sequences. There are many crucial refinements, for example, including a distantly related sequence to indicate

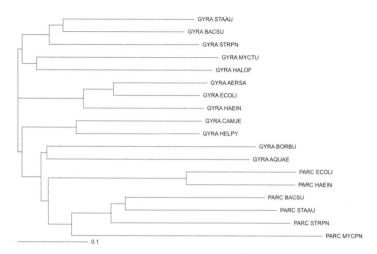

Figure 4. Phylogenetic tree of protein sequences. Unrooted phylogenetic tree of a series of GyrA and ParC sequences produced by the TreeView program, based on a Clustal multiple sequence alignment.

the position of the 'root' of the tree (that is, the position of the most likely common ancestor of all the sequences), correcting for the possibility of multiple substitutions at one position, and a variety of statistical tests of the validity of the branching of the tree.

Structural bioinformatics

The methods described in this section so far have been concerned with proteins as sequences, and perhaps with specific motifs or active site residues. Of course, crucial to protein function are the details of the three-dimensional structures that they adopt. We do have structural information from X-ray crystallography and NMR spectrometry (Section A5) but, although the rate of acquiring new structures has increased dramatically with technological advances, it has not kept pace with the industrial-scale sequencing of DNA. Whilst the number of known protein sequences is of the order of several million (depending on how you count them), the protein data bank (PDB), the primary repository of protein structural information, contains only (!) 80 000 structures (late 2011). To get around this mismatch, a number of approaches has been made to the determination of structure directly from sequence information.

The first and oldest such method is **secondary structure prediction**, in which we determine the likelihood that a particular part of a sequence forms one of the main secondary structure features, α-helices, β-sheets, and coil or loop regions (Section A3). The original methods relied solely on the fact that specific amino acids have a varying **propensity** to form part of an α-helix or β-sheet, so that a run of strongly helix-forming residues would be predicted to form a helix (and likewise for a β-sheet). These propensities are by no means absolute, however, with almost every amino acid able to adopt any conformation, and these methods are somewhat unreliable. More recently, the reliability has been improved by factoring in other information. One example is the tendency of many α-helices to be **amphipathic**, i.e. to have a hydrophobic and a hydrophilic face. This results in specific patterns of sequence that can be used to recognize α-helix-forming regions. Multiple sequence alignments in particular can be very helpful in the assignment of secondary structure, since in general the structural features of a protein (the overall fold, and the locations of the secondary structural features) are more highly conserved than the sequence itself. Hence the tendency to form, say, an α-helix will be conserved across the varying sequences in a multiple alignment. Other aspects of multiple alignments help to delineate structure. One example is the tendency of insertions and deletions between related proteins to occur in loops connecting secondary structural features rather than within the features themselves, where they are less likely to disrupt the overall structure of a protein. In addition, if a test sequence has sequence similarity to a protein with known structure, then the alignment between the two can give very powerful clues to the likely secondary structure of different regions. Tests of the most sophisticated prediction programs using all the above methods have a success rate of around 70–80% for assigning an individual residue to the correct structural type.

In fact, the use of multiple sequence alignments and known structures of homologous proteins can be taken much further. In the procedure known as **comparative** or **homology modeling**, a known structure can be used as the starting point to produce a complete three-dimensional structure of a new homologous protein. The structures are aligned, and positions of the core peptide backbone atoms of the new sequence are placed in the related positions in the framework of the known protein. Again, the tendency of insertions and deletions to be accommodated in loops between more defined structural features is a relevant factor, and the likely conformation of loops of different sizes must be predicted. When a new model has been built, it can be refined by adjusting the positions of the amino acid side chains to produce an optimized arrangement.

In related methods, it is possible to predict the function of specific regions of new protein sequence with good efficiency, for example the identification of particular types of domains or protein families (as enumerated in the **PFAM** database), or the identification of **signal peptides** (Section L4) and hence the likely cellular location of a protein, or the location of **transmembrane domains** of membrane-spanning proteins (Section A4), or domains that can form **fibrillar networks**, such as those involved in Alzheimer's disease and other degenerative conditions (Section A3).

Structure prediction bioinformatic methods are also being used to address the possible structure and function of nonglobular proteins, which are less amenable than proteins with well-defined globular structures to experimental structural determination. One focus of recent attention has been on **intrinsically disordered proteins**, which are proteins or domains with no native secondary or tertiary structure, but which are increasingly recognized as containing short motifs that can form a defined structure when interacting with a binding partner. Such **short linear motifs** may be involved in relatively weak, but functionally important protein–protein, or protein–DNA interactions.

There are many studies aimed at the complete (so-called *ab initio*, from first principles) prediction of protein structure from sequence. This is a theoretical possibility since many proteins are known to fold into their correct conformation entirely on their own, implying that the final structure is determined directly by the sequence (Section A3). Computational approaches to this problem can be broadly divided into two: **physics-based methods**, which attempt to derive structure from the physical interactions that govern protein folding, and **statistics-based methods**, which work with the statistical tendencies of short sequences to exhibit specific conformations. *Ab initio* modeling has so far proved reasonably successful at 'predicting' the fold of small proteins, whose structure is known by experimental methods, and it is the focus of ongoing research. These modeling efforts are computationally intensive, and have been the subject of **distributed computing** projects, such as **Rosetta@home**.

7 Systems and synthetic biology

Key Notes	
Systems biology	Systems biology is the opposite of traditional reductionist biology. Rather than considering the role of each gene or protein alone, systems biology involves consideration of a set of biological molecules, processes or entities together in an integrated way to establish how they combine to lead to biological function.
Network biology	Following the development of genome sequencing and 'omics technologies, large data sets have been constructed. Network biology seeks to integrate these data sets into a more complete understanding of the organization of biological networks. Networks are analyzed using statistical inference and commonly represented graphically as nodes, which are connected by edges.
Network motifs	Network motifs are simple connectivity patterns that are found to recur in biological networks. In gene expression networks, specific network motifs may be responsible for specific dynamic characteristics in gene expression.
Quantitative biology	Over the past few years, there has been an increasing trend towards more quantitative measurements in biology. This involves the measurement of absolute concentrations of key molecules and the rates of specific reactions.
Quantitative mathematical models	Mathematical modeling is used as a major tool in systems biology. Deterministic mathematical models use ordinary differential equations. Noise may be simulated in stochastic models using Monte Carlo methods and stochastic differential equations. Spatial models use partial differential equations.
Integration across biological scales	A major goal in systems biology is to be able to integrate information across scales from molecules to cells to organs to organisms to populations. This represents the greatest challenge, as it is simply impossible to derive models of organ function at the molecular level. This will require new tools to produce more workable models that can be integrated together.
Synthetic biology	Synthetic biology combines science and engineering to design and construct new biological functions and systems that are not found in nature. It may be used as a tool to

understand (and perhaps simplify) a biological system as part of a systems biology approach or to manipulate important processes in organisms to engineer a new biosynthetic process whose product may be of value.

Related topics   (Section O) Gene manipulation
(Section P) Cloning vectors
(Section R) Analysis and uses of cloned DNA
(S1) Introduction to the 'omics

(S2) Global gene expression analysis
(S3) Proteomics
(S4) Cell and molecular imaging
(S6) Bioinformatics

Systems biology

Over the past 30 years, molecular biology has led to great advances in our understanding of cell biology. By studying biology in broken down parts – the **reductionist** approach – we have built up a partial picture of how cells and organisms work, molecule by molecule and gene by gene. In the past 15 years, our ability to acquire large sets of data (genomic, transcriptomic, proteomic, metabolomics, etc.) has increased dramatically (Section S1). Organisms do not function as a set of separate parts, but rather in an integrated manner, where all parts or components are functionally connected. **Systems biology** seeks to study biology in an integrated manner by considering how all parts of the system contribute to the overall behavior of the system. Systems biology therefore has to tackle the **complexity** that is inherent to biology. An essential component of systems biology is the integration of experimental and theoretical biology. Here, mathematics and computer science approaches allow the handling of the large data sets and the assembly of mathematical models to make experimentally testable predictions. The loop between experimental validation and theoretical prediction can be seen as being a form of traditional generation and testing of hypotheses, which has been a standard mode of science for generations (Figure 1).

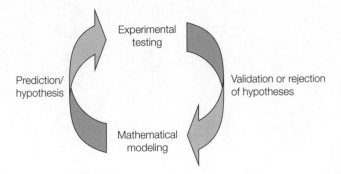

Figure 1. The iterative loop of model prediction and experimental validation that underlies systems biology.

Network biology

Many types of biological networks exist, but few have been characterized in detail, even in the simplest organisms. The structure of the network is referred to as the **network topology**. The development of 'omics technologies (Section S1) has provided large data sets that are often not fully quantitative, but provide a great deal of new biological information. One aim of systems biology is the assembly and visualization of network diagrams that show network topology. The challenge here is the **statistical inference** of the network structure. This often involves bioinformatic approaches (Section S6). This inference of network topology relies on searching for patterns of partial correlation between network components. This often involves statistical approaches such as **Bayesian networks**, which are graphical models that display the most probable network structure. The network can be described from graph theory as a series of **nodes** with a series of **edges** between the nodes.

- In **transcriptional networks**, genes are the nodes. If a gene expresses a transcriptional activator or inhibitor, then it may have a positive or negative connection to the genes that it controls. These are the edges.

- In **signaling networks**, proteins are the nodes and the edges are the directed reactions between them. Examples might be phosphorylation and dephosphorylation

- In **metabolite networks**, metabolites are the nodes and the edges are the enzymatic or chemical reactions that convert them.

Such biological networks operate at all scales up to intra- and inter-species communication networks.

Network motifs

Network motifs are particular patterns of connectivity in a network that recur more frequently than would be expected at random. Different types of network seem to be composed of varying network motifs that may occur again and again. These motifs can be seen as small circuits from which the network is composed. The concept of network motifs in transcriptional circuits was first proposed by Uri Alon from studies of transcriptional networks in *E. coli* (and subsequently in other organisms). Examples of simple network motifs in transcription networks are shown in Figure 2.

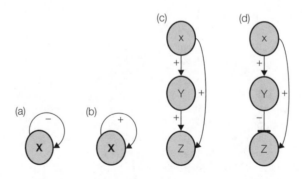

Figure 2. Common network motifs: (a) negative autoregulation; (b) positive autoregulation; (c) one example of a coherent feed-forward loop; (d) one example of an incoherent feed-forward loop.

(i) **Negative autoregulation**, where a transcription factor represses its own transcription. This can accelerate the transcriptional response and buffer against noise (Figure 2a).

(ii) **Positive autoregulation**, where a transcription factor activates its own transcription. This form of amplification can slow down responsiveness and lead to bimodal behavior where the system can be stable in two states: off or on (Figure 2b).

(iii) **Feed forward loops**, where a target gene Z is activated by two transcription factors X and Y, where Y is also regulated by X. This may be: coherent, where all interactions are positive, or incoherent, where for example X activates Y and Z, but Y represses Z. A **coherent feed-forward loop** may result in a situation where both X and Y are required for Z activation, or a situation where either is required (Figure 2c). An **incoherent feed-forward loop** may give rise to pulsatile behavior and accelerate the response (Figure 2d).

Quantitative biology

The ultimate aim of systems biology is to provide a quantitative and dynamic model of a biological system of interest. For many years molecular cell biology has relied on experimental methods that measure relative levels of biological molecules at fixed times but which do not provide absolute numbers. Most of the global 'omics technologies do not provide absolute quantification either. Quantitative mathematical models require absolute concentrations and rates of reactions. Therefore, there is an increasing requirement for new experimental methods that quantify biological concentrations and processes. In this respect, cell and molecular imaging is becoming increasingly important as a source of dynamic single cell data (Section S4). There is also a growing requirement for improved methods for quantitative measurement of key components, such as RNA (e.g. from RNA-Seq, Section S2) and protein concentrations (e.g. from quantitative proteomics, Section S3), as well as protein interactions and modifications and the concentrations of key metabolites.

The component reactants and products in a given biological reaction are often referred to as **variables**. The rates of reactions are referred to as **parameters**. It is generally true that the variables are far easier to measure than the parameters. Therefore, the parameters often have to be inferred by **fitting** of the mathematical model. It has generally only been possible so far to develop quantitative models for smaller subsystems that consist of only a subset of components.

Quantitative mathematical models

Mathematical models may be constructed based on **ordinary differential equations**, which describe the rates of, for example, chemical reactions. This generally simulates the average of all cells in a population, assuming a single rate for each reaction in the network. These are referred to as **deterministic mathematical models**. When the numbers of molecules involved in a reaction (such as gene transcription) are low, there may be significant variation due to the chance of small numbers of molecules colliding. In order to assess and predict this variation, **stochastic mathematical models** are used that are often based on **Monte Carlo methods**, which use random variation in repeated cycles of calculation, inspired in part by the mathematics of games of chance, hence the name. Stochastic models often use **stochastic differential equations**.

Even in relatively simple systems some parameters (rates) have to be estimated from fitting to experimental data. In these cases, the quality of dynamic experimental data is key to the success of the model. As a general rule, it is relatively easy to fit a model to data, but the model will often be wrong. The key to systems biology is to iteratively test the model in order to refine and improve it. Thus, one function of a useful model is its ability to provide new and interesting hypotheses. The other main aim of a model may be to allow the visualization of a core network in order to allow a better understanding of its function.

Integration across biological scales

A major long-term aim in systems biology is the modeling of biological systems across scales (e.g. from single molecules to cells to tissues to organisms to populations). This represents the principal challenge, since biological function emerges from the interactions of system components (such as the genes and proteins) that cannot be predicted by reductionist approaches. A key concept in multi-level modeling is therefore that information does not arise only from the genes, but arises from the interactions of all components of the system. This implies that information is not one-way and may work in any direction across scales.

A classic example of the application of multi-scale modeling to systems biology is the development of a model of the heart by Dennis Noble and Peter Hunter, first published in 1960. This originated from earlier classical work by Alan Hodgkin and Andrew Huxley on modeling of ion transport and gating in the squid giant axon (for which they received a Nobel prize in 1963). The heart model simulates a spatially organized, electrically connected set of cells anatomically similar to the heart. The model can simulate many aspects of heart function that are directly relevant to disease. However, it has not been possible to include genes in this model. The information required to compute this system in molecular detail would be greater than the capability of all the computers on the planet! Therefore, the ultimate goal of developing a model of a virtual physiological human will necessarily depend on new mathematical methods to simplify models and to link them across scales.

Synthetic biology

Synthetic biology combines science and engineering to design and construct new biological functions and systems that are not found in nature. It arises from the powerful tools of molecular biology for the sequencing, manipulation, and mutagenesis of genomes. Synthetic biology is related to systems biology in that the ability to manipulate cells and genomes and then to use a model to analyze and predict the resulting function is a good way in which to understand them. Synthetic biology may also be used to create useful bio-manufacturing processes, such as the development of bio-fuels.

An important early development in synthetic biology was the synthesis of the **repressilator**. The repressilator is a synthetic genetic regulatory network that was synthesized in *E. coli* to exhibit oscillations in the expression of green fluorescent protein (Section S4) (Figure 3). This system was found experimentally to act like an electrical oscillator system with fixed periods. The repressilator was an early milestone of synthetic biology that tested how a simple artificial network controlled the dynamic behavior of the system. This experiment gives us a new appreciation of the stability and function of other biological oscillators.

An example of the power of synthetic biology is the construction of the world's first synthetic organism by J. Craig Venter and colleagues in 2010. They designed a 1.08-Mb chromosome of a modified *Mycoplasma mycoides* genome in the computer, which was

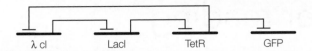

Figure 3. The repressilator: in this synthetic circuit λ cl represses Lacl, which in turn represses TetR. TetR represses both λ cl and GFP. The resulting circuit results in oscillations in GFP expression.

then synthesized in the laboratory and transplanted into a recipient cell denuded of its own chromosome to produce a self-replicating cell controlled by the synthetic genome. Initially the synthetic genome was built in blocks of DNA, which were assembled in yeast into the final genome, which was grown as a yeast artificial chromosome (Section P2). The new genome was introduced into a different species of *Mycoplasma* and initially, it was found to be nonfunctional. Combinations of synthetic and natural genome fragments were then combined to allow different sections of the genome to each be tested for biological function. Correction of a single base pair deletion in an essential gene then gave rise to biological function from the synthetic genome. There has inevitably been considerable ethical debate about these developments and their many possible applications, which include the creation of artificial organisms able to produce new medicines, fuels, and other products.

Further reading

There are many comprehensive textbooks covering molecular biology and no single book can satisfy the needs of every student. Different readers subjectively prefer different textbooks and so we do not feel that it would be particularly helpful to recommend one book over another. Instead, we have listed some of the leading textbooks that we know from experience have served their student readers well.

General reading

Alberts, B., Johnson, A., Lewis, J., Raff, M., Roberts, K, and Walter, P. (2008) *Molecular Biology of the Cell*, 5th Edn. Garland Science, New York and Oxford.

Brown, T.A. (2006) *Genomes 3*. Garland Science, New York and Oxford.

Krebs, J.E., Goldstein, E.S., and Kilpatrick, S.T. (2011) *Lewin's Genes X*. Jones and Bartlett Publishers Inc., Sudbury.

Liljas, A., Liljas, L., Piskur, J., Lindblom, G., Nissen, P., and Kjeldgaard, M. (2009) *Textbook of Structural Biology*. World Scientific Publishing, Hackensack, New Jersey.

Lodish, H., Berk, A., Kaiser, C.A., Krieger, M., Scott, M.P., Bretscher, A., Ploegh, H., and Matsudaira, P. (2007) *Molecular Cell Biology*, 6th Edn. W.H. Freeman, New York.

Strachan, T. and Read, A. (2010) *Human Molecular Genetics*. Garland Science, New York and Oxford.

Watson, J.D. (2004) *DNA: The Secret of Life*. Arrow Books, London.

Watson, J.D., Baker, T.A., Bell, S.B., Gann, A., Levine, M., and Losick, R. (2008) *Molecular Biology of the Gene*, 6th Edn. Pearson Education, New York.

More advanced reading

The following selected books and articles are recommended to readers who wish to know more about specific subjects. Some of the review articles may be too advanced for first year students but are very useful sources of information for subjects that may be studied in greater depth in later years.

Section A Informational macromolecules

Calladine, C.R., Drew, H.R., Luisi, B.F., and Travers, A.A. (2004) *Understanding DNA: The Molecule and How it Works*, 3rd Edn. Elsevier, London.

Cavanagh, J., Fairbrother, W., Palmer, A.G.III, Rance, M., and Skelton, N.J. (2006) *Protein NMR Spectroscopy: Principles and Practice*, 2nd Edn. Academic Press, Burlington, VT.

Chiu, W., Baker, M.L., and Almo, S.C. (2006) Structural biology of cellular machines. *Trends Cell Biol.* **16**, 144–150.

Cox, M.M. and Phillips, G.N.Jr. (Eds.) (2008) *Handbook of Proteins: Structure, Function and Methods*, Vols. 1 and 2. Wiley-Blackwell, Chichester.

Demain, A.L. and Preeti, V. (2009) Production of recombinant proteins by microbes and higher organisms. *Biotechnol. Adv.* **27**, 297–306.

Ellis, R.J. (2006) Molecular chaperones: assisting assembly as well as folding. *Trends Biochem. Sci.* **31**, 395–401.

Esposito, D. and Chatterjee, D.K. (2006) Enhancement of soluble protein expression through the use of fusion tags. *Curr. Opin. Biotechnol.* **17**, 353–358.

Lesk, A.M. (2004) *Introduction to Protein Science*. Oxford University Press, Oxford.

Neidle, S. (2002) *Nucleic Acid Structure and Recognition*. Oxford University Press, Oxford.

Rupp, B. (2009) *Biomolecular Crystallography: Principles, Practice and Applications to Structural Biology*. Garland Science, New York and Oxford.

Twyman, R.M. (2004) *Principles of Proteomics*. Garland Science, New York and Oxford.

Whitford, D. (2005) *Proteins – Structure & Function*. John Wiley & Sons, Chichester.

Section B Properties of nucleic acids

Bates, A.D. and Maxwell, A. (2005) *DNA Topology.* Oxford University Press, Oxford.

Blackburn, G.M., Gait, M., Loakes, D., and Williams, D.M. (Eds.) (2005) *Nucleic Acids in Chemistry and Biology*. Royal Society of Chemistry, London.

Calladine, C.R., Drew, H.R., Luisi, B.F., and Travers, A.A. (2004) *Understanding DNA: The Molecule and How it Works*, 3rd Edn. Elsevier, London.

Creighton, T.E. (2010) *The Biophysical Chemistry of Nucleic Acids and Proteins*. Helvetian Press, London.

Deweese, J.E., Osherhoff, M.A., and Osherhoff, N. (2009) DNA topology and topoisomerases. *Biochem. Mol. Biol. Edu.* **37**, 2–10.

Neidle, S. (2008) *Principles of Nucleic Acid Structure*. Academic Press, London.

Wang, J.C. (2002) Cellular roles of DNA topoisomerases: a molecular perspective. *Nat. Rev. Mol. Cell Biol.* **3**, 430–440.

Section C Prokaryotic and eukaryotic chromosome structure

Cockerill, P.N. (2011) Structure and function of active chromatin and DNase I hypersensitive sites. *FEBS J.* **278**, 2182–2210.

Henikoff, S. and Ali, S. (2011) Histone modification: cause or cog? *Trends Genet.* **27**, 389–396.

Klose, K.J. and Bird, A.P. (2006) Genomic DNA methylation: the mark and its mediators. *Trends Biochem. Sci.* **31**, 89–97.

Rando, O.J. and Chang, H.Y. (2009) Genome-wide views of chromatin structure. *Annu. Rev. Biochem.* **78**, 245–271.

Reyes-Lamothe, R., Wang, X., and Sherratt, D. (2008) *Escherichia coli* and its chromosome. *Trends Microbiol.* **16**, 238–245.

Shapiro, J.A. and von Sternberg, R. (2005) Why repetitive DNA is essential to genome function. *Biol. Rev.* **80**, 227–250.

Thanbichier, M., Viollier, P.H., and Shapiro, L. (2005) The structure and function of the bacterial chromosome. *Curr. Opin. Genet. Dev.* **15**, 153–162.

Tremethick, D.J. (2007) Higher-order structures of chromatin: the elusive 30 nm fiber. *Cell* **128**, 651–654.

Weiner, A.M. (2002) SINEs and LINEs: the art of biting the hand that feeds you. *Curr. Opin. Cell Biol.* **14**, 343–350.

Woodcock, C.L. and Dimitrov, S. (2001) Higher order structure of chromatin and chromosomes. *Curr. Opin. Genet. Dev.* **11**, 130–135.

Section D DNA replication

Bebenek, K. and Kunkel, T.A. (2004) Functions of DNA polymerases. *Adv. Protein. Chem.* **69**, 137–165.

DePamphilis, M.L. (Ed.) (2006) *DNA Replication and Human Disease. Cold Spring Harbor Monograph Series 46*. Cold Spring Harbor Laboratory Press, New York.

Johansson, E. and MacNeill, S.A. (2010) Eukaryotic replicative DNA polymerases take shape. *Trends Biochem. Sci.* **35**, 339–347.

Johnson, A. and O'Donnell, M. (2005) Cellular DNA replicases: components and dynamics at the replication fork. *Annu. Rev. Biochem.* **74**, 283–315.

Masai, H., Matsumoto, S., You, Z.Y., Yoshizawa-Sugata, N., and Oda, M. (2010). Eukaryotic DNA replication: where, when and how? *Annu. Rev. Biochem.* **79**, 89–130.

McEachern, M.J., Krauskopf, A., and Blackburn, E.H. (2000) Telomeres and their control. *Annu. Rev. Genet.* **34**, 331–358.

Mott, M.L. and Berger, J.M. (2007) DNA replication initiation: mechanisms and regulation in bacteria. *Nature Rev. Microbiol.* **5**, 343–354.

Pomerantz, R.T. and O'Donnell, M. (2007) Replisome mechanics: insights into a twin DNA polymerase machine. *Trends Microbiol.* **15**, 156–164.

Scholefield, G., Veening, J.W., and Murray, H. (2011) DnaA and ORC: more than just DNA replication initiators. *Trends Cell Biol.* **21**, 188–194.

Section E DNA damage, repair, and recombination

Friedberg, E.C., Walker, G.C., Siede, W., Wood, R.D., Schultz, R.A., and Ellenberger, T. (2006) *DNA Repair and Mutagenesis*, 2nd Edn. American Society for Microbiology, Washington, DC.

Grindley, N.D.F., Whiteson, K.L., and Rice, P.A. (2006) Mechanisms of site-specific recombination. *Annu. Rev. Biochem.* **75**, 567–605.

Jiricny, J. (2006) The multifaceted mismatch-repair system. *Nat. Rev. Mol. Cell Biol.* **7**, 335–346.

Kazazian, H.H.Jr., (2004) Mobile elements: drivers of genome evolution. *Science* **303**, 1626–1632.

Ljungman, M. (2010) The DNA damage response – repair or despair. *Environ. Mol. Mutagen.* **51**, 879–889.

Persky, N.S. and Lovett, S.T. (2008) Mechanisms of recombination: lessons from *E. coli*. *Crit. Rev. Biochem. Mol. Biol.* **43**, 347–370.

Prakash, S., Johnson, R.E., and Prakash, L. (2005) Eukaryotic translesion DNA polymerases: specificity of structure and function. *Annu. Rev. Biochem.* **74**, 317–353.

Robertson, A.B. Klungland, A., Rognes, T., and Leiros, I. (2009) DNA repair in mammalian cells – base excision repair: the long and short of it. *Cell. Mol. Life Sci.* **66**, 981–993.

Roos, W.P. and Kaina, B. (2006) DNA damage-induced cell death by apoptosis. *Trends Mol. Med.* **12**, 440–450.

San Filippo, J., Sung, S., and Klein, K. (2008) Mechanism of eukaryotic homologous recombination. *Annu. Rev. Biochem.* **77**, 229–257.

Sancar, A., Lindsey-Boltz, L.A., Ünsai-Kaçmaz, K., and Linn, S. (2004) Molecular mechanisms of mammalian DNA repair and the DNA damage checkpoints. *Annu. Rev. Biochem.* **73**, 39–85.

Siede, W., Kow, Y.W., and Doetsch, P.W. (Eds.) (2005) *DNA Damage Recognition*. Taylor & Francis, New York.

Wyman, C. and Kanaar, R, (2006) Double-strand break repair: all's well that ends well. *Annu. Rev. Genet.* **40**, 363–383.

Section F Transcription in bacteria

Borukhov, S. and Nudler, E. (2008) RNA polymerase: the vehicle of transcription. *Trends Microbiol.* **16**, 126–134.

Merino, E. and Yanofsky, C. (2005) Transcription attenuation: a highly conserved regulatory strategy used by bacteria. *Trends Genet.* **21**, 260–264.

Paget, M.S.B. and Helmann, J.D. (2003) The σ^{70} family of sigma factors. *Genome Biology* **4**, 203.

Roberts, J.W., Shankar, S., and Filter, J.J. (2008) RNA polymerase elongation factors. *Annu. Rev. Microbiol.* **62**, 211–233.

Wagner, R. (2000) *Transcription Regulation in Prokaryotes.* Oxford University Press, Oxford.

Young, B.A., Gruber, T.M., and Gross, C.A. (2002) Views of transcription initiation. *Cell* **109**, 417–420.

Section G Regulation of transcription in bacteria

Elliot, D. and Ladomery, M. (2010) *Molecular Biology of RNA.* Oxford University Press, Oxford.

Merino, E., Jensen, R.A., and Yanofsky, C. (2008) Evolution of bacterial trp operons and their regulation. *Curr. Opin. Microbiol.* **11**, 78–86.

Gottesman, S. (2005) Micros for microbes: non-coding regulatory RNAs in bacteria. *Trends Genet.* **21**, 399–404.

Gruber, T.M. and Gross, C.A. (2003) Multiple sigma subunits and the partitioning of bacterial transcription space. *Annu. Rev. Microbiol.* **57**, 441–466.

Müller-Hill, B. (1996) *The* lac *Operon: a Short History of a Genetic Paradigm.* W. de Gruyter, Berlin.

Müller-Hill, B. (1998) Some repressors of bacterial transcription. *Curr. Opin. Microbiol.* **1**, 145–151.

Österberg, S., del Peso-Santos, T., and Shingler, V. (2011) Regulation of alternative sigma factor use. *Annu. Rev. Microbiol.* **65**, 37–55.

Rutberg, B. (1997) Antitermination of transcription of catabolic operons. *Mol. Microbiol.* **23**, 413–421.

Wagner, R. (2000) *Transcription Regulation in Prokaryotes.* Oxford University Press, Oxford.

Section H Transcription in eukaryotes

Cramer, P., Armache, K.-J., Baumli, S. *et al.* (2008) Structure of eukaryotic RNA polymerases. *Annu. Rev. Biophys.* **37**, 337–352.

Darzacq, X., Yao, J., Larson, D.R. *et al.* (2009) Imaging transcription in living cells. *Annu. Rev. Biophys.* **38**, 173–196.

Dieci, G., Fiorino, G., Castelnuovo, M., Teichmann, M., and Pagano, A. (2007) The expanding RNA polymerase III transcriptome. *Trends Genet.* **23**, 614–622.

Egloff, S. and Murphy, S. (2008) Cracking the RNA polymerase II CTD code. *Trends Genet.* **24**, 280–288.

Hahn, S. (2004) Structure and mechanism of the RNA polymerase II transcription machinery. *Nat. Struct. Mol. Biol.* **11**, 394–403.

Juven-Gershon, T., Hsu, J.-Y., Theisen, J.W.M., and Kadonaga, J.T. (2008). The RNA polymerase core promoter – the gateway to transcription. *Cur. Opin. Cell Biol.* **20**, 253–259.

Latchman, D.S. (2005) *Gene Regulation: A Eukaryotic Perspective,* 5th Edn. Taylor & Francis, New York and Oxford.

Latchman, D.S. (2007) *Eukaryotic Transcription Factors,* 5th Edn. Academic Press, London.

Lee, T.I. and Young, R.A. (2000) Transcription of eukaryotic protein-coding genes. *Annu. Rev. Genet.* **34**, 77–137.

Muñoz, M., de la Mata, M., and Kornblihtt, A.R. (2010) The carboxy-terminal domain of RNA polymerase II and alternative splicing. *Trends Biochem. Sci.* **35**, 497–504.

Selth, L.A., Sigurdsson, S., and Svejstrup, J.Q. (2010) Transcript elongation by RNA polymerase II. *Annu. Rev. Biochem.* **79**, 271–293.

Smale, S.T. and Kadonaga, J.T. (2003) The RNA polymerase II core promoter. *Annu. Rev. Biochem.* **72**, 449–479.

Thomas, M.C. and Chiang, C.M. (2006) The general transcription machinery and general cofactors. *Crit. Rev. Biochem. Mol. Biol.* **41**, 105–178.

White, R.J. (2001) *Gene Transcription: Mechanisms and Control*. Blackwell Science, Oxford.

White, R.J. (2008) RNA polymerases I and III, non-coding RNAs and cancer. *Trends Genet.* **24**, 622–629.

Section I Regulation of transcription in eukaroytes

Almeida, R. and Allshire, R.C. (2005) RNA silencing and genome regulation. *Trends Cell Biol.* **15**, 251–258.

Bai, L. and Morozov, A.V. (2010) Gene regulation by nucleosome positioning. *Trends Genet.* **26**, 476–483.

Brodersen, P. and Voinnet, O. (2006) The diversity of RNA silencing in plants. *Trends Genet.* **22**, 268–280.

Carninici, P., Yasuda J., and Hayashizaki, Y. (2008) Multifaceted mammalian transcriptome. *Curr. Opin. Cell Biol.* **20**, 274–280.

Elliot, D. and Ladomery, M. (2010) *Molecular Biology of RNA*. Oxford University Press, Oxford.

Francis, G.A., Fayard, E., Picard, F., and Auwerx, J. (2003) Nuclear receptors and the control of metabolism. *Annu. Rev. Physiol.* **65**, 261–311.

Handel, A.E., Ebers, G.C., and Ramagopalan, S.V. (2010) Epigenetics: molecular mechanisms and implications for disease. *Trends Mol. Med.* **16**, 7–16.

Kouzarides, T. (2007) Chromatin modifications and their function. *Cell* **128**, 693–705.

Latchman, D.S. (2005) *Gene Regulation: A Eukaryotic Perspective*, 5th Edn. Taylor & Francis, New York and Oxford.

Latchman, D.S. (2007) *Eukaryotic Transcription Factors*, 5th Edn. Academic Press, London.

Latchman, D.S. (2010) *Gene Control*. Garland Science, New York and Oxford.

Meister, G. (2011) *RNA Biology: An Introduction*. Wiley VCH, Weinheim.

Mellor, J. (2006) Dynamic nucleosomes and gene transcription. *Trends Genet.* **22**, 320–329.

Nilsen, T.W. (2007) Mechanisms of microRNA-mediated gene regulation in animal cells. *Trends Genet.* **23**, 243–249.

Privalsky, M.L. (2004) The role of corepressors in transcriptional regulation by nuclear hormone receptors. *Annu. Rev. Physiol.* **66**, 315–360.

Rando, O.J. and Chang, H.Y. (2009) Genome-wide views of chromatin structure. *Annu. Rev. Biochem.* **78**, 245–271.

Sharp, P.A. (2009) The centrality of RNA. *Cell* **136**, 577–580.

Warren, A.J. (2002) Eukaryotic transcription factors. *Curr. Opin. Struct. Biol.* **12**, 107–114.

White, R.J. (2001) *Gene Transcription: Mechanisms and Control*. Blackwell Science, Oxford.

Wutz, A. (2007) *Xist* function: bridging chromatin and stem cells. *Trends Genet.* **23**, 457–464.

Section J RNA processing and RNPs

Bentley, D.L. (2005) Rules of engagement: co-transcriptional recruitment of pre-mRNA processing factors. *Curr. Opin. Cell Biol.* **17**, 251–256.

Black, R. (2003) Mechanisms of alternative pre-messenger RNA splicing. *Annu. Rev. Biochem.* **72**, 291–336.

Elliot, D. and Ladomery, M. (2010) *Molecular Biology of RNA*. Oxford University Press, Oxford.

Evans, D., Marquez, S.M., and Pace, N.R. (2006) RNAse P: interface of the RNA and protein worlds. *Trends Biochem. Sci.* **31**, 333–341.

Gerbi, S.A. and Borovjagin, A.V. (2004) Pre-ribosomal RNA processing in multicellular organisms. In: *The Nucleolus* (ed. M.O.J. Olson), pp. 170–198. Kluwer Academic, New York.

Han, J., Xiong, J., Wang, D., and Fu, X.-D. (2011) Pre-mRNA splicing: where and when in the nucleus. *Trends Cell Biol.* **21**, 336-343.

Jackowiak, P., Nowacka, M., Strozycki, P.M., and Figlerowicz, M. (2011) RNA degradome – its biogenesis and functions. *Nucleic Acids Res.* **39**, 7361–7370.

Licatalosi, D.D. and Darnell, R.B. (2010) RNA processing and its regulation: global insights into biological networks. *Nat. Rev. Genet.* **11**, 75–87.

Meister, G. (2011) *RNA Biology: An Introduction.* Wiley VCH, Weinheim.

Muñoz, M., de la Mata, M., and Kornblihtt, A.R. (2010) The carboxy-terminal domain of RNA polymerase II and alternative splicing. *Trends Biochem. Sci.* **35**, 497–504.

Warf, M.B. and Berglund, J.A. (2010) Role of RNA structure in regulating pre-mRNA splicing. *Trends Biochem. Sci.* **35**, 169–178.

Wu, J. (2012) *Post-transcriptional Gene Regulation: RNA Processing in Eukaryotes.* Wiley VCH, Weinheim.

Section K The genetic code and tNRA

Cropp, T.A. and Schultz, P.G. (2004) An expanding genetic code. *Trends Genet.* **20**, 625–630.

Di Giulio, M. (2005) The origin of the genetic code: theories and their relationships, a review. *Biosystems* **80**, 175–184.

Elliot, D. and Ladomery, M. (2010) *Molecular Biology of RNA.* Oxford University Press, Oxford.

Ibba, M. and Söll, D. (2000) Aminoacyl-tRNA synthesis. *Annu. Rev. Biochem.* **69**, 617–650.

Ibba, M. and Söll, D. (2004) Aminoacyl-tRNAs: setting the limits of the genetic code. *Genes Dev.* **18**, 731–738.

Meister, G. (2011) *RNA Biology: An Introduction.* Wiley VCH, Weinheim.

Soll, D. and Raj Bhandary, T.L. (1995) *Molecular Biology of tRNA.* American Society for Microbiology, Washington, DC.

Section L Protein synthesis

Bashan, A. and Yonath, A. (2008) Correlating ribosome function with high-resolution structures. *Trends Microbiol.* **16**, 326–335.

Elliot, D. and Ladomery, M. (2010) *Molecular Biology of RNA.* Oxford University Press, Oxford.

Fabian, M.F., Sonenberg, N., and Filipowicz, W. (2010) Regulation of mRNA translation and stability by microRNAs. *Annu. Rev. Biochem.* **79**, 351–379.

Hernandez, G., Altmann, M., and Lasko, P. (2010) Origins and evolution of the mechanisms regulating translation initiation in eukaryotes. *Trends Biochem. Sci.* **35**, 63–73.

Jackson, R.J., Hellen, C.U.T., and Pestova, T.V. (2010) The mechanism of eukaryotic translation initiation and principles of its regulation. *Nat. Rev. Mol. Cell Biol.* **11**, 113–127.

Kapp, L.D. and Lorsch, J.R. (2004) The molecular mechanics of eukaryotic translation. *Annu. Rev. Biochem.* **73**, 657–704.

Kawamata, T. and Tomari, Y. (2010) Making RISC. *Trends Biochem. Sci.* **35**, 368–376.

Korostelev, A. and Noller, H.F. (2007) The ribosome in focus: new structures bring new insights. *Trends Biochem. Sci.* **32**, 434–441.

Liljas, A. (2004) *Structural Aspects of Protein Synthesis.* World Scientific Publishing, Singapore.

Meister, G. (2011) *RNA Biology: An Introduction.* Wiley VCH, Weinheim.

Nilsen, T.W. (2007) Mechanisms of microRNA-mediated gene regulation in animal cells. *Trends Genet.* **23**, 243–249.

Rodnina, M., Beringer, M., and Wintermeyer, W. (2007) How ribosomes make peptide bonds. *Trends Biochem. Sci.* **32**, 20–26.

Schmeing, T.M. and Ramakrishnan, V. (2009) What recent ribosome structures have revealed about the mechanism of translation. *Nature* **461**, 1234–1242.

Section M Bacteriophages and eukaryotic viruses

Birge, E.A. (2006) *Bacterial and Bacteriophage Genetics*, 5th Edn. Springer, New York.

Boss, I.W., Plaisance, K.B., and Renne, R. (2009) Role of virus-encoded microRNAs in herpesvirus biology. *Trends Microbiol.* **17**, 544–553.

Cann, A.J. (2011) *Principles of Molecular Virology*, 5th Edn. Academic Press, London.

Damania, B. (2007) DNA tumor viruses and human cancer. *Trends Microbiol.* **15**, 38–44.

DePamphilis, M.L. (Ed.) (2006) *DNA Replication and Human Disease. Cold Spring Harbor Monograph Series 46*. Cold Spring Harbor Laboratory Press, New York.

Dickson, A.M. and Wilusz, J. (2011) Strategies for viral RNA stability: live long and prosper. *Trends Genet.* **27**, 286–293.

Dimmock, N.J., Easton, A.J., and Leppard, K.N. (2006) *Introduction to Modern Virology*, 6th Edn. Blackwell Publishing, Oxford.

Knobler, C.M. and Gelbart, W.M. (2009) Physical chemistry of DNA viruses. *Annu. Rev. Phys. Chem.* **60**, 367–383.

Oppenheim, A.B., Kobiler, O., Stavans, J., Court, D.L., and Adhya, S. (2005) Switches in bacteriophage lambda development. *Annu. Rev. Genet.* **39**, 409–429.

Section N Cell cycle and cancer

Clarke, P.R. and Allan, L.A. (2009) Cell-cycle control in the face of damage – a matter of life or death. *Trends Cell Biol.* **19**, 89–98.

Damania, B. (2007) DNA tumor viruses and human cancer. *Trends Microbiol.* **15**, 38–44.

Green, D.R. (2010) *Means to an End: Apoptosis and Other Cell Death Mechanisms*. Cold Spring Harbor Laboratory Press, New York.

Hanahan, D. and Weinberg, R.A. (2000) The hallmarks of cancer. *Cell* **100**, 57–70.

Macdonald, F., Ford, C.H.J., and Casson, A.G. (2004) *Molecular Biology of Cancer*, 2nd Edn. Garland Science, New York and Oxford.

Morgan, D. (2006) *The Cell Cycle: Principles of Control*. New Science Press, London.

Knowles, M.A. and Selby, P.J. (Eds.) (2005) *Introduction to the Cellular and Molecular Biology of Cancer*, 4th Edn. Oxford University Press, Oxford.

Macleod, K. (2000) Tumor suppressor genes. *Curr. Opin. Genet. Dev.* **10**, 81–93.

Potten, C. and Wilson, J. (2004) *Apoptosis: The Life and Death of Cells*. Cambridge University Press, Cambridge.

Riedl, S.J. and Shi, Y. (2004) Molecular mechanism of caspase regulation during apoptosis. *Nat. Rev. Mol. Cell Biol.* **5**, 897–907.

Roos, W.P. and Kaina, B. (2006) DNA damage-induced cell death by apoptosis. *Trends Mol. Med.* **12**, 440–450.

Wang, C. and Youle, R.J. (2009) The role of mitochondria in apoptosis. *Annu. Rev. Genet.* **43**, 95–118.

Weinberg, R.A. (2006) *The Biology of Cancer*. Garland Science, New York and Oxford.

White, R.J. (2008) RNA polymerases I and III, non-coding RNAs and cancer. *Trends Genet.* **24**, 622–629.

Section O Gene manipulation, Section P Cloning vectors, and Section Q Gene libraries and screening

Brown, T.A. (2010) *Gene Cloning & DNA Analysis: An Introduction*, 6th Edn. Wiley-Blackwell, Oxford.

Dale, J.W., von Schantz, M., and Plant, N. (2012) *From Genes to Genomes: Concepts and Applications of DNA Technology*, 3rd Edn. Wiley-Blackwell, Oxford.

Hartley, J.L. (2006) Cloning technologies for protein expression and purification. *Curr. Opin. Biotechnol.* **17**, 359–366.

Lundstrom, K. (2003) Latest development in viral vectors for gene therapy. *Trends Biotechnol.* **21**, 117–122.

Primrose, S.B. and Twyman, R.M. (2006) *Principles of Gene Manipulation and Genomics*, 7th Edn. Blackwell Publishing, Massachusetts, USA.

Sambrook, J. and Russell, D. (2001) *Molecular Cloning: a Laboratory Manual*, 3rd Edn. Cold Spring Harbor Laboratory Press, New York.

Section R Analysis and uses of cloned DNA

Bustin, S.A., Benes, V., Nolan, T., and Pfaffl, M.W. (2005) Quantitative real-time PCR – a perspective. *J. Mol. Endocrinol.* **34**, 597–601.

Bustin, S.A. (Ed.) (2009) *The PCR Revolution: Basic Technologies and Applications.* Cambridge University Press, Cambridge.

Dale, J.W., von Schantz, M., and Plant, N. (2012) *From Genes to Genomes: Concepts and Applications of DNA Technology*, 3rd Edn. Wiley-Blackwell, Oxford.

Davies, K. (2010) *The $1,000 Genome: The Revolution in DNA Sequencing and the New Era of Personalized Medicine.* Free Press, New York.

Kircher, M. and Kelso, J. (2010) High throughput DNA sequencing: concepts and limitations *BioEssays* **32**, 524–536.

Mardis, E.R. (2008) The impact of next-generation sequencing technologies on genetics. *Trends Genet.* **24**, 133–141.

McPherson, M.J. and Moller, S.G. (2006) *PCR: The Basics*, 2nd Edn. Taylor & Francis, New York and Oxford.

Metzker, M. (2010) Sequencing technologies – the next generation. *Nat. Rev. Genet.* **11**, 31–46.

Suter-Crazzolara, C., Klemm, M., and Reiss, B. (1995) Reporter genes. *Meth. Cell Biol.* **50**, 425–438

Section S Functional genomics and the new technologies

Alon, U. (2006) *An Introduction to Systems Biology: Design Principles of Biological Circuits.* CRC Press, Boca Raton, CA.

Amabile, G. and Meissner, A. (2009) Induced pluripotent stem cells: current progress and potential for regenerative medicine. *Trends Mol. Med.* **15**, 59–68.

Avison, M. (2006) *Measuring Gene Expression.* Taylor & Francis, New York and Oxford.

Bialik, S., Zalckvar, E., Ber, Y., Rubinstein, A.D., and Kimchi, A. (2010) Systems biology analysis of programmed cell death. *Trends Biochem. Sci.* **35**, 556–564.

Bruggeman, F.J. and Westerhoff, H.V. (2007) The nature of systems biology. *Trends Microbiol.* **15**, 45–50.

Carninci, P. (2006) Tagging mammalian transcriptome complexity. *Trends Genet.* **22**, 501–510.

Chait, B.T. (2011) Mass spectrometry in the postgenomic era. *Annu. Rev. Biochem.* **80**, 239–246.

Collas, P. (2010) The current state of chromatin immunoprecipitation. *Mol. Biotechnol.* **45**, 87–100.

Darzacq, X., Yao, J., Larson, D.R. et al. (2009) Imaging transcription in living cells. *Annu. Rev. Biophys.* **38**, 173–196.

Giacca, M. (2010) *Gene Therapy.* Springer-Verlag Italia, Milan.

Gibson, D.G. et al. (2010) Creation of a bacterial cell controlled by a chemically synthesized genome. *Science* **329**, 52–56.

Giepmans, B.N.G., Adams, S.R., Ellisman, M.H., and Tsien, R.Y. (2006) The fluorescent toolbox for assessing protein location and function. *Science* **312**, 217–224.

Goncalves, M.A.F.V. (2005) A concise peer into the background, initial thoughts and practices of human gene therapy. *Bioessays* **27**, 506–517.

Hodgman, T.C., French, A., and Westhead, D.R. (2009) *Instant Notes in Bioinformatics*. 2nd Edn. Taylor & Francis, New York and Oxford.

Honore, B., Ostergaard, M., and Vorum, H. (2004) Functional genomics studied by proteomics. *Bioessays* **26**, 901–915.

Khalil, A.S. and Collins, J.J. (2010) Synthetic biology: applications come of age. *Nature Rev. Genet.* **11**, 367–379.

Lanza, R., Gearhart, J., Hogan, B., Melton, D., Pederson, R., Thomas, E.D., Thomson, J., and Wilmut, I. (2009) *Essentials of Stem Cell Biology*. Academic Press, Burlington, VT.

Lesk, A.M. (2007) *Introduction to Genomics*. Oxford University Press, Oxford.

Noble, D. (2008) *The Music of Life: Biology Beyond the Genome*. Oxford University Press, Oxford.

Pevsner, J. (2009) *Bioinformatics and Functional Genomics*. Wiley-Blackwell, Hoboken New Jersey.

Pop, M. and Salzberg, S.L. (2008) Bioinformatics challenges of new sequencing technology. *Trends Genet.* **24**, 142–149.

Reece, R.J. (2012) *Analysis of Genes and Genomes*, 2nd Edn. Wiley-Blackwell, Chichester.

Smith, A.G. (2008) Embryo-derived stem cells: of mice and men. *Stem Cells* **1**, 435–462.

Snap, E.L. (2009) Fluorescent proteins: a cell biologist's user guide. *Trends Cell Biol.* **19**, 649–655.

Stephens, D. (Ed.) (2006) *Cell Imaging: Methods Express*. Scion Publishing, Oxford.

Stephens, D.J. and Allan, V.J. (2003) Light microscopy techniques for live cell imaging. *Science* **300**, 82–86.

Stoughton, R.B. (2005) Applications of DNA microarrays in biology. *Annu. Rev. Biochem.* **74**, 53–82.

Twyman, R.M. (2004) *Principles of Proteomics*. Garland Science, New York and Oxford.

Walther, T.H. and Mann, M. (2010) Mass spectrometry-based proteomics in cell biology. *J. Cell Biol.* **190**, 491–500.

Zhou, J., Thompson, D.K., Xu, Y., and Tiedje, J.M. (2004) *Microbial Functional Genomics*. John Wiley & Sons, Hoboken New Jersey.

Abbreviations

α-TIF	α-trans-inducing factor	dGDP	deoxyguanosine 5′-diphosphate
ADP	adenosine 5′-diphosphate	dGTP	deoxyguanosine 5′-triphosphate
AIDS	acquired immune deficiency syndrome		
ALV	avian leukosis virus	DNA	deoxyribonucleic acid
AMP	adenosine 5′-monophosphate	DNAse I	deoxyribonuclease I
AP	apurinic or apyrimidinic	dNTP	deoxynucleoside 5′-triphosphate
ARS	autonomously replicating sequence	DOP-PCR	PCR using degenerate oligonucleotide primers
ATP	adenosine 5′-triphosphate	DSB	double-strand break
BAC	bacterial artificial chromosome	DSCAM	Down syndrome cell-adhesion molecule
BER	base excision repair	dsDNA	double-stranded DNA
bHLH	basic HLH	dsRNA	double-stranded RNA
BLAST	Basic Local Alignment Search Tool	dTTP	deoxythymidine 5′-triphosphate
bp	base pairs	EDTA	ethylenediamine tetra-acetic acid
BRF	TFIIB-related factor		
BSE	bovine spongiform encephalopathy	EF	elongation factor
		eIF	eukaryotic initiation factor
BUdR	bromodeoxyuridine	ENU	ethylnitrosourea
bZIP	basic leucine zipper	ER	endoplasmic reticulum
cAMP	cyclic AMP	eRF	eukaryotic release factor
CAP	catabolite activator protein	ES	embryonic stem
CDK	cyclin-dependent kinase	ESI	electrospray ionization
cDNA	complementary DNA	EST	expressed sequence tag
CFTR	cystic fibrosis transmembrane conductance regulator	EtBr	ethidium bromide
		ETS	external transcribed spacer
CHEF	contour clamped homogeneous electric field	FADH	flavin adenine dinucleotide
		FASTA	Fast-All
ChIP	chromatin immunoprecipitation	FIGE	field inversion gel electrophoresis
CJD	Creutzfeldt–Jakob disease	FISH	fluorescent in situ hybridization
CMP	cytidine 5′-monophosphate		
CRP	cAMP receptor protein	GFP	green fluorescent protein
CSF-1	colony-stimulating factor-1	GMO	genetically modified organism
CTD	carboxy-terminal domain	GST	glutathione-S-transferase
Da	Daltons	GTP	guanosine 5′-triphosphate
dATP	deoxyadenosine 5′-triphosphate	HAT	histone acetyltransferase
		HDAC	histone deacetylase
dCTP	deoxycytidine 5′-triphosphate	HDL	high-density lipoprotein
ddNTP	dideoxynucleotide 5′-triphosphate	HIV	human immunodeficiency virus

HLH	helix–loop–helix	MMTV	mouse mammary tumor (retro) virus
hnRNA	heterogeneous nuclear RNA	mRNA	messenger RNA
hnRNP	heterogeneous nuclear ribonucleoprotein	MS	mass spectrometry
HR	homologous recombination	NAD$^+$	nicotinamide adenine dinucleotide
HSP	heat-shock protein	ncRNA	noncoding RNA
HSV-1	herpes simplex virus-1	NER	nucleotide excision repair
HSVTK	herpes simplex virus thymidine kinase	NHEJ	nonhomologous end joining
ICC	immunocytochemistry	NMD	nonsense mediated decay
ICE	interleukin-1β converting enzyme	NMN	nicotinamide mononucleotide
IF	initiation factor	NMR	nuclear magnetic resonance
IgG	immunoglobulin G	nt	nucleotides
IHC	immunohistochemistry	NTP	nucleoside 5′-triphosphates
Int	integrase	NTPase	nucleoside triphosphatase
IP	immunoprecipitation	OE-PCR	overlap extension PCR
iPS	induced pluripotent stem	OMIM®	Online Mendelian Inheritance in Man database
IPTG	isopropyl-β-D-thiogalactopyranoside	ORC	origin recognition complex
IRE	iron response element	ORF	open reading frame
IRES	internal ribosome entry site	PABI	poly(A) binding protein I
IR$_L$	internal repeat (long segment)	PABII	poly(A) binding protein II
IR$_S$	internal repeat (short segment)	pADPR	poly(ADP-ribose)
IS	insertion sequence	PAGE	polyacrylamide gel electrophoresis
ISH	*in situ* hybridization	PAP	poly(A) polymerase
ISP	iron sensing protein	PARP1	poly(ADP-ribose) polymerase-1
ITS	internal transcribed spacer	PCNA	proliferating cell nuclear antigen
kb	kilobase pairs in duplex nucleic acid, kilobases in single-stranded nucleic acid	PCR	polymerase chain reaction
		PDB	Protein Data Bank
kDa	kiloDaltons	PDGF	platelet-derived growth factor
LINES	long interspersed elements	PFGE	pulsed field gel electrophoresis
lncRNA	long noncoding RNA	piRNA	piwi-associated RNA
LTR	long terminal repeat	PMF	peptide mass fingerprint
LUCA	last universal common ancestor	PPi	pyrophosphate
MALDI	matrix-assisted laser desorption/ionization	pre-mRNA	mRNA precursor
		pri-miRNA	primary miRNA
MBP	maltose-binding protein	PTH	phenylthiohydantoin
MCS	multiple cloning site	qPCR	quantitative PCR
MDa	mega Daltons	RACE	rapid amplification of cDNA ends
Met-tRNA	methionyl-tRNA	RB1	retinoblastoma gene
MFC	multifactor complex		
miRNA	microRNA	RBS	ribosome-binding site
MMS	methylmethane sulfonate	RF	release factor

RFLP	restriction fragment length polymorphism	SSB	single-strand break
RFP	red fluorescent protein	SSCP	single stranded conformational polymorphism
RISC	RNA-induced silencing complex	ssDNA	single-stranded DNA
RITS	RNA-induced transcriptional silencing	SSLP	simple sequence length polymorphism
RNA	ribonucleic acid	STR	simple tandem repeat
RNA Pol I	RNA polymerase I	SV40	simian virus 40
RNA Pol II	RNA polymerase II	TAF	TBP-associated factor
RNA Pol III	RNA polymerase III	TAFI	TAFs for RNA Pol I transcription
RNAi	RNA interference	TBP	TATA-binding protein
RNase A	ribonuclease A	TdT	terminal deoxynucleotidyl transferase
RNP	ribonucleoprotein		
ROS	reactive oxygen species	TLS	translesion DNA synthesis
RP-A	replication protein A	T_m	melting temperature
RRF	ribosome recycling factor	tmRNA	transfer-messenger RNA
rRNA	ribosomal RNA	TOF	time-of-flight
RT	reverse transcriptase	Tris	tris(hydroxymethyl) aminomethane
RT-PCR	reverse transcriptase-polymerase chain reaction	TR_L	terminal repeat (long segment)
SCID	severe combined immunodeficiency	tRNA	transfer RNA
SCNT	somatic cell nuclear transfer	TR_S	terminal repeat (short segment)
SDS	sodium dodecyl sulfate	UBF	upstream binding factor
SDS-PAGE	sodium dodecyl sulfate polyacrylamide gel electrophoresis	UCE	upstream control element
		U_L	long unique section
SECIS	selenocysteine insertion sequence	URE	upstream regulatory element
		U_S	short unique section
SILAC	stable isotope labeling with amino acids in cell culture	UTP	uridine 5′-triphosphate
		UTR	untranslated region
SINEs	short interspersed elements	UV	ultraviolet
siRNA	short interfering RNA	VNTR	variable number tandem repeat
SL1	selectivity factor 1		
snoRNP	small nucleolar ribonucleoprotein particle	X-gal	5-bromo-4-chloro-3-indolyl-β-D-galactopyranoside
SNP	single nucleotide polymorphism	*Xist*	X-inactive specific transcript
snRNA	small nuclear RNA	XP	xeroderma pigmentosum
snRNP	small nuclear RNP	XP-V	xeroderma pigmentosum variant
sRNA	small RNA	YAC	yeast artificial chromosome
SRP	signal recognition particle	YEp	yeast episomal plasmid
Ssb	single-stranded binding protein		

Index

ab initio structure prediction, 32, 343, 349
Abasic site, 79
Abortive initiation, 106
Acanthamoeba, 218
Acetylation, 57, 147, 208, 329
Actin, 21, 23
Adenine, 5, 8, 11
Adenosine, 6, 175, 189
Adenosine deaminase, 175, 189
Adenovirus, 218–219, 281, 340
S-adenosylmethionine (SAM), 155
A-DNA, 9
ADP-ribose, 87, 100
Aequoria victoria, 336
Aequorin, 336
Affinity chromatography, 27
Agarose, 62, 246, 255–257, 259, 262, 265, 275, 285, 287, 294–295, 311, 322, 330
Agrobacterium tumefaciens, 46, 246, 280
AIDS virus, 2, 150–152, 223–225,
Alkaline phosphatase, 248, 260–262, 271, 289, 294
Alkylating agent, 80, 84, 87
Allele, 62–63, 236–238
Allolactose, 112
Alu element, 60–61, 95, 153
Alzheimer's disease, 18, 209, 349
α-Amanitin, 124
Amino acids, 13–21, 24–25, 28, 31
Aminoacyl-adenylate, 184, 186
Aminoacyl-tRNA, 177, 184–186
  delivery to ribosome, 197
Aminoacyl-tRNA synthetases, 184, 191
  proofreading, 186
Aminopeptidase, 208
Amphipathic helix, 348
Ampicillin, 249, 262, 266, 275
α-Amylase gene, 172
Amyloid, 18
Antennapedia, 143, 146, 152
Anti-conformation, 10
Anti-terminator, 117
Antibodies, 21, 23, 27, 30–32, 76, 247, 291–292, 295, 324, 330–332
  diversity, 94
  primary, 31, 334–335
  secondary, 31, 335
Anticodon, 162, 177, 182–186, 189–191, 193
Antigen, 21, 30, 31, 211, 330, 334
Antiparallel, 8, 17, 188
Antisense RNA, 96, 153, 180, 205, 221, 295
Antisense strand, 96, 106–109, 124, 180, 310
AP-1, 235
AP endonuclease, 88–89
AP site, *see* Apurinic site
Apolipoprotein, 24, 174
Apoprotein, 19
Apoptosis, 43, 153, 186, 241–243
  and cancer, 235–238, 243
  apoptotic bodies, 241–242
  bax, 243
  bcl-2, 242–243
  caspases, 243
  ICE proteases, 243
  in *C. elegans*, 241–242
Apurinic (AP) site, 34, 79, 88

Archaea, 3, 14, 76, 124, 138, 178, 193
Argonaute, 206
Arylating agent, 80, 84
Ataxia telangiectasia, 90
att sites, 94, 268–269
Attenuation, 116–118
AU-rich elements, 205
Autoradiography, 292, 295, 298, 310–311, 322
Autoregulation, 353
Auxotroph, 276
Avian leukosis virus, 224

Bacillus subtilis, 121
Bacteria, 3
Bacterial artificial chromosome (BAC), 277, 285, 292
Bacteriophage, 211–217
  DNA purification, 293
  φX174, 70, 180, 215
  integration, 214, 264
  lambda (λ) 94, 143, 214–217, 245, 256, 267, 271
  M13, 70, 211, 214–215, 245, 274, 295, 298
  Mu, 214, 217
  P1, 94
  packaging, 214, 272, 275, 284–285
  recombination, 94
  replication, 66, 69, 214
  repressor, 143, 215, 217
  RNA polymerases, 100, 121, 248, 295
  σ factors, 121
  SP6, 248, 267, 295
  SPO1, 121
  T3, 100, 248, 295
  T4, 70, 100, 121, 211, 248, 259, 285
  T7, 70, 100, 121, 267, 295
  transposition, 217
  vectors, 245, 271–275
Base
  discriminator, 185
  pairs, 8–9, 78, 83–89
  structures, 5
  tautomers, 34–35, 83–84
Bax, 243
Bayesian networks, 352
Bcl-2, 242–243
B-DNA, 8, 9
Bioinformatics, 341–349
  databases, 343
  structural, 348–349
  tools, 343–346
Biotinylated dUTP, 294
BLAST, 345,– 346
Blastocyst, 326
Blot
  northern, 295–296, 308, 322
  Southern, 62, 295–296, 322
  western, 31, 329
  zoo, 296
Bovine spongiform encephalopathy (BSE), 18
Branchpoint sequence, 168–169
BRCA1 & 2, 92
Bromodeoxyuridine (BUdR), 76
Buoyant density, 36, 61, 65
bZIP proteins, 144–145

Caenorhabdidtis elegans, 241, 327, 343

cAMP, *see* cyclic AMP
cAMP receptor protein, *see* Catabolite activator protein
Camptothecin, 44
Cancer, 31, 44, 63, 77, 84–85, 88, 90, 92, 172, 206, 209, 212, 224, 229–238, 243, 319, 330, 340
Cap (mRNA), 134, 166–167, 170, 190, 202
Carboxy-terminal domain (CTD), 125, 138, 140, 146, 151, 169–170
Carboxypeptidase, 208
Carcinogenesis, 78, 83, 235
Caspases, 243
Catabolite activator protein (CAP), 104, 113, 115, 120, 143, 267
Catalytic RNA, 162
CCAAT box, 135, 220
CDK, *see* cyclin-dependent kinase
cDNA, 27–28, 175, 267, 290, 293–302, 306–307, 309–316, 322–323
cDNA library, 166, 246–247, 286–292, 307, 332, 342
Cell
    determination, 151
    differentiation, 58, 94, 121, 151, 229, 241, 338
    transformation, 212, 235, 279
Cell cycle, 49, 55, 57, 75, 140, 152, 226–230, 319
    activation, 229
    anaphase, 228
    checkpoints, 228, 238, 243
    cyclin-dependent kinases (CDKs), 75, 140, 152, 228–230
    cyclins, 228–239
    DNA synthesis phase, 228
    E2F, 229
    G0 phase, 228
    G1 phase, 227
    G2 phase, 228
    gap phase, 227–228
    inhibition, 229–230, 241, 243
    interphase, 49, 55, 60, 126, 228
    M phase, 228
    metaphase, 49, 228
    mitogen, 228, 234–235
    mitosis, 54, 57, 127, 228, 236
    p53, 230, 238, 243
    phases, 227–228
    prophase, 228
    quiescence, 228
    restriction point (R point), 228
    retinoblastoma protein (Rb), 229, 238
    S phase, 228
Cell imaging, 334–336
Central dogma, 1–3
Centromere, 54–55, 61, 228, 275–276
Cesium chloride, 36, 251
CFTR, *see* Cystic fibrosis transmembrane conductance regulator
Chaperones, 18, 27, 208
Charge dipoles, 16, 34
CHEF, *see* Electrophoresis
ChIP, *see* Chromatin immunoprecipitation
    ChIP-chip, 324–325
    ChIP-Seq, 324–325
Chloramphenicol acetyltransferase, 312
Chloroplast, 2, 124, 159, 190, 199, 208, 284, 318
Chromatid, 54–55, 91, 228
Chromatin, 24, 49–52, 55–58, 74, 228, 241
    30 nm fiber, 52, 55–57
    chromatosome, 51
    CpG methylation, 57, 147, 285
    DNase I hypersensitivity, 56–57, 103
    euchromatin, 56, 75, 147
    heterochromatin, 55, 75, 153

higher order structure, 52
histones, 16, 24, 49–52, 57–58, 75–76, 87, 147, 153, 208, 219, 324
    immunoprecipitation (ChIP), 324
    linker DNA, 51–52
    modification, 57, 87, 147, 153
    nuclear matrix, 52, 55, 76
    nucleosomes, 24, 29, 49–52, 56–57, 75, 147, 241
    remodeling, 147
    solenoid, 52
Chromatosome, 51
Chromosome
    bacterial artificial (BAC), 277, 285, 292
    centromere, 54–55, 61, 228, 275–276
    DNA domains, 46–47
    DNA loops, 46
    Escherichia coli, 45–47
    eukaryotic, 36, 41, 49–58, 60–63, 67
    interphase, 49, 55, 60, 126, 228
    kinetochore, 55, 61
    microtubules, 23, 55
    mitotic, 54
    nuclear matrix, 52, 55, 76
    prokaryotic, 40, 45–47, 66
    scaffold, 55, 76
    spindle, 23, 54–55, 61, 228
    supercoiling, 40–44, 70–71, 104, 106, 219
    telomere, 55, 76, 275
    translocation, 88
    X, 55, 153
    yeast artificial (YAC), 275–277, 284–285
Chymotrypsin, 20–21
Cilia, 23
Circular dichroism, 28
Cis-acting sequence, 121, 153, 174, 206, 223
Cistron, 111, 170, 205, 212
Cleavage and polyadenylation specificity factor, 166
Cleavage stimulation factor, 166
Clone, *see* DNA clone
Cloning, *see* DNA cloning
Cloning vectors, 210, 215, 265–282
    2μ plasmid, 279–280
    bacterial artificial chromosome (BAC), 277, 285, 292
    bacteriophage, 246, 271–275, 289, 291, 295, 298
    baculovirus, 220, 246, 281–282
    cosmid, 245, 275, 284–285
    destination vector, 267
    expression vector, 27, 219, 267–268, 274, 289, 292, 334
    Gateway® vector, 94, 264, 267–268
    λ (lambda), 245, 271–274, 291
    λ replacement vector, 271–272, 285
    λ gt11, 289
    M13, 214, 245, 274, 295, 298
    plasmid, 245–246, 249, 253, 259, 262, 265–268, 285, 289, 291, 314
    retroviral, 223, 281, 340
    shuttle vector, 279–281
    SV40, 219, 246, 279, 281
    Ti plasmid, 46, 246, 280–281
    viral, 281–282, 327, 340
    yeast artificial chromosome (YAC), 246, 275–277, 284–285
    yeast episomal plasmid (YEp), 246, 279–280
Closed circular DNA, 40–42, 45, 71, 215, 251, 314
Clustal, 346
Coactivator, 149
Cockayne syndrome, 89
Coding strand, 96
Codon, 27, 83, 177–179,
    initiation (start), 172, 177–178, 190, 193, 197, 200–201, 267

Codon – *continued*
  termination (stop), 83, 172, 174, 177–178, 198, 203, 225, 309
    synonymous, 177–179
    usage bias, 27, 179, 184, 289
Codon–anticodon interaction, 182–183, 185–186, 188–190
Cohesive ends, 215, 248, 254, 259, 271
Coiled coil structure, 145,
Collagen, 14, 17, 21, 208
Colony hybridization, 291
Colony-stimulating factor-1 (CSF-1), 234
Comparative genomic microarray, 324
Comparative modeling, 343, 349
Competent cells, 261–262, 272, 316
Complementation, 212, 223, 225, 292
Concatamers, 163, 215, 220, 272
Confocal microscope, 335
Conjugated protein, 19, 23–24
Consensus sequence, 75, 103–104, 113, 120, 135, 149–150, 174, 200
Copolymer, 177
Corepressor, 115, 149
cos sites (ends), 215–216, 271–272, 275
Cosmid vector, 245, 275, 284–285
CpG, 57, 147, 285
Cre-loxP, 94, 326
Cre recombinase, 94
Creutzfeld-Jakob disease (CJD), 18
Cro gene and protein, 143, 215–217
Crown gall tumor, 280
Cryo-electron microscopy, 29
CTD, *see* Carboxy-terminal domain
Cyclic AMP, 113
Cyclin, 228–230
Cyclin-dependent kinase (CDK), 75, 140, 152, 228–230
Cystic fibrosis, 275, 326, 340
Cystic fibrosis transmembrane conductance regulator (CFTR), 275, 340
Cytidine, 6
Cytosine, 5, 8, 11
Cytoskeleton, 19, 21–23
  microfilaments, 23
  microtubules, 19, 22

Dcp2, 170
Decapping, 170, 203
Degradome, 320
Denaturation
  DNA, 34–36, 39, 63, 80, 251, 295, 298, 304–305
  protein, 18, 29, 208, 250–251
  RNA, 34–35, 39
Density gradient centrifugation, 36, 61, 65, 67, 251
2′-Deoxyribose, 6, 10, 34–35, 79
Deoxyribonuclease, *see* Nuclease
Deoxyribonucleic acid, *see* DNA
Deoxyribonucleotide, 6–7, 229
Deoxythymidine, 6
Dephosphorylation, 289, 294, 352
Depurination, 79
Dicer, 163, 206
Dideoxynucleotides, 298
Differentiation, 58, 94, 121, 151, 229, 241, 338
Dihydrouridine, 182
Dimethyl sulfate, 311
DNA,
  A_{260}, 37–39
  A_{260}/A_{280} ratio, 38–39
  adaptors, 287-288, 294
  A-DNA, 9
  B-DNA, 8–9

ΔLk, 41–42, 46–47
annealing, 39, 92, 109, 254, 275, 288, 295, 298, 302, 304, 306–307, 314–315
antiparallel strands, 8, 67
automated synthesis, 289
axial ratio, 36
base pairs, 8–9, 78, 83–89
base stacking, 34, 38–39
bases, 1, 5, 8
bending, 47, 113, 138
binding proteins, 16, 46–47, 97, 127, 132, 139, 142–146, 149–152, 324
buoyant density, 36, 61, 65
catenated, 43, 73
chloroplast, 2, 284, 318
closed circular, 40–42, 45, 71, 215, 251, 314
complementary strands, 8, 34, 39, 64
complexity, 59–63
CpG methylation, 57, 147, 285
crosslink, 80, 90
denaturation, 34–36, 39, 63, 80, 251, 295, 298, 304–305
dispersed repetitive, 61, 95, 153
double helix, 7–9, 34, 41, 71
effect of acid, 34
effect of alkali, 34
end labeling, 294, 298, 310–311
end repair, 284
ethidium bromide binding, 42, 251–252, 256–257, 307
fingerprinting, 61–62, 304
G+C content, 36, 39, 61, 296, 306
highly repetitive, 60–61
hybridization, 39, 247, 291, 295, 304, 323–325, 334
hydrogen bonding, 8, 33–35, 39
hypervariable, 61
intercalators, 42, 84
lesions, 78–80, 84–85, 87–89, 92
libraries, *see* DNA libraries
ligation, 71–72, 88–89, 92, 246, 254, 257, 259–263, 265, 289, 300, 302, 306
linking number (Lk), 41
$Lk°$, 41, 46–47
long interspersed elements (LINES), 61, 95
major groove, 8–10, 92, 115, 143–144
measurement of purity, 38–39
melting, 39
melting temperature (T_m), 39
methylation, 11, 57, 80, 89, 285
microsatellite, 61
minisatellite, 61
minor groove, 8–10, 138
mitochondrial, 2, 178, 180, 284, 318
moderately repetitive, 60–61
modification, 11, 57, 253–254, 260, 285
nicked, 89, 92, 214, 252, 256, 260, 263, 295, 314
noncoding, 60, 79, 83–85, 96, 246, 318, 344
open circular, 256
partial digestion, 285
probes, 28, 39, 62, 247, 267, 289–296, 302, 304, 308, 310 322–323, 334
quantitation, 38–39
reassociation kinetics, 60
relaxed, 41–42
renaturation, 39, 70, 251
replicative form, 70, 214, 274
ribosomal, 60, 343–344
satellite, 55, 60–61
sequencing, 32, 34, 60, 63, 179, 245, 295, 297–302, 304, 324, 342–344, 346, 354
shearing, 36, 54, 284, 324
short interspersed elements (SINES), 61, 95, 153
sonication, 36, 284, 324

stability, 33–35
supercoiling, *see* DNA supercoiling
thermal denaturation, 39
topoisomerases, 43–44, 71, 73, 94, 346
torsional stress, 10, 42
twist (*Tw*), 41–44, 47
unique sequence, 60, 62
UV absorption, 37–38
viscosity, 36
writhe (*Wr*), 42, 50
Z-DNA, 9–10
DNA clone, 244–247
  cDNA, 166, 246–247, 286–292, 306–307, 332, 342
  characterization, 293–296
  gene polarity, 293, 309
  genomic, 283–285
  identification, 290–292
  insert orientation, 259, 263–264, 294
  mapping, 247, 294, 310–312
  mutagenesis, 313–316
  organization, 309–312
DNA cloning
  alkaline lysis, 250–251
  antibiotic resistance, 249, 262, 268, 324
  β-galactosidase, 266, 289, 292, 312, 331
  blue-white screening, 266–267, 332
  chips, 322, 324
  cohesive ends, 215, 248, 254, 259, 271
  colonies, 261–262, 264–265, 284, 291–292
  colony hybridization, 291
  competent cells, 261–262, 272, 316
  direct DNA transfer, 279
  double digest, 264, 294
  electroporation, 260–261, 279
  ethanol precipitation, 251–252
  expression screening, 292
  expression vectors, 267, 279–282
  fragment orientation, 259, 263–264, 294
  functional screening, 292
  gene gun, 279
  glycerol stock, 263
  his-tag, 27, 267
  host organism, 27, 245–246, 260, 267, 274, 279–282,
    314
  IPTG, *see* Isopropyl-β-D-thiogalactopyranoside
  isolation of DNA fragments, 255–257,
  λ lysogen, 267, 271, 274
  λ packaging, 272–273, 275, 284–285
  lacZ, 266, 289, 292
  lacZ', 266–267, 292
  ligation, 246, 254, 257, 259–263, 265, 289, 300, 306
  microarrays, 319, 322–325
  microinjection, 279, 338
  minipreparation, 250–251, 255, 262, 265, 277
  multiple cloning site (MCS), 266–267
  packaging extract, 272
  phenol extraction, 251
  phenol-chloroform, 251, 284, 294
  plaque hybridization, 291
  plaque lift, 292
  plaques, 273, 284, 304
  recombinant DNA, 3, 245
  replica plating, 291
  restriction digests, 62, 246, 253–257
  restriction fragments, 254–257
  selectable marker, 245, 275–277, 279–280, 339
  selection, 246, 262, 266, 275–277, 279, 324
  shuttle vectors, 279–281
  sticky ends, 215, 248, 254, 259, 271
  subcloning, 246, 250, 262, 264
  T-DNA, 280–281

  Ti plasmid, 46, 246, 280–281
  transfection, 232, 279, 281, 336, 338
  transformation, 246, 260–263, 277, 324
  vectors, *see* Cloning vectors
  X-gal, 266, 312, 322
DNA chip, 322, 324
DNA cloning enzymes, 248
DNA crosslink, 80
DNA damage
  abasic site, 79, 88
  aflatoxin, 80
  alkylation, 80, 84, 87
  O^4-alkylthymine, 87
  apurinic site, 79, 88
  benzo[a]pyrene adduct, 80–81
  cytosine deamination, 78, 84
  depurination, 79
  3-methyladenine, 80
  7-methylguanine, 80
  O^6-methylguanine, 80, 87
  oxidative, 79, 90
  8-oxoguanine, 79
  pyrimidine dimers, 80–81, 84–89
DNA fingerprinting, 61–62, 304
DNA glycosylases, 79, 87–88
DNA gyrase, 43, 71, 106, 345–346
DNA helicases, 70–73, 75, 91, 215, 220
DNA lesion, 78–80, 84–85, 87–89, 92
DNA library, *see* Gene library
DNA ligase, 68, 71, 76, 87–89, 92, 95, 215, 248, 259, 285,
    289, 302
DNA methylase, 253
DNA modification
  6-*N*-methyladenine, 11, 253, 314
  4-*N*-methylcytosine, 11
  5-methylcytosine, 57, 79, 253, 285
DNA photolyase, 86
DNA polymerase, 67, 83, 85, 217, 220, 298, 315
  Klenow, 248, 284, 298
  Pfu, 307
  pol I, 71, 89, 92, 95, 248, 295
  pol III holoenzyme, 71, 72, 76
  pol IV, 85
  pol V, 85
  pol α (alpha), 75
  pol β (beta), 89
  pol δ (delta), 75, 89
  pol ε (epsilon), 75, 89
  pol η (eta), 85
  pol ι (iota), 85
  pol κ (kappa), 85
  pol ζ (zeta), 85
  polymerase-primase, 75
  T7, 298
  Taq, 248, 306–307
DNA primase, 70–72, 75, 220
DNA recombination
  branch migration, 92
  Cre-LoxP, 94, 326
  Cre recombinase, 94
  D-loop, 92
  general, 91
  Holliday junction, 92
  homologous, 88, 91–93, 280, 324, 331
  illegitimate, 94–95
  in DNA repair, 88
  meiotic, 83, 89, 92
  site-specific, 94, 264, 267, 271, 280, 326, 339–340
DNA repair
  adaptive response, 87
  alkyltransferase, 87

DNA repair – *continued*
  base excision repair (BER), 88–89
  double-strand break, 88
  error-prone, 85
  mismatch, 89
  non-homologous end joining (NHEJ), 88
  nucleotide excision repair (NER), 88–89
  PARP, *see* Poly(ADP)ribose polymerase
  photoreactivation, 86–87
  single-strand break, 87–88
  transcription-coupled, 89
  translesion DNA synthesis (TLS), 85
  XRCC1, 88
DNA repair defects, 89–90
DNA replication, 64–77
  autonomously replicating sequences (ARS), 75, 276
  bacteriophage, 69–70, 213–217
  bidirectional, 66, 70, 215
  Cdc6 protein, 75
  Cdt1 protein, 75
  clamp loader, 71
  concatamers, 215, 220, 272
  DNA gyrase, 43, 71
  DNA ligase, 68, 71, 76
  DNA polymerase I, 71, 73
  DNA polymerase III holoenzyme, 71–72
  DNA primase, 70–72, 75, 220
  DNA topoisomerase IV, 43, 73
  Dna2 endonuclease, 76
  DnaA protein, 70–71
  DnaB helicase, 70–73, 215
  dNTPs, 6, 65–66
  euchromatin, 56, 75
  eukaryotic DNA polymerases, 75
  Fen1 endonuclease, 76
  fidelity, 68, 71, 83–85
  heterochromatin, 55, 75
  initiation, 66, 70, 75
  lagging strand, 67–68, 70–71, 75
  leading strand, 67–68, 71–73, 75–77, 84
  licensing factor, 75
  MCM proteins, 75
  minichromosomes, 70
  Okazaki fragments, 67–68, 71, 76
  oriC, 70
  origin recognition complex (ORC), 75–76
  origins, 66–67, 70–71, 75–76
  pre-priming complex, 70
  proliferating cell nuclear antigen (PCNA), 76
  proofreading, 71, 75, 83–85
  replication forks, 65–68, 70–76, 84, 88, 92
  replication protein A (RP-A), 75
  replicons, 66, 75
  replisome, 19, 72
  RNA primers, 70–73, 75–76, 84
  RNAse H, 76, 224
  rolling circle, 215
  Saccharomyces cerevisiae, 75
  semi-conservative, 65
  semi-discontinuous, 67
  simian virus 40 (SV40), 75, 212, 219
  single-stranded binding protein, 70–71
  sliding clamp, 71
  telomerase, 24, 55, 76–77
  telomeres, 55, 76
  template, 11, 64–66, 71
  termination, 66, 72–73, 76–77
  unwinding, 71
  viral, 75, 218–221
  Xenopus laevis, 75
DNA sequencing, 34, 245, 293, 298–302, 342–344, 354
  chain terminator, 298
  cycle sequencing, 298–299
  dye terminator, 299
  emulsion PCR, 300
  fluorescent reversible terminator, 300
  high throughput, 63, 245, 298, 300, 324
  ligation, 302
  next generation, 302
  primer, 298, 300, 302
  pyrosequencing, 300, 302
  Sanger, 298
  shotgun, 299–300
DNA supercoiling, 40–44
  ΔLk, 41–42, 46–47
  constrained and unconstrained, 41, 46–47, 50
  in eukaryotes, 50, 219
  in nucleosomes, 49–50
  in prokaryotes, 41, 46–47, 70–71, 251
  in transcription, 46, 104, 106
  linking number (Lk), 41
  Lk^o, 41, 46–47
  on agarose gels, 256
  positive and negative, 41, 50, 70–71
  topoisomer, 41
  topoisomerases, 43–44, 71, 73, 94, 346
  torsional stress, 10, 42
  twist (*Tw*), 41–44, 47
  writhe (*Wr*), 42, 50
DNA synthesis, *see* DNA replication
DNA topoisomerases
  DNA gyrase, 43, 71, 106, 345–346
  topoisomerase IV, 43, 73
  type I, 43–44, 94
  type II, 43–44, 71, 73, 346–347
DNA viruses, 218–221
  adenoviruses, 218–219, 281, 340
  baculoviruses, 220, 246, 281–282
  Epstein–Barr, 132, 212, 220
  genomes, 212
  herpesviruses, 145, 211-212, 217–221, 281, 339
  papovaviruses, 212, 218–219
  SV40, 75, 135, 212, 219, 246, 281
  vaccinia, 220, 281, 337
DNase I footprinting, 311
DNase I hypersensitivity, 56
Docking protein, 207
Domain (DNA), 46–47, 52, 75
Domain (protein), 19, 25, 125, 331, 343–344, 349
  activation, 142, 145–147, 149, 331
  basic, 144
  dimerization, 142, 152
  DNA-binding, 138, 142–146, 149, 152, 331
  glutamine-rich, 146, 149
  helix-loop-helix, 144–145
  helix-turn-helix, 143–144, 152
  homeodomain, 143, 145, 152
  leucine zipper, 144
  ligand-binding, 142–143, 146, 149
  proline-rich, 146
  repressor, 146
  zinc finger, 143–144
Dominant negative effect, 238
Down syndrome cell adhesion molecule (DSCAM), 174
Drosha, 163
Drosophila melanogaster, 61, 95, 143, 146, 152, 172, 174, 343
  P element transposase, 172
DSCAM, *see* Down syndrome cell adhesion molecule
Dystrophin, 165

E2F, 229–230
EcoRI methylase, 253
Edman degradation, 28, 342
eIF4E binding proteins, 202, 205
Electrophoresis
    agarose gel, 62, 246, 255–257, 259, 262–265, 275, 285, 287, 294–295, 311, 322
    contour clamped homogeneous electric field (CHEF), 275
    field inversion gel (FIGE), 275
    polyacrylamide gel (PAGE), 29, 31, 287, 311, 322
    pulsed field gel (PFGE), 275
    two dimensional gel, 329
Electroporation, 260–261, 279
Electrospray ionization (ESI), 29–30
EMBL database 343, 345
EMBL3, 271, 285
Embryonic stem cells, 325–326, 338–339
End labeling, 294
End-product inhibition, 115, 117
Endonuclease, 50, 88, 155, 166, 170, 214
    AP, 88–89
    Dna2, 76
    DNase I, 56–57, 103, 295, 311
    Fen1, 76
    MutH, 89
    restriction, 246, 253–254, 314
    RNase A, 248, 251
    RNAses, *see* Nuclease
    RuvC, 92
    UvrABC, 88
Endoplasmic reticulum, 95, 207, 335
Enhancer, 135–136, 142, 146, 153, 174, 230, 233–234
Epigenetics, 57, 147, 153
Episomal, 246, 279, 281
Epitope(s), 31, 292
    tag, 27, 31, 331, 334–335
Epstein–Barr virus, 132, 212, 220
Equilibrium density gradient centrifugation, 36, 61, 65, 251
ErbA oncogene, 235
ErbB oncogene, 233
ES cells, *see* Embryonic stem cells
ESI, *see* Electrospray ionization
EST, *see* Expressed sequence tag
Ethanol precipitation, 251
Ethidium bromide, 42, 251–252, 256–257, 307
Ethylnitrosourea, 80
Etoposide, 44
Euchromatin, 56, 75, 147
Eukaryotes, 3
Evolutionary relationship, 342–343, 354–347
Excisionase, 94
Exons, 19, 168–169, 172–174
    alternative, 172
    combinatorial use, 174
    mutually exclusive, 174
    splicing enhancer, 174
Exonuclease, 50, 71, 88, 155, 157, 166, 168, 170, 203, 295
    proofreading, 71, 83, 307
    RNAses, *see* Nuclease
Exosome, 157, 170, 205
Expect (E) value, 345
Expressed sequence tag (EST), 342, 344
Expression microarray, 324
Expression profiling, 319
Expression screening, 292
Extein, 208
External transcribed spacer (ETS), 157

F factor, 277

Fanconi anemia ,90
FASTA, 345
Ferritin, 21, 206
Fibroblast, 151–152, 232, 235, 238, 338–339
Fibroblast growth factor (FGF), 234
Firefly luciferase, 312, 336
FISH, *see* Fluorescence *in situ* hybridization
Flagella, 23
Fluorescence, 31, 42, 76, 247, 256, 295, 299–302, 307–308, 323, 331, 334–335
Fluorescence *in situ* hybridization, 334
Fluorescent proteins, 312, 336, 354
Fms oncogene, 234
Formamide, 35, 296, 310
N-Formylmethionine, 190–191, 200
5-Formyluracil, 79
Fos oncogene, 235, 238
Frameshift, 83, 197, 225, 234
Functional complementation, 292
Functional genomics, 317–349
Functional screening, 292
Fusion protein, 27–28, 142, 267, 289, 291–292, 312, 330–331, 335–336

G418, *see* Geneticin
G-protein, 234
Gain-of-function phenotype, 326
β-Galactosidase, 111–112, 266, 289, 292, 312, 331
Ganciclovir, 339
Gel electrophoresis, *see* Electrophoresis
Gel retardation (gel shift), 310–311
GenBank database, 343
Gene cloning, *see* DNA cloning
Gene clusters, 60
Gene expression, 2
Gene gun, 279
Gene knockdown, 206, 327
Gene knockout, 32, 94, 324–327
Gene library, 283–289
    cDNA, 166, 246–247, 286–292, 307, 332, 342
    genomic, 246, 283–285, 287
    representative, 283–285
    screening of, 290–292
    size of, 284
    subtracted, 287
Gene synthesis, 289
Gene therapy, 3, 94, 225, 245, 281, 339–340
Genes, overlapping, 180
Genetic code, 2, 14, 28, 176–179, 189, 290, 306
    ambiguity, 177–178
    codon usage, 27, 179, 184, 289
    codons, 27, 83, 177–179
    deciphering, 177
    degeneracy, 177–178
    modifications, 14, 178
    overlapping, 180
    reading frame, 177, 179
    redundancy, 177, 189, 290
    synonymous codons, 177–179, 184
    table, 178
    universality, 177–178
Genetic engineering, 3, 245, 278, 338
Genetic polymorphism, 62–63, 83, 245, 342
    restriction fragment length polymorphism (RFLP), 63
    simple sequence length polymorphism (SSLP), 63
    single nucleotide polymorphism (SNP), 63, 324
    single stranded conformational polymorphism (SSCP), 63
Genetically modified organism (GMO), 3, 32, 94, 292, 337–339
    gene knockout, 32, 94, 324–327

Genetically modified organism (GMO) – *continued*
  nuclear transfer, 338
  transgenic, 3, 32, 94, 292, 337–339
Geneticin (G418), 279
Genome, 1–3, 11, 318
  ambisense, 212
  Carsonella ruddi, 45
  complexity, 59–63
  Drosophila, 61, 95
  Escherichia coli, 45–47, 60, 155,196
  eukaryotic, 48–52, 61, 95, 153
  human, 61, 172, 178, 206, 275, 284, 299, 319, 323, 339
  mimivirus, 218
  Mycoplasma mycoides, 3, 354–355
  negative sense, 212
  organellar, 2, 175, 178, 318
  Ostreococcus tauri, 60
  positive sense, 212
  RNA, 2, 11, 222–225
  Saccharomyces cerevisiae, 32, 75, 95, 324–326
  Sorangium cellulosum, 45
  viral, 2, 75, 135, 180, 210–225
Genome sequencing, 3, 28, 32, 63, 179, 233, 245, 297–302, 275, 299, 339
Genome-wide analysis, 322
Genomic library, 246, 283–285, 287
Genomics, 245, 318
  functional genomics, 318
  structural genomics, 318
GFP, *see* Green fluorescent protein
Globin, 18, 20–21, 287
Glucocorticoid receptor, 145, 149
Glutathione, 27, 331
Glutathione-S-transferase, 27, 330
Glycerol stock, 263
Glycomics, 320
Glycosylation, 23, 208, 329
Glycosylic (glycosidic) bond, 6, 79
GMO, *see* Genetically modified organism
Green fluorescent protein, 312, 334–336
Growth factors
  epidermal (EGF), 233
  fibroblast (FGF), 234
  platelet-derived (PDGF), 234
GST, *see* Glutathione-S-transferase
GTPase, 234
Guanine, 5, 8, 34, 79–80
Guanosine, 6, 163
Guide RNAs, 175

H-NS protein, 47
Hairpin structure, 10, 98, 108–109, 116–117, 157, 163, 168, 327
Heat shock, 120
  gene, 120, 312
  promoters, 104, 121, 312
  proteins, 120, 149
Hemimethylation, 57, 89, 314
Hemoglobin, 18–21
Heparin, 100
Herpesviruses, 145, 211–212, 219–221, 281, 339
Heterochromatin, 55, 61, 75, 153
Heteroduplex, 92
Heterogeneous nuclear RNA, *see* hnRNA
Heteromer, 18
Hfq protein, 122
High performance liquid chromatography, 319, 329
his operon, 118
Histidine tag (His-tag), 27, 267
Histone-like proteins, 47, 76

Histone(s), 16, 24, 49, 75–76, 87, 208, 219
  acetylation, 57, 147, 153, 324
  core, 49
  deacetylation, 147
  genes, 60–61
  H1, 49, 51–52
  H5, 58
  methylation, 57, 147, 153, 324
  octamer core, 50–51
  phosphorylation, 57, 324
  mRNA, 168, 170
  variants, 57
Histone downstream element, 168
HIV, 2, 223
  gene expression, 150–151
  Rev protein, 225
  TAR sequence, 150–151
  Tat protein, 150–151, 216, 225
hnRNA (pre-mRNA), 11, 134, 152, 165–170, 172, 174–175, 174
hnRNP, 153, 165, 174
Holoenzyme, 19, 71–72, 99–101, 106–107, 120–121
Homeobox, 143, 152
Homeodomain, 143, 145, 152
Homeotic genes, 152
Homoeostasis, 210, 241
Homologous recombination, 88, 91–94, 280, 324, 331
Homology, 20, 247, 344, 346
Homology box, 349
Homology modeling, 343, 349
Homomer, 18
Homopolymer, 177
Housekeeping genes, 57, 149
Human growth hormone, 21, 245
Human immunodeficiency virus, *see* HIV
Huntington's disease, 63, 209
Hybridoma, 31
Hybridization, 39, 247, 287, 290, 295–296, 319, 322–324
  colony, 291–292
  in situ, 334
  plaque, 291–292
  probe, 62, 247, 267, 290, 304, 310
  stringency, 295–296
Hydrogen bonds, 8–10, 15–18, 21, 33–35, 184
Hydrophobicity, 15–16, 18–19, 21, 24–25, 34–35, 38, 144–145, 207–208, 348
Hyperchromicity, 38
Hypochromicity, 38
Hypoxanthine, 84, 182, 189

ICE proteases, 243
Identity elements, 185–186
IHF, *see* Integration host factor
Imaging (cell), 333–336
Immunochemical, 292
Immunocytochemistry, 31, 334
Immunodeficiency, 2, 243, 340
Immunofluorescence, 31, 76, 331, 335
Immunoglobulin, 30–31, 94, 172, 230, 235, 330
Immunohistochemistry, 334
Immunoprecipitation (IP), 32, 330
  chromatin (ChIP), 324
In vitro transcription, 140, 267, 295
Inclusion body, 27
Induced pluripotent stem (iPS) cells, 339
Initiator element, 135, 140
Initiator tRNA, 190–191, 193, 197, 200–202
Inosine, 175, 182–183, 189
Insertion sequences (IS), 95
Insertional inactivation, 266
Insulin, 15, 21, 28, 208, 245

int-2 gene, 234
Integrase, 94–95, 223–224, 280
Integration host factor (IHF), 47, 94
Intein, 208
Interactome, 319, 330
Intercalating agents, 42, 84
Interferon, 150, 202
Internal ribosome entry site (IRES), 202, 223
Internal transcribed spacer (ITS), 157
Intron, 60–61, 157, 161, 168–169, 172–174, 206, 274
    processing, 157, 161–163, 168–169, 172–174
    splicing enhancer, 174
    splicing silencer, 174
Inverted repeat, 111, 113, 220
IPTG, see Isopropyl-β-D-thiogalactopyranoside
IRES, see Internal ribosome entry site
Iron response element (IRE), 206
Iron sensing protein (ISP), 206
Isopropyl-β-D-thiogalactopyranoside (IPTG), 112–113, 266–267, 274
Isopycnic centrifugation, 36, 61, 65, 251
Isothermal titration calorimetry, 29
ITS, see Internal transcribed spacer

JAK kinase, 150
Jun oncogene, 146, 235

Kanamycin, 324
Keratin, 15, 21
Kinase
    ATM, 90
    cyclin-dependent, 75, 140, 152, 228–230
    mTor, 205
    polynucleotide, 248, 294
    protein, 15, 21, 140, 150–151, 205, 320
    thymidine, 220, 339
Kinetochore, 55, 61
Kinome, 320
Klenow polymerase, 248, 284, 298
Knockdown (RNAi), 206, 327
Knockout
    gene, 94, 324–327
    mice, 32, 94, 325–326
Kozak sequence, 200

*L*1 element, 61
Lac inducer, 112
Lac operon, 102, 110–113
Lac repressor, 111–112, 115, 120, 266
lacI gene, 111
β-Lactamase, 95, 249, 262
Lactose, 111–113
lacZ gene, 111, 266, 289, 292
lacZ' gene, 266–267, 292
lacZYA genes, 111–113
Lambda, see Bacteriophage lambda
Lariat, 168
Last Universal Common Ancestor (LUCA), 3
Latency, 221
Leishmania, 175
Lentiviruses, 223, 225
Lesion, 78–80, 84–85, 87–89, 92
Leucine zippers, 144
LexA repressor, 142
LINES, see Long interspersed elements
Linking number (*Lk*), 41–43
Lipidomics, 320
Lipids, 15–16, 19, 23–25, 320
Lipoproteins, 15, 19, 24
Liposome, 279

lncRNA, 11, 61, 134, 147, 153, 165
Long interspersed elements (LINES), 61, 95
Long non-coding RNA, see lncRNA
Long terminal repeat (LTR), 95, 224–225
Loss-of-function phenotype, 326
Loss of heterozygosity, 237
LTR, see Long terminal repeat
Luciferase, 312, 336
Luciferin, 336
Lysogenic infection, 214–217, 271, 274
Lytic infection, 214–217, 271

MALDI, see Matrix-assisted laser desorption/ionization
Malignancy, 235
Maltose-binding protein (MBP), 27
Mass spectrometry, 28–30, 320, 329–330, 342, 344
Mathematical models, 351–354
Matrix-assisted laser desorption/ionization, 29–30
MCM helicase, 75
MCS, see Multiple cloning site
Meiosis, 91
Melting temperature (T_m), 39, 296, 306
Membrane,
    endoplasmic reticulum, 95, 207, 335
    lipids, 25
    nitrocellulose, 31, 291, 295
    nuclear, 205, 221
    plasma, 23, 25, 149, 234, 241, 279
    proteins, 25, 29, 46
Messenger RNA, see mRNA
Metabolomics, 319–320, 343, 351
Metaphase, 49, 55, 91, 228
Methyl methanesulfonate (MMS), 80, 84
Methylation,
    CpG, 57, 147, 285
    DNA bases, 11, 57, 80, 89, 253, 314
    histone, 147, 153
    protein, 208
    RNA, 155, 157, 165–166, 170
6-*N*-Methyladenine, 11, 253, 314
4-*N*-Methylcytosine, 11
5-Methylcytosine, 57, 79
Methylguanines, 80, 87
7-Methylguanosine, 166
2'-*O*-Methylribose, 157
Microarray, 319, 322–325
Micrococcal nuclease, 50, 52
Microinjection, 279, 292, 338
Microfilaments, 23
microRNA, see miRNA
Microtubules, 19, 22–23, 55
miRNA, 11, 60, 124, 133, 153, 163, 170, 175, 206, 221
Mitochondria, 2, 124, 159, 175, 178–180, 190, 199, 208, 284, 318
Mitochondrial DNA, 2, 178, 180, 284, 318
Mitosis, 54, 57, 127, 228, 236
MMS, see Methyl methanesulfonate
Molecular barcode, 325
Molecular clock, 347
Monocistronic mRNA, 205
Monoclonal antibody, 31
Monte Carlo modeling, 353
Motif,
    DNA, 136, 142, 150, 344, 346
    network, 352
    protein, 19, 21, 143–145, 149, 343, 346, 348–349
Mosaic, 193, 326, 339
mRNA, 2, 11, 47, 318–319, 322–323, 342
    alternative processing, 172–175
    cap(ping), 134, 165–166, 167, 170, 190, 202
    circularization, 202

mRNA – *continued*
  decapping, 170, 203
  degradation, 157, 170, 203, 206
  editing, 174–175
  export, 168–169, 225
  fractionation, 287
  guanylyltransferase, 166
  isolation, 287
  masked, 205
  methylation, 170
  monocistronic, 205
  nonpolyadenylated, 168–170
  polyadenylated, 168, 224, 287
  polyadenylation, 166–168, 170, 174
  poly(A) tail, 166, 168, 170, 202–203, 219, 287, 307
  polycistronic, 111, 202, 205, 212
  polysomes, 191
  pre-mRNA, 124, 134, 165–170
  processing, 165–170, 172–175, 225
  ribosome binding site (RBS), 121, 190, 200, 205
  secondary structure, 116, 205
  splicing, 166, 168–169, 172–174
  start codon, 177–178, 191, 193, 200
  stop codon, 177–179, 198
  surveillance, 169, 203
  synthetic, 177
  untranslated regions (UTRs), 121, 147, 150, 153, 172, 193, 205–206, 235
mTor kinase, 205
Mucoproteins, 23
Multifactor complex, 200
Multimer, 18
Multiple cloning site (MCS), 266–267
Multiple sequence alignment, 343, 346–349
Mutagenesis
  deletion, 315
  direct, 78, 84
  indirect, 78, 80, 84–85
  insertional inactivation, 32, 266, 276, 315, 324
  PCR, 289, 306, 314–316
  Quikchange™, 314
  random, 315–316
  site-directed, 27, 274, 313, 336
  targetted, 85
  terminal transferase, 94
  translesion DNA synthesis (TLS), 84–85
  transposon, 32, 95
  untargetted, 85
Mutagens
  alkylating agents, 80, 84, 87
  arylating agents, 80, 84
  base analogs, 84
  intercalating agents, 42, 84
  nitrous acid, 84
  radiation, 84
Mutation, 32, 63, 78, 89, 92, 100, 179, 223, 225, 339, 342
  cancer, 231, 233–4, 236–8, 243
  deletion, 83–84, 88, 92, 94, 233, 235, 238, 324–326, 338, 355
  fixation, 84
  frameshift, 83, 234
  gain-of-function, 326
  insertion, 83–84, 238, 349
  loss-of-function, 326–327
  microarray, 324
  missense, 83
  nonsense, 83
  point, 79, 83, 337
  recessive, 63, 236–237
  spontaneous, 83, 89, 261
  transition, 83–84, 179

  transversion, 83–84, 179
Myc oncogene, 234–235, 238, 243
Mycoplasma, 3, 179, 289, 354–355
MyoD, 145, 151–152, 230
Myosin, 21, 23
Myosin gene, 172

Nascent transcript, 98
ncRNA, 11, 60–61, 121, 131, 152–153, 163, 165, 319
Necrosis, 241
NEDD8, 208
Neomycin phosphotransferase, 279
Network biology, 352
Neutron diffraction, 158
NHEJ, *see* Nonhomologous end joining
Nick translation, 295
NIH-3T3 cells, 232, 235–236
Nitrocellulose membrane, 31, 291, 295
NMR, *see* Nuclear magnetic resonance
Noncoding RNA, *see* ncRNA
Nonhomologous end joining, 88
Nonsense mediated mRNA decay, 174
Nonstop mRNA decay, 203
Northern blotting, 295–296, 308, 322
Nuclear localization signal, 149, 208
Nuclear magnetic resonance (NMR), 29, 318, 320, 342–343, 348
Nuclear matrix, 52, 55, 76
Nuclear oncogene, 234–235
Nuclear receptor, 149, 235
Nuclease,
  Bacillus cereus RNase, 302
  DNase I, 56–57, 103, 295, 311
  endonuclease, 50, 88, 155, 166, 170, 214
  exonuclease, 50, 71, 88, 155, 157, 166, 168, 170, 203, 295
  micrococcal nuclease, 50
  proofreading, 71, 83, 307
  restriction enzymes, 246, 253–254, 314
  RNase III, 155, 163
  RNase A, 248, 251
  RNase D, 161
  RNase E, 161
  RNase F, 161
  RNase H, 76, 224
  RNase M16, 155
  RNase M23, 155
  RNase M5, 155
  RNase P, 11, 122, 161–162, 182
  RNase Phy M, 302
  RNase T1, 302
  RNase U2, 302
  S1, 248, 310
  XRN1, 170
Nucleic acid
  3′-end, 7
  5′-end, 7
  annealing, *see* hybridization
  apurinic, 29
  base tautomers, 34–35, 83–84
  bases, 5
  denaturation, 34–36, 39, 63, 80, 251, 295, 298, 304–305
  effect of acid, 34
  effect of alkali, 34–35
  end labeling, 294, 298, 310–311
  hybridization, 39, 247, 291, 295, 304, 323–325, 334
  hypochromicity, 38
  λ_{max}, 37
  nucleosides, 6
  nucleotides, 6–7
  phosphodiester bond, 7

probe, 28, 39, 62, 247, 267, 289–296, 302, 304, 308, 310, 322–323, 334
  quantitation, 38–39,
  stability, 33–35
  strand-specific labeling, 295
  sugar-phosphate backbone, 25, 35, 37, 39, 248
  uniform labeling, 294
  UV absorption, 37–38
Nucleocapsid, 211
Nucleoid, 45–47
Nucleolar organizer, 127
Nucleolus, 60, 124, 127, 156–157, 166
Nucleoproteins, 19, 24, 25, 49, 55, 155, 165
Nucleoside, 6
Nucleosomes, 24, 29, 49–52, 56–57, 75, 147, 241
Nucleotide, 6–7
  anti-conformation, 10
  syn-conformation, 10
Nucleus, 1, 3, 49, 55
Nylon membrane, 291, 295, 322

Oligo(dA), 170
Oligo(dC), 307
Oligo(dG), 288
Oligo(dT), 287, 307, 309
Oligomer, 18
Oligonucleotide, 38, 88, 289, 322
  adaptor-primers, 287–288
  primers, 287, 298–302, 304–306
  probes, 28, 247, 289–290
OMIM® database, 339
Oncogene, 224, 230–235, 238, 242–243
  nuclear, 234–235
  viral, 219, 224, 232
Open reading frame, 179, 219–220, 309, 318
Operator, 111–113, 115–116, 118, 142
Operon, 110–118, 121, 155, 161, 205
ORF, *see* Open reading frame
Ortholog, 21, 346
Overlapping genes, 180

p21, 152, 230
p53, 153, 230, 238, 241, 243, 340
PABI & PABII, *see* Poly(A) binding proteins
Packaging extract, 272
Palindrome, 111, 115–116, 254
PANDA, 153
Papovaviruses, 212, 218–219
Paralog, 20, 92, 346
PARP, *see* Poly(ADP)ribose polymerase
Partial digest, 285, 311
PCR, *see* Polymerase chain reaction
PDB, *see* Protein Data Bank
Peptide mass fingerprint (PMF), 30, 329
Peptidyl transferase, 163, 197–198
PFAM database, 346, 349
PFGE, *see* Electrophoresis
Pfu, *see* DNA polymerase,
Phage, *see* Bacteriophage
Phase extraction, 284
Phenol extraction, 251
Phenol-chloroform, 251, 284, 294
Phenotype, 32, 292, 320, 324, 326
Phosphodiester bond, 7
Phosphoproteins, 15, 208
Phosphoproteome, 320
Phosphoramidite, 289
Phosphorylation,
  DNA, 294
  protein, 19, 57, 125, 140, 146, 149–151, 170, 202, 205, 208, 228–229, 329, 352

Phosphotyrosine, 43
Photinus pyralis, 336
Phylogenetic tree, 343, 347–348
Pichia pastoris, 280
piRNA, 11, 163, 206
Piwi-associated RNA, *see* piRNA
Plaque, 273, 284, 304
  hybridization, 291
  lift, 292
Plasmid, 40, 45–46, 69–70, 75, 95, 165, 180
  2μ, 279–280
  miniprep, 250
  purification, 250–252
  Ti, 46, 246, 280–282
  vectors, 245–246, 249, 253, 255–268, 284–285, 289, 314, 325
  yeast episomal (Yep), 279–280
Pluripotency, 325, 339
PMF, *see* Peptide mass fingerprint
Poliovirus, 211, 223
Poly(A)
  alternative processing, 172, 174
  binding proteins, 168
  cleavage factors, 166
  nuclease, 170
  polymerase, 166–167
  tail, 166, 168, 170, 202–203, 219, 287, 307
Polyadenylation, 166–168, 170, 174
Poly(ADP)ribose polymerase (PARP), 87–89, 92
Polycistronic mRNA, 111, 202, 205, 212
Polyclonal antibody, 31
Polymerase chain reaction (PCR), 245, 247, 264, 285, 288–291, 300, 303–308, 314
  annealing temperature, 304, 306–307
  cycle, 304, 308
  degenerate oligonucleotide primers (DOP), 306
  emulsion, 300
  enzymes, 306–307
  error-prone, 315
  inverse, 306
  magnesium concentration, 307
  multiplex, 308
  mutagenesis, 313–316
  nested, 307
  optimization, 307
  overlap extension, 315
  Pfu polymerase, 307
  primary, 315
  primers, 62, 304–305
  probes, 304, 308
  quantitative (qPCR), 307–308, 325
  rapid amplification of cDNA ends (RACE), 307
  real time, 307–308, 322, 324
  reverse transcriptase (RT)-PCR, 307, 322
  secondary, 315
  Taq polymerase, 248, 306
  template, 304–305
  threshold cycle, 308
Polynucleotide kinase, 248, 294
Polynucleotide phosphorylase, 177
Polyprotein, 206–208, 225
Polypyrimidine tract, 168
Polyribosomes, *see* Polysomes
Polysomes, 191, 287
Post-transfer editing, 187
Post-translational modifications, 23, 27–28, 207–209, 282, 318–319, 329
Pre-mRNA, *see* hnRNA
Pre-miRNA, 163
Pre-transfer editing, 186
Pribnow box, 103

Primer,
    adaptor-primer, 287–288, 300
    cDNA, 287
    degenerate, 306
    extension, 310
    forward and reverse, 306
    gene-specific, 307
    mutagenic, 314–315
    oligo(dT), 287
    PCR, 62, 304–305
    RNA, 70–73, 75–76, 84, 223
    sequencing, 289, 298
Pri-miRNA, 163
Prion, 18
Processivity, 140, 150–151
Programmed cell death, *see* Apoptosis
Prokaryotes, 3
Promoter, 97, 102–104
    -10 sequence, 103, 106, 120
    -35 sequence, 103, 106, 120
    alternative, 132, 172
    analysis, 311–312, 336
    bacteriophage, 214–216
    clearance, 107
    efficiency, 103–104
    Escherichia coli, 100, 102–104, 106
    heat shock, 120–121
    lac, 104, 111–113
    RNA Pol I, 127
    RNA Pol II, 135–136, 138–140, 147, 149–151
    RNA Pol III, 131–132
    trp, 115
    viral, 219–220, 224
    vector, 264, 266–267, 274, 279, 282, 295, 326, 331, 338
Proofreading,
    DNA replication, 71, 75, 83–85
    exonuclease, 71, 75, 83, 307
    protein synthesis, 186–187
Prophage, 94, 215, 217
PROSITE database, 346
Prosthetic groups, 15, 19, 86
Protease complex, 209
Protease digestion, 284
Proteasome, 209
Protein
    3$_{10}$-helix, 17
    α-helix, 17, 19, 348
    array, 332
    β-sheet, 17–19, 336, 348
    C-terminus, 15
    chip, 332
    circular dichroism, 28
    conjugated, 19, 23
    degradation, 208–209
    domains, 19, 25, 331, 343–344, 349
    engineering, 316
    families, 19–20, 234, 346, 349
    fibrous, 15, 17
    globular, 15, 17, 19, 28–29
    hydrogen bonds, 15–18
    hydrophobic forces, 15–16, 18–19, 21, 24, 27
    identification, 30, 32, 247, 329
    interactions, 16, 29, 32, 128, 208, 319, 330–332
    intrinsically disordered, 349
    mass determination, 29–30
    mass spectrometry, 29–30, 320, 329, 342, 344
    motifs, 19, 21, 343, 346, 348–349
    N-terminus, 15
    NMR spectroscopy, 29, 318, 320, 342–343, 348
    noncovalent interactions, 16, 18–19, 21, 24
    peptide bond, 15

    primary structure, 15–16
    purification, 27, 331
    quaternary structure, 18–19
    secondary structure, 16–17
    secretion, 23, 95, 207–208
    sequencing, 28, 329, 342
    splicing, 208
    supersecondary structure, 19–20
    tertiary structure, 17–18
    triple helix, 17
    X-ray crystallography, 28–29, 342
Protein A, 330
Protein Data Bank (PDB), 343, 348
Protein disulfide isomerase, 208
Protein HU, 46
Protein kinase, 15, 21, 228, 320
Protein-protein interactions, 16, 29, 32, 128, 208, 319, 330–332
Protein synthesis, 11, 188–209
    30S initiation complex, 193
    70S initiation complex, 197
    80S initiation complex, 200, 202
    cell-free systems, 287
    elongation, 197, 202
    elongation factors, 193, 197, 200, 207
    frameshifting, 197, 225
    initiation, 190, 193, 197, 200–202
    initiation factors, 193, 200
    multifactor complex, 200
    post-transfer editing, 187
    pre-transfer editing, 186
    proofreading, 186–187
    release factors, 198, 203
    ribosome, 24, 158–159
    ribosome binding site (RBS), 190, 193, 267
    scanning, 200
    termination, 198, 203
Proteoglycans, 23–24
Proteome, 319, 329
Proteomics, 319, 328–332, 343
Proto-oncogene, 231, 233, 236, 242
Protoplast, 279
Provirus, 223–224, 232–234
Pseudogene, 92, 95
Pseudouridine, 157, 182
Pull-down assay, 330
Purification tag, 27, 267, 331
Purine, 5–6
Pyrimidine, 5–6
Pyrimidine dimer, 80, 84, 86–88
Pyrrolysine, 14, 178

qPCR, *see* Quantitative PCR
Quadrupole analyzer, 29–30
Quantitative biology, 353
Quantitative PCR (qPCR), 307–308
Quasispecies, 223
Quaternary structure, 18–19
Quikchange™, 314

Rad51 protein, 92
Ras genes, 231, 234–235, 238
Rb, *see* Retinoblastoma
RBS, see Ribosome binding site
rDNA, see Ribosomal DNA
Reading frame, 177
Reassociation kinetics, 60
RecA protein, 92, 94, 217
RecBCD protein, 92
Receptor,
    cell-surface, 21, 233–234

nuclear, 144–146, 149, 235
ribosome, 207
signal recognition particle, 207
transferrin, 206
Recessive mutation, 63, 89, 236–237
Recombinant DNA, 3, 32, 245, 259, 265, 273, 275, 281, 289, 293
Recombinant protein, 27, 31, 267, 280, 326, 330, 335
Recombination, *see* DNA recombination
Red fluorescent protein, 336
Reductionism, 351
Regulator gene, 111, 216
Release factors, 198, 203
Repetitive DNA, 60–61, 92, 95, 153, 283
Replica plating, 291
Replication, *see* DNA replication
Replicative form, 70, 214, 274
Replisome, 19, 72
Repressilator, 354
Repression, 51, 57, 77, 111, 115, 147, 153, 205, 216, 327, 335, 353
Repressor, 111–112, 115, 120, 143, 146, 149, 216, 266
Reporter genes, 311–312, 331, 336
Restriction enzyme, 62, 246, 248, 253–257
   BamHI, 255–256, 271, 285, 294
   blunt ends, 248, 254, 259, 285
   cohesive ends, 254, 259
   double digest, 259, 294
   DpnI, 314
   EcoRI, 254–257, 259, 263, 294
   mapping, 247, 294
   palindromic sequence, 254
   partial digestion, 284–285
   recognition sequence, 61, 254, 264, 266, 277, 287, 306
   Sau3A, 285
   sticky ends, 248, 254, 285, 288
Restriction fragment, 61, 257, 263
Restriction fragment length polymorphism, 63
Restriction mapping, 247, 294
Restriction site, 61, 254, 264, 266, 277, 287, 306
Reticulocyte lysate, 287
Retinoblastoma, 237
   gene (*RB1*), 238
   protein (Rb), 229, 238
Retinoic acid receptor, 149
Retrotransposons, 92, 95, 223
Retroviruses, 2, 95, 212, 223–25, 232–233, 246, 281, 340
   avian leukosis virus, 224
   env gene, 223–224
   frameshifting, 225
   gag gene, 224–225
   genome, 224
   HIV, 2, 150–151, 216, 223, 225
   integrase, 223
   mutation rates, 225
   pol gene, 223–225
   provirus, 223, 232
   replication, 184, 212
   rev protein, 225
   reverse transcriptase, 76, 212, 223, 248, 287, 322
Rev protein, 225
Reverse transcriptase, 76, 212, 223, 248, 287, 322
Reverse transcriptase-PCR, 307, 322
Reversed phase chromatography, 329
RFLP, *see* Restriction fragment length polymorphism
RFP, *see* Red fluorescent protein
Rho factor, 98, 109, 215
Ribonuclease, *see* Nuclease
Ribonuclease protection assay, 322
Ribonucleic acid, *see* RNA

Ribonucleoprotein (RNP), 24, 154–175
Ribonucleotide, 6
Ribose, 6, 34
Ribosomal DNA, 60, 343–344
Ribosomal proteins, 158–159
Ribosomal RNA, *see* rRNA
Ribosome, 24, 158–159
   30S subunit, 24, 158–159, 193, 205
   50S subuint, 24, 158–159, 197
   A, P and E sites, 191, 197
   bacterial, 24, 158
   eukaryotic, 159
   peptide bond formation, 197
   protein components, 158–159
   recycling factor, 198
   RNA components, 24, 158
   structural features, 158–159
Ribosome binding site, 116, 190, 193, 267
Ribosome receptor proteins, 207
Riboswitch, 121
Ribothymidine, 182
Ribozyme, 11, 24, 157, 162–163, 197
Rifampicin, 100
RISC, *see* RNA-induced silencing complex
RITS, *see* RNA-induced transcriptional silencing
RNA
   7SL, 11, 95, 207
   acid, effect of, 34
   alkali, effect of, 34
   antisense strand, 295
   bases, 5
   binding experiments, 157
   chain elongation, 107
   chain initiation, 106
   chain termination, 107–108
   genes, 2, 11, 60, 124, 126–127, 131–132, 134, 153, 156, 161, 163, 165, 206, 342
   hairpin, 10, 98, 108–109, 116–117, 157, 163, 168, 327
   hydrogen bonding, 10
   hydrolysis in alkali, 34
   interference, 206, 327
   knockdown, 206, 327
   long non-coding, *see* lnc RNA
   masked, 205
   micro, *see* miRNA
   modification, 11, 157, 161
   non-coding, 11, 61, 121, 134, 147, 152–153, 163, 165, 206, 235, 319
   piwi-associated, *see* piRNA
   processing, 154–175
   replication, 2, 223
   ribosomal, *see* rRNA
   RNA-Seq, 324
   sense strand, 295
   sequencing, 302, 324
   short interfering, *see* siRNA
   silencing, 206, 327
   stability, 33–35
   stem-loop, 10, 98, 150, 155, 188, 205–206
   structure, 7, 10
   transfer, *see* tRNA
   transfer-messenger, *see* tmRNA
   UV absorption, 38
RNA-DNA hybrid, 75–76, 98, 107, 109, 310
RNA editing, 3, 28, 174–175, 302
RNA-induced silencing complex, 153, 163, 206
RNA-induced transcriptional silencing, 153
RNA interference, 206, 327
RNA Pol I, 124, 126, 156
   promoters, 127
   transcription factors, 127–129, 138

RNA Pol II, 124, 134, 153, 163, 165–168, 206
  basal transcription factors, 138–140
  CTD, 125, 138, 140, 146, 151, 169–170
  enhancers, 135
  promoters, 135, 138, 172
RNA Pol III, 124, 131, 156, 166
  promoters, 132
  termination, 132
  transcription factors, 132, 138, 143
RNA polymerase, 97–98
  α subunit, 100
  β subunit, 100
  β′ subunit, 100
  bacteriophage SP6, 248, 267, 295
  bacteriophage T3, 100, 248, 295
  bacteriophage T7, 100, 121, 248, 267, 274, 295, 298
  core enzyme, 99, 120
  E. coli, 98–101, 124
  eukaryotic, *see* RNA Pol I, RNA Pol II, RNA Pol III
  holoenzyme, 99–100
  sigma (σ) factor(s), 99–102, 104, 106–107, 120–121
RNA processing, 154–175
RNA replication, 2, 223
RNA synthesis, *see* Transcription
RNA viruses, 223–225
  HIV, 2, 224–225
  poliovirus, 223
  retroviruses, 223–225
RNAi, *see* RNA interference
RNAse, *see* Nuclease
RNAse protection, 157
RNP, *see* Ribonucleoprotein
RppH, 170
rRNA, 11, 24, 126–127, 155–157
  16S, 24, 155, 158, 190
  18S, 60, 126, 156–158, 190
  23S, 24, 155–156, 158, 163, 197
  28S, 60, 126, 156–157, 159, 163, 197
  45S, 60, 124, 126–129, 156
  5.8S, 60, 126, 156–157, 159
  5S (bacterial), 24, 155, 158
  5S (eukaryotic), 124, 131–132, 156, 159
  genes, 60, 126, 132, 342
  methylation, 157
  peptidyl transferase activity, 163, 197–198
  processing, 155–157
  self-splicing, 163
RT, *see* Reverse transcriptase
RT-PCR, *see* Reverse transcriptase-PCR
RuvA, B, C proteins, 92

S1 nuclease, 248, 310, 322
Satellite DNA, 55, 60–62
Scanning model, 200
SCID, *see* Severe Combined Immunodeficiency
Scrapie, 18
SDS, *see* Sodium dodecyl sulfate
SDS-PAGE, *see* Electrophoresis (PAGE)
Second messenger, 234
Secondary structure prediction, 32, 343, 348–349
Sedimentation coefficient, 24, 155
Selectivity factor, 129, 138
Selenocysteine, 14, 178
Self-splicing, 163
Sense strand, 97, 106, 135, 223
Sequence
  alignment, 343, 345–349
  annotation, 343, 346
  databases, 342–346
  gap, 344–346
  identity, 94, 344–345

  mismatch, 85, 89, 344, 348
  similarity, 20, 213, 343–349
  similarity searching, 343–344, 347
Sequencing, *see* DNA sequencing & RNA sequencing
Severe Combined Immunodeficiency, 340
7SL RNA, 11, 95, 207
Shaker-2, 292
Shine–Dalgarno sequence, 121, 190
Short interfering RNA, *see* siRNA
Sigma (σ) factor(s), 99–102, 104, 106–107, 120–121
Signal
  nuclear localization, 149, 208
  peptidase, 208
  peptide, 207
  polyadenylation, 166, 170
  recognition particle (SRP), 95, 207
  sequence, 207–208
  transduction, 149
SILAC, 329
Simple sequence length polymorphism, 63
Simple tandem repeat, 61
SINES, 61, 95, 153
Single nucleotide polymorphism, 63, 324
Single-stranded conformational polymorphism, 63
siRNA, 11, 32, 134, 163, 206, 327, 337
Site-directed mutagenesis, *see* Mutagenesis
SL1, *see* Selectivity factor
Sm proteins, 166
Small nuclear ribonucleoprotein, *see* snRNP
Small nuclear RNA, *see* snRNA
Small RNA, *see* sRNA
Smith–Waterman algorithm, 344–345
SNP, *see* Single nucleotide polymorphism
snRNA, 11, 124, 131–132, 134, 165–166, 169
snRNP, 162, 166, 168
Sodium dodecyl sulfate, 29, 31, 251, 291
Somatic cell nuclear transfer, 338
Southern blotting, 62, 295–296, 322
SP1, 135, 144, 146, 149
SP6 RNA polymerase, 248, 267, 295
Spliceosome, 168–169
Splicing, 133, 166, 168–170
  alternative, 172–174, 219, 319, 329
  exon, 19, 168, 172–174
  intron, 60–61, 157, 161–163, 168–170, 172, 174, 179, 206, 287, 309
  self, 163
SPO1 phage, 121
Sporulation, 121
sRNA, 122
SRP receptor, 207
SSCP, *see* Single-stranded conformational polymorphism
SSLP, *see* Simple sequence length polymorphism
Start codon, 172, 177–178, 180, 190–191, 193, 200, 202, 267, 309
STAT proteins, 149–150
Stem cells, 325, 338–340
Stem-loop, 10, 98, 150, 155, 188, 205–206
Steroid hormones, 149
  receptors, 143–144, 146, 149–150
  response elements, 149
Sticky ends, 215, 248, 254, 259, 271
Stop codon, 27, 83, 116, 172, 174, 177–179, 190, 193, 197–198, 200, 203, 205, 225, 309
STR, *see* Simple tandem repeat
Streptavidin, 295
Streptolydigins, 100
Structural gene, 110, 206, 318
Structure prediction, 32, 343, 348–349
Subcloning, 246, 250, 264, 267, 298

Subtilisin, 20–21
Sucrose gradients, 67, 285, 287
SUMO, 208
Supercoiled DNA, *see* DNA supercoiling
Surface plasmon resonance, 29
SV40, 75, 135, 212, 219, 246, 279, 281
Swiss-Prot, 343, 345
Syn-conformation,10
Synchrotron radiation, 29
Synonymous codons, 178–179, 184
Synthetic biology, 3, 354–355
Systems biology, 351

T-antigen, 219
T3 RNA polymerase, 100, 248, 295
T4 DNA ligase, 248, 259, 285, 289
T7 DNA polymerase, 298
T7 RNA polymerase, 100, 121, 248, 267, 274, 295, 298
TAF$_I$s, 129
TAF$_{II}$s, 138, 146
Tandem gene clusters, 60
TAP-tag, 331
Taq DNA polymerase, 248, 306–307
Targeted gene disruption, 324, 339
Targeted insertional mutagenesis, 324
Tat protein, 140, 150, 216, 225
TATA binding protein, *see* TBP
TATA box, 132, 135, 138–140, 220
TBP, 129, 132, 135, 138, 140, 149
TBP-associated factors, *see* TAF$_I$s *and* TAF$_{II}$s
Telomerase, 11, 24, 76–77
Telomere, 55, 75–77, 275–276
Template strand, 11, 64, 96
Terminal (deoxynucleotidyl) transferase, 94, 248, 288,
    294, 307
Terminator sequence, 98, 107–109, 116–117, 267
Tetracycline, 249
Tetrahymena, 157, 163
TFII transcription factors, 89, 138–140, 146, 149
TFIII transcription factors, 132, 138, 143, 151
Thioredoxin, 27
Thymidine, 6, 67
Thymidine kinase, 220, 339
Thymine, 5, 8, 57
Thymine dimer, 80–81
Thyroid hormone receptor, 143, 146, 149, 235
Time-of-flight, 30
TLS, *see* Translesion DNA synthesis
T$_m$, *see* Melting temperature
tmRNA, 198
Tn transposon, 95
TOF, *see* Time-of-flight
Topoisomer, 41–42
Topoisomerases, *see* DNA topoisomerases
trans-acting, 122, 153, 206, 220, 223
Transcription, 2, 8, 96–153
    abortive initiation, 106
    anti-terminator, 117
    bubble, 107
    closed complex, 106
    complex, 98, 107, 109, 138, 140, 142, 146, 149–150,
        168, 331
    elongation (bacterial), 98, 100, 107
    elongation (eukaryotic), 140, 150
    enhancers, 135–136, 142, 146, 153, 174, 230, 233–234
    in vitro, 267, 295
    initiation (bacterial), 97, 99–100, 102–103, 106, 120
    initiation (eukaryotic), 124–125, 129, 131–133, 135,
        138–140, 151
    open complex, 106
    profiling, 319

promoter, 97, 102–104
promoter clearance, 107
start site, 103, 106, 127, 131–132, 135, 150, 220,
    310–311
stop signal, 98, 107–109, 116–117, 267
termination (bacterial), 98, 107–109, 116
termination (eukaryotic), 132
terminator sequence, 98, 107–109, 116–117, 267
ternary complex, 107
unit, 97, 107, 111, 115, 127, 131
Transcription factors, 21, 47, 124, 138–147
    activation domains, 142, 145–146, 149, 331
    Antennapedia, 143, 146, 152
    AP2, 146
    Bicoid, 143
    binding sites, 106, 111, 113, 115, 128, 132, 142, 146,
        149, 220, 314
    BRF, 132
    cJun, 146, 235
    coactivator, 149
    corepressor, 115, 149
    DNA-binding domains, 142–146, 149, 152, 331
    domain swap experiments, 142, 146
    domains, 142–146
    Gal4, 142, 145–146
    Gcn4, 142, 145
    glucocorticoid receptor, 145, 149
    homeodomain proteins, 152
    Lac repressor, 111
    LexA repressor, 142
    MyoD, 145, 151–152, 230
    nuclear hormone receptors, 143–144, 146, 149–150,
        235
    Oct2, 146
    phosphorylation, 140, 146, 149, 151
    sigma (σ) factor(s), 99–102, 104, 106–107, 120–121
    SL1, 129, 138
    SP1, 144, 146, 149
    STAT, 149–150
    TAF$_I$s, 129
    TAF$_{II}$s, 138, 146
    Tat, 140, 150, 216, 225
    TBP, 129, 132, 135, 138, 142, 149
    TFII factors, 89, 138–140, 146, 149
    TFIII factors 132, 138, 143, 151
    Trp repressor, 115
    UBF, 127–129
    WT1, 146
Transcriptome, 318–319, 322–323
Transcriptomics, 318–319, 343, 346, 351
Transduction, 281, 340
Transfection, 232, 235, 279, 281, 336, 338, 340
Transfer-messenger RNA, *see* tmRNA
Transferrin, 21, 206
Transfer RNA, *see* tRNA
Transformation
    cellular, 212, 235, 279
    DNA, 246, 260–262, 277, 324
Transformylase, 191
Transgene, 279, 338–339
Transgenic organism, 3, 32, 94, 292, 337–339
Transition mutation, 83, 179
Translation, *see* Protein synthesis
Translesion DNA synthesis, 84–85, 89
Translocase, 197
Translocation
    chromosomal, 88, 230, 234
    protein, 207, 335–336
    ribosome, 197–198, 200
    telomerase, 77
Translocon, 207

Transposase, 91, 95, 172, 173, 217
Transposition, 61, 94–95
  Alu element, 60–61, 95, 153
  bacteriophage, 213–214, 217
  insertion sequences (IS), 95
  L1 element, 61
  mutagenesis, 32, 95
  retrotransposons, 92, 95, 223
  Tn transposon, 95
  Ty element, 95
Transposon, *see* Transposition
Transversion mutation, 83, 179
TrEMBL database, 343
Triplex, 92
tRNA, 11, 182–187
  amino acid acceptor stem, 182
  CCA end, 161–162, 182
  cloverleaf, 182–183
  D-loop, 131, 182–183
  function, 184, 193–197
  genes, 131–132, 161
  initiator, 190–191, 193, 197, 200, 202
  invariant nucleotides, 182–184
  isoaccepting, 186
  modified bases, 182
  nucleotidyl transferase, 161
  primary structure, 182
  processing, 161–162
  proofreading, 186–187
  secondary structure, 182–183
  semi-variant nucleotides, 183
  T-loop, 183
  tertiary structure, 184
  transcription, 131–132
  variable arm, 183
  wobble, 178, 189–191
Trp
  attenuator, 116
  leader peptide, 116
  leader RNA, 116
  operon, 115–118
  repressor, 115
Tubulin, 22–23
Tumor suppressors, 146, 153, 229–230, 236–238, 243,
  340
Tumor viruses, 212, 219–220
Twist (*Tw*), 41–44, 47
Two-hybrid analysis, 331–332, 343
Ty retrotransposon, 95

U6 snRNA, 124, 132, 166, 168
U7 snRNP, 168
UBF, *see* Upstream binding factor
Ubiquitin, 208–209, 329
UniProt database, 343, 345
Untranslated region (UTR), 121, 147, 150, 153, 172, 193,
  206
Upstream, 97
Upstream binding factor (UBF), 127–128
Upstream control element (UCE), 127
Upstream Regulatory Element (URE), 135, 142
Uracil, 5, 79
Urea, 27, 35, 298
Uridine, 6
UTR, *see* Untranslated region

UV absorption, 15, 28, 37–38
UV irradiation, 84, 157, 256, 292

van der Waals forces, 16, 18, 21
Variable number tandem repeat, 61–62
Vector, *see* Cloning vector
Virion, 210
Virus
  budding, 211
  capsid, 211, 214–215
  complementation, 212, 223, 225
  enhancer, 135, 233–234
  envelope, 211, 219, 221, 223–225
  genomes, 212
  helper, 212
  infection, 211–212, 215, 220, 223
    latent, 221
    lysogenic, 214, 217, 271, 274
    lytic, 214, 271
  matrix, 220
  nucleocapsid, 211
  receptors, 211
  virulence, 212, 223, 281
Virus types
  adenoviruses, 218–219, 281, 340
  avian leukosis, 224
  baculoviruses, 220, 246, 281–282
  Epstein–Barr, 132, 212, 220
  hepatitis C, 2
  herpes simplex, 220–221, 239
  herpesviruses, 145, 211-212, 217–221, 281, 339
  influenza, 2
  mimivirus, 218
  mouse mammary tumor, 234
  papovaviruses, 212, 218–219
  poliovirus, 211, 223
  poxviruses, 220
  retroviruses, 2, 95, 212, 223–25, 232–233, 246, 281,
    340
  rhabdovirus, 211
  SV40, 75, 135, 212, 219, 246, 281
  vaccinia, 220, 281, 337
  varicella zoster, 219
VNTR, *see* Variable number tandem repeat

Western blot, 31, 329
Wheat germ extract, 287
Wilms tumor gene, 146
Wobble, 178, 189–191
Writhe (*Wr*), 42–50

X chromosome inactivation, 153
Xenotransplantation, 338
Xeroderma pigmentosum, 89
X-gal, 266, 312, 332
Xist, 153
X-ray crystallography, 28–29, 50, 318, 342–343, 348
X-ray diffraction, 28, 158
XRCC1, 88

Yeast artificial chromosome (YAC), 275, 277, 284–285

Z-DNA, 10
Zinc finger domains, 143–144, 149
Zoo blot, 296